ASTROPHYSICS IN ANTARCTICA

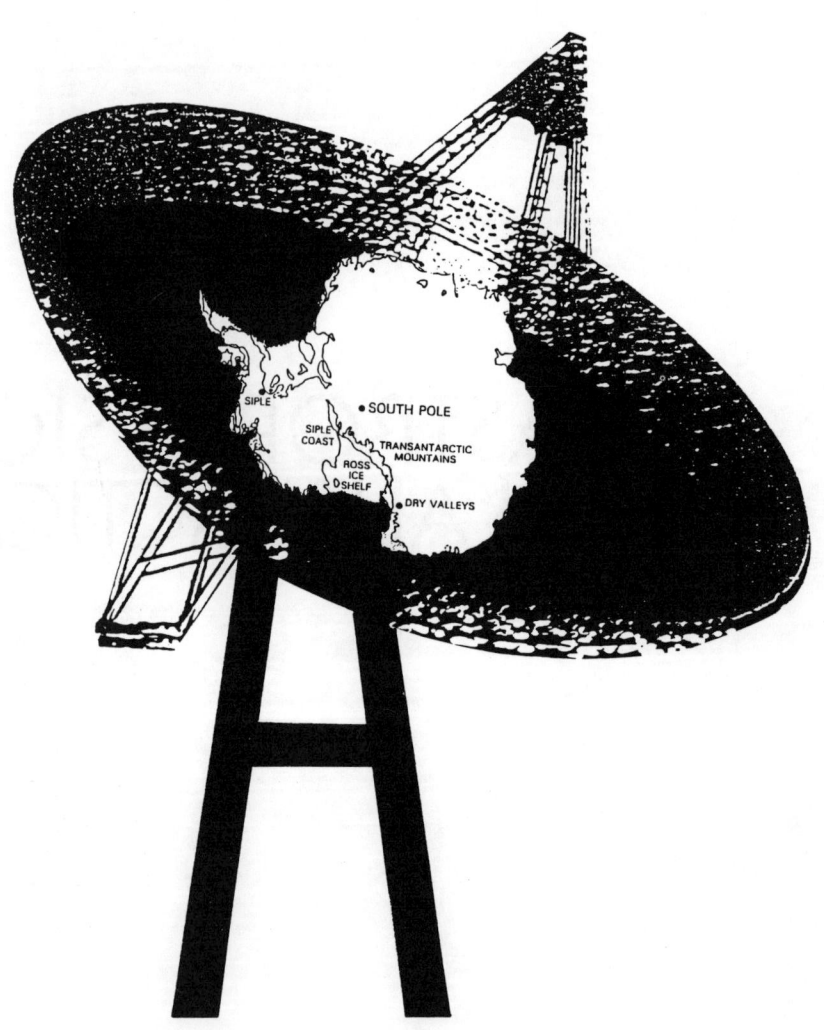

AIP CONFERENCE PROCEEDINGS 198

RITA G. LERNER
SERIES EDITOR

ASTROPHYSICS IN ANTARCTICA

NEWARK, DELAWARE 1989

EDITORS:
DERMOTT J. MULLAN,
MARTIN A. POMERANTZ, &
TODOR STANEV

BARTOL RESEARCH INSTITUTE
UNIVERSITY OF DELAWARE

AIP
American Institute of Physics   New York

Authorization to photocopy items for internal or personal use, beyond the free copying permitted under the 1978 US Copyright Law (see statement below), is granted by the American Insitute of Physics for users registered with the Copyright Clearance Center (CCC) Transactional Reporting Service, provided that the base fee of $3.00 per copy is paid directly to CCC, 27 Congress St., Salem, MA 01970. For those organizations that have been granted a photocopy license by CCC, a separate system of payment has been arranged. The fee code for users of the Transactional Reporting Service is: 0094-243X/87 $3.00.

Copyright 1989 American Institute of Physics.

Individual readers of this volume and non-profit libraries, acting for them, are permitted to make fair use of the material in it, such as copying an article for use in teaching or research. Permission is granted to quote from this volume in scientific work with the customary acknowledgment of the source. To reprint a figure, table or other excerpt requires the consent of one of the original authors and notification to AIP. Republication or systematic or multiple reproduction of any material in this volume is permitted only under license from AIP. Address inquiries to Series Editor, AIP Conference Proceedings, AIP, 335 E. 45th St., New York, NY 10017.

L.C. Catalog Card No. 89-46421
ISBN 0-88318-398-6
DOE CONF 8906234

Printed in the United States of America.

Contents

Preface ix

List of Attendees xii

GROUND-BASED GAMMA-RAY ASTRONOMY, NEUTRINO ASTRONOMY, AND NEW DIRECTIONS

Very High Energy Gamma Ray Astronomy 3
    Trevor C. Weekes
The South Pole as a Site for Monitoring 100 TeV Cosmic
Gamma Rays by Means of an Air Shower Array 12
    A. M. Hillas
Experimental Test of the Anomalous Density Spectrum
of Cosmic Ray Showers, as an Adjunct to the South Pole
Air Shower Experiment? 16
    A. M. Hillas
Design Considerations for a TeV Telescope
at the South Pole 20
    K. Harris, M. A. Pomerantz, P. T. Reynolds, G. Vacanti,
    A. Walker, and T. C. Weekes
A South Pole Facility to Observe Very High Energy Gamma Ray Sources 24
    Bartal, Purdue, Smithsonian, Wisconsin Collaboration
Prospects for Ground-based Gamma Ray Astronomy
at the South Pole: Working Group Report 35
    A. A. Watson
Neutrino Astronomy 39
    F. Halzen, J. Learned, and T. Stanev
ICEMANd: Microwave Detection of Ultra-high Energy
Neutrinos in Ice 52
    J. P. Ralston and D. W. McKay
New Ideas in South Pole Experiments 61
    D. Seckel

MILLIMETER, SUB-MILLIMETER, AND INFRARED ASTRONOMY

Tilting and Vibration of the "Skylab" Building
at the South Pole 67
    A. A. Stark and J. Gress
Low-frequency Measurements of the CMB Spectrum 71
    A. Kogut, M. Bensadoun, G. De Amici, S. Levin, M. Limon,
    G. Smoot, G. Sironi, M. Bersanelli, and G. Bonelli

High-altitude Measurements of Fluctuations in the CMB 75
    R. D. Davies
CBR Anisotropy and Galactic Emission Observations
from Antartica 79
    P. Andreani, G. Dall'Oglio, M. DiBari, L. Martinis,
    L. Piccirillo, L. Pizzo, and C. Santillo
Cosmic Background Anisotropy Studies at 10° Angular Scales
with a HEMT Radiometer.. 84
    T. Gaier, J. Schuster, and P. Lubin
South Pole Studies of the Anisotropy of the Cosmic
Background Radiation at One Degree... 88
    P. R. Meinhold, P. M. Lubin, A. O. Chingcuanco, J. A. Schuster, and M. Seiffert
First Sub-millimeter Observations from the South Pole:
The Integrated Galactic Emission 93
    F. Pajot, R. Gispert, J. M. Lamarre, J. L. Puget,
    M. A. Pomerantz, and R. Peyturaux
South Pole Sub-millimeter Isotropy Measurements
of the Cosmic Microwave Background.. 97
    M. Dragovan, S. R. Platt, R. J. Pernic, and A. A. Stark
Atmospheric Transparency over Antarctica from the Mid-infrared
to Centimeter Wavelengths... 100
    J. Bally
On the Future of Sub-millimeter Astronomy at the South Pole 106
    A. A. Stark
Millimeter and Sub-millimeter Photometry from Antarctica... 116
    J. B. Peterson
Infrared Astronomy in Antarctica.. 123
    D. A. Harper

GEOSPACE PLASMAS

Geospace Plasmas... 133
    R. L. Arnoldy
Ground-level Detection of LF and MF High Latitude
Terrestrial Emissions.. 145
    R. F. Benson
Geospace Plasmas: Report of the Working Group 149
    T. J. Rosenberg

ANTARCTIC BALLOONING

Cosmic and Solar Gamma-ray, X-ray, and Particle Measurements
from High-altitude Balloons in Antarctica... 157
    R. P. Lin

Long-duration Ballooning at Mid-latitudes and in Anarctica 166
    W. V. Jones
New Doors Opened by the Polar Patrol Balloon—Cosmic
Quark Matter, Micrograins, and Gamma Rays 176
    Y. Fukada, Y. Hatano, T. Saito, R. Fujii, H. Oda,
    and T. Yanagita
Search for Antimatter at the 10^{-7} Level with the Polar
Patrol Balloon 180
    Y. Fukuda, Y. Hatano, T. Saito, R. Fujii, H. Oda,
    and I. Yamamoto
A Measurement of the Isotopic Composition of Iron-group
Elements in the Galactic Cosmic Rays Using Balloon-borne
Track-recording Detectors in Antarctica 184
    A. J. Westphal, P. B. Price, and D. P. Snowden-Ifft
Zero Pressure Balloon Behavior in Antarctica 190
    J. R. Ground
ROBFIT: A General Purpose Spectral Analysis Package 194
    G. J. Bamford, R. L. Coldwell, and A. C. Rester
JACEE Long-duration Balloon Flights 198
    R. J. Wilkes *et al.*

SOLAR AND STELLAR ASTRONOMY

Solar and Stellar Observations from the South Pole 205
    J. L. Linsky
Driving of the Solar *P*-modes by Radiative Pumping
in the Upper Photosphere 218
    J. M. Fontenla, A. G. Emsli, and R. L. Moore
A Solar Tracking Platform for Use at the South Pole 222
    S. M. Jefferies, R. Pfeiffer, M. A. Pomerantz, L. Schulman,
    and W. Ball
Solar Observing Conditions at the South Pole 227
    J. Harvey
Full-disk Helioseismology in the Antarctic 231
    E. Fossat, B. Gelly, G. Grec, and F.-X. Schmider
Infrared Physics from the South Pole 234
    D. Deming
Photometry of Selected Variable Stars at the South Pole 238
    K. Y. Chen
Solar-stellar Astronomy Working Group Summary 242
    J. Harvey

ANTARCTIC ENVIRONMENT AND EXPERIENCE

A Proposal for an International Station in Antarctica 249
    J. T. Lynch

The Role of the Winter-over Scientist
in Antarctic Astrophysics 253
    W. B. Gail

Field Operations in Antarctica from the Winter-over
Scientist's Perspective 256
    D. P. Clements

Astrophysical Experimentation in Antarctica:
A Winter-over Scientist's View 260
    L. E. Kay

A Graduate Student's Appreciation of the "Winter-over"
Experience 264
    N. J. T. Smith

Preface

The idea of performing optical/infrared astronomy in Antarctica was apparently given serious consideration for the first time in print almost 20 years ago, in a report issued by the U.S. National Academy of Sciences in 1970. That report dealt with a variety of issues concerning research in Antarctica. The article dealing with Antarctic Astronomy in the N.A.S. report was written by Arne A. Wyller, then Professor at Bartol Research Foundation. Wyller's article was based in part on semi-quantitative evaluations of seeing conditions at South Pole. These evaluations were performed by various Bartol observers who had been wintering over since the inception of Martin A. Pomerantz's neutron monitor program in 1964. In the course of their winter stays, several of these observers used a small telescope to make rough visual estimates of the quality of the solar image and of stellar images. In view of the visual results, the N.A.S. report was optimistic that observing programs for both solar and stellar astronomy could profitably be mounted at South Pole. Rough estimates were also available on the precipitable water vapor at South Pole, indicating that the infrared observing conditions at the site should be of high quality.

Despite the favorable coverage in the N.A.S. report of the possibilities for astronomical research in Antarctica, no systematic effort has so far been forthcoming at the national level to exploit the unique advantages of South Pole for astronomy. Up to the present, the work which has been performed has been more or less piecemeal. Examples of these programs are described in the next three paragraphs: the first two paragraphs refer to summer-only observing, whereas the third includes both summer and winter data collection.

Almost ten years were to pass following the N.A.S. report before the first optical research program was performed at South Pole. Towards the end of 1979, a team consisting of E. Fossat and G. Grec (Observatoire de Nice) and M. A. Pomerantz took a sodium vapor cell coupled to a small telescope (provided by the Swedish Solar Observatory, where A. A. Wyller was then Director), to South Pole and obtained an unbroken run of 120 hours of observations for purposes of measuring solar oscillations. The data enabled the frequencies of hundreds of solar eigenmodes to be determined with unprecedentedly high precision. A second program aimed at solar oscillations was subsequently developed in a collaboration between J. Harvey (National Solar Observatory), T. Duvall (Goddard Space Flight Center), and M. A. Pomerantz: this program established that information could be obtained on solar features as small as a few seconds of arc. Results from these two programs (which have both returned several times to Pole) have dramatically substantiated the optimistic predictions of the 1970 N.A.S. report regarding solar work in Antarctica.

It was not until well into the 1980's that programs were begun to take advantage of the splendid infrared observing conditions which had been pointed out in the 1970 N.A.S. report. These programs resulted from collaborations between M. A. Pomerantz and, on the one hand, J. Puget and colleagues (L.P.S.P., Paris) and, on the other hand with M. Dragovan and colleagues (University of Chicago and AT&T Bell Laboratories). The first of these collaborations was directed at a study of infrared emission from the Galaxy. The second collaboration involved studies of the cosmic microwave background and its anisotropy.

Meanwhile, the neutron monitor program of M. A. Pomerantz has continued since 1964 to study energetic charged particles by exploiting the advantages of Antarctica (high magnetic latitude and, at South Pole, low atmospheric pressure). Another aspect of high-energy astrophysics has become the object of a recent Antarctic collaboration between A. A. Watson (University of Leeds) and Bartol: an air shower array has been established at South Pole in an attempt to identify sources of high-energy photons. The only winter-over program in optical astronomy which has so far been performed at South Pole is that from the University of Florida (K. Y. Chen, Principal Investigator).

The fact that individual collaborations have led to successful experiments in various areas of astronomy in Antarctica suggests that it is now timely to consider the performance of research in a more organized manner. Accordingly, when the National Science Foundation announced its new program of Science and Technology Centers (STC), it seemed appropriate to consider an STC dedicated to a formal program of research in astrophysics in Antarctica. A Proposal for an STC Planning Grant entitled "South Pole Astrophysics Research Center" (Principal Investigator: Martin A. Pomerantz) was funded by the NSF in December 1988 to explore the possibility of establishing such a Center. The central facet of the Planning Grant was the convening of a conference which would draw together researchers with interests in all areas of astrophysics that might profit from the observing conditions in Antarctica. An Advisory Committee was invited to assist in the preparations for the conference. Members of this Committee (as well as members of the Local Organizing Committee) are listed below.

The present volume contains the proceedings of the Astrophysics in Antarctica Conference, which was held at the University of Delaware June 8–10, 1989. The contents of these Proceedings, including as they do past, present, and future research, testify to the active interest among the astrophysics community in performing research in Antarctica.

The organizers gratefully acknowledge the support provided for the Astrophysics in Antarctica Conference by the National Science Foundation, the University of Delaware, and the Bartol Research Institute.

D. Mullan, A. Pomerantz,
and T. Stanev

ADVISORY COMMITTEE

C. A. Beichman, Caltech
J. W. Harvey, NSO, Tucson
T. J. Rosenberg, Maryland
J. C. Learned, Hawaii
P. B. Price, Berkeley
A. A. Starck, Bell Labs
C. H. Townes, Berkeley
A. A. Watson, Leeds
D. T. Wilkinson, Princeton

LOCAL ORGANIZING COMMITTEE

M. Pomerantz (chairman)
J. Bieber
P. Evenson
T. Gaisser
W. Matthaeus
D. Mullan
N. Ness
J. Perrett
S. Owocki
Q. Shafi
T. Stanev

ASTROPHYSICS IN ANTARCTICA CONFERENCE
JUNE 8–10, 1989
LIST OF ATTENDEES

Name	Address
Morris L. Aizenman	National Science Foundation, Washington, D.C.
Mark Ander	National Laboratory, Los Alamos
Roger L. Arnoldy	University of New Hampshire
John Bally	AT&T Bell Labs
Gary Bamford	University of Florida
Thomas M. Bania	Boston University
Stephen Barr	Bartol Research Institute
Laura P. Bautz	National Science Foundation, Washington, D.C.
James R. Benbrook	University of Houston
Robert F. Benson	NASA/GSFC
John Bieber	Bartol Research Institute
Alessandro Cacciani	University "La Sapienza" Roma
Kwan-Yu Chen	University of Florida
J. Chen	Bartol Research Institute
Erick Chiang	National Science Foundation
Dave Clements	Bartol Research Institute
Carol Jo Crannel	NASA/GSFC
Giorgio Dall'Oglio	University of Rome
R. D. Davies	University of Manchester, U.K.
Drake Deming	GSFC, Greenbelt
Brian R. Dennis	NASA/GSFC
M. Dragovan	Princeton University
W. C. Erickson	University of Maryland
Paul Evenson	Bartol Research Institute
Vollie Fields	Raytheon Company, Burlington, MA
Juan Fontenla	University of Alabama in Huntsville
J. A. Gaidos	Purdue University
T. Gaier	University of California, Santa Barbara
William B. Gail	The Aerospace Corp., Los Angeles, CA
Thomas Gaisser	Bartol Research Institute

Reinhard Genzel	Max-Planck Institüt für Physik ünd Astrophysik, Monchen, Germany
Dan Gezari	NASA/GSFC
Richard Gispert	Institut d'Astrophysique Spatiale, France
Francis Halzen	University of Wisconsin
D. A. Harper	Yerkes Observatory
John Harvey	National Solar Observatory, AZ
Dick Herr	University of Delaware
Anthony Michael Hillas	University of Leeds, U.K.
Bernard V. Jackson	UCSD, California
Matthew Jaworski	University of Wisconsin
Stuart Jefferies	Bartol Research Institute
W. Vernon Jones	NASA Hq., Washington, D.C.
Hilary Kane	GSFC, Greenbelt
John P. Katsufrakis	Stanford University, CA
Laura Kay	Lick Observatory, CA
Almus Kenter	University of Wisconsin
Al Kogut	University of California, Berkeley
T. Kostiuk	GSFC, Greenbelt
J. Learned	University of Hawaii at Manoa
Robert P. Lin	University of California, Berkeley
Craig Lindberg	AT&T Bell Labs
David Lindley	Nature Magazine (Press)
Jeffery L. Linsky	University of Colorado, Boulder
K. Y. Lo	University of Illinois
Robert F. Loewenstein	Yerkes Observatory, WI
Douglas M. Lowder	University of California, Berkeley
John Lynch	National Science Foundation, Washington, D.C.
James MacDonald	University of Delaware
Eugene Maier	NASA/GSFC
Robert H. March	University of Wisconsin
Ron Moore	NASA/MSFC, Huntsville, AL
Robert M. Morse	University of Wisconsin
Dermott Mullan	Bartol Research Institute
Michael Mumma	GSFC, Greenbelt

Z. E. Musielak	University of Alabama, Huntsville
James Neff	NASA/GSFC
Norman Ness	Bartol Research Institute
Stan Owocki	Bartol Research Institute
Lyman A. Page	MIT, Cambridge, MA
Francois Pajot	Institute d'Astrophysique Spatiale, France
Vernon Pankonin	National Science Foundation, Washington, D.C.
Benjamin F. Peery, Jr.	National Science Foundation, Washington, D.C.
Robert J. Pernic	Yerkes Observatory, WI
Jay Perrett	Bartol Research Institute
Jeffrey Peterson	Princeton University
Gabriella Pizzo	University of Rome
Stephen R. Platt	Yerkes Observatory, WI
Martin Pomerantz	Bartol Research Institute
Marius Potgieter	Potchefstroom University, South Africa
P. B. Price	University of California
A. C. Rester	University of Florida
Theo. J. Rosenberg	University of Maryland
Takeshi Saito	University of Tokyo
Robert Schaefer	Ohio State University
David Seckel	Bartol Research Institute
Qaisar Shafi	Bartol Research Institute
Harry Shipman	University of Delaware
Walter Siegmund	University of Washington
David H. Smith	Sky Publishing Corp., MA
Nigel Smith	University of Leeds, U.K.
Todor Stanev	Bartol Research Institute
Antony A. Stark	AT&T Bell Labs
Robin Stebbins	University of Colorado
J. R. Storey	University of Auckland, New Zealand
Robert E. Streitmatter	NASA/GSFC
Charles H. Townes	University of California
Chang-Hua Tsao	Bartol Research Institute
Evelyn Tuska	Bartol Research Institute

Alan A. Watson	University of Leeds, U.K.
Trevor C. Weekes	Whipple Observatory, AZ
Andrew Westphal	University of California
R. Jeffrey Wilkes	University of Washington
David T. Wilkinson	Princeton University
Peter E. Wilkniss	National Science Foundation, Washington, D.C.
Barbara Williams	University of Delaware
Robert W. Wilson	AT&T Bell Labs
Thomas Winch	University of Wisconsin
James P. Wright	National Science Foundation, Washington, D.C.
Jonas Zmuidzinas	University of Illinois

Ground-based gamma-ray astronomy, neutrino astronomy and new directions

VERY HIGH ENERGY GAMMA RAY ASTRONOMY

Trevor C. Weekes

Whipple Observatory,
Harvard-Smithsonian Center for Astrophysics,
Box 97, Amado, Az 85645-0097.

ABSTRACT

Ground-based research in gamma-ray astronomy is rapidly developing; the South Pole offers much promise as the location for telescopes in the TeV and PeV energy range.

INTRODUCTION

Unlike some of the astrophysical disciplines under review at this workshop Very High Energy Gamma Ray Astronomy is a relatively new and hence immature discipline. Although the basic techniques were established 30 years ago it is only in the last five years or so that consistent results have been presented; on the basis of these it is now possible to think of TeV gamma-ray astronomy as a legitimate astronomical window[1]. There is even evidence for the existence of a few sources of PeV (and even one of EeV) gamma rays.

As is to be expected with any new discipline there is a fair share of controversial results; this is to be expected and is probably the result of a combination of poor statistics, improperly understood detection techniques, transient source emission, unusual astrophysics and possibly even some new physics. It is too soon yet to see the real fires through all the smoke but at least we can say that there is sufficient smoke that there must be some fires, that some of these fires are not where we might have expected them to be and that we are not going to have a clear picture of what is going on until we have improved detection techniques.

In this review I will not address specifically the role of Antarctica in gamma ray research but rather will highlight the overall status of the field upon which subsequent discussion of new and existing South Pole experiments can be based. It will be apparent that experiments at this location compliment existing and planned gamma-ray facilities in a unique way. In this context I note that the upcoming missions of Gamma-1 (French-U.S.S.R) and GRO (U.S.A.-Europe) will ensure that this coming decade will be marked by more activity in gamma-ray astronomy at all wavelengths than ever before. The four gamma-ray telescopes on GRO (the Gamma Ray Observatory) will cover four decades of the gamma-ray spectrum (from 1 MeV to 30 Gev) with an improvement in flux sensitivity of a factor of 20.

TECHNIQUES

At energies above 10 GeV the expected gamma-ray fluxes are so small that they are beyond the range of detectability in the current generation of satellites. Fortunately detection is possible by indirect means using the electromagnetic cascade that the gamma ray initiates in the earth's atmosphere. The earth's atmosphere acts as both a detection medium and an amplifier and provides an economical method of doing gamma-ray astronomy from the ground. This approach has the inherent disadvantage that it is not possible to have a veto to reject the much more numerous cosmic ray background and it is difficult to uniquely identify any shower as coming from a gamma-ray primary.

The basic principles involved in the two main methods of detection are illustrated in Figure 1. The two methods are compared in Table 1 where it is seen that they are quite different and are

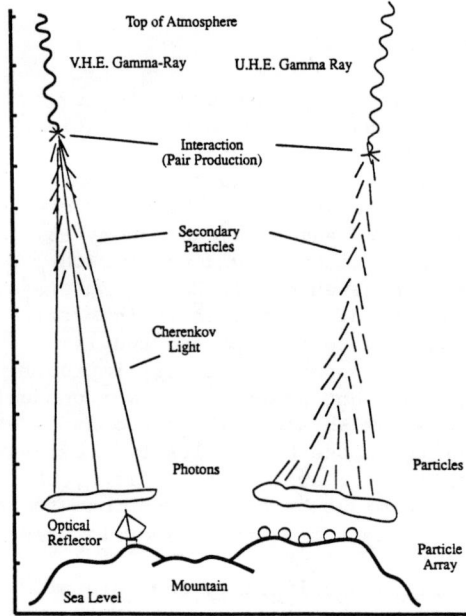

Figure 1. The two principal methods of detecting gamma-ray air showers from the ground.

complimentary since there is no overlap in energy sensitivity. The basic element in each detection technique is shown in Figure 2. It is clear that the enclosed scintillator element has an inherent advantage compared with the optical detector which must be exposed to the dark night-sky.

In the first generation of detectors no attempt was made to discriminate gamma-ray showers (electromagnetic cascades) from the more numerous background cosmic ray showers. Various methods have been suggested (Table 2). It has long been known that at energies greater than 100 TeV it is possible to select gamma-ray showers on the basis of their low muon content. However to date this has not been demonstrated to be a viable technique and where sources have been claimed at PeV energies the muon content has been similar to the background. In fact the selection by shower age has apparently been more effective although simulations show it should not be an

Table 1

Detection Techniques

	VHE	UHE
Energy range	TeV (0.1-10 TeV)	PeV (0.1-1000 PeV)
Technique	Atmospheric Cherenkov	Air Shower Array
Field of view	Target; 1-2°	Zenith; ±30°
Angular resolution	0.5°	0.5°
Collection Area	2×10^4 m^2	10^3 m^2
Duty cycle	<10%	100%
Detector element	Optical (mirror+pmt)	Scintillator
Motion	Steerable	Fixed
Background discrimination	Difficult	Easier

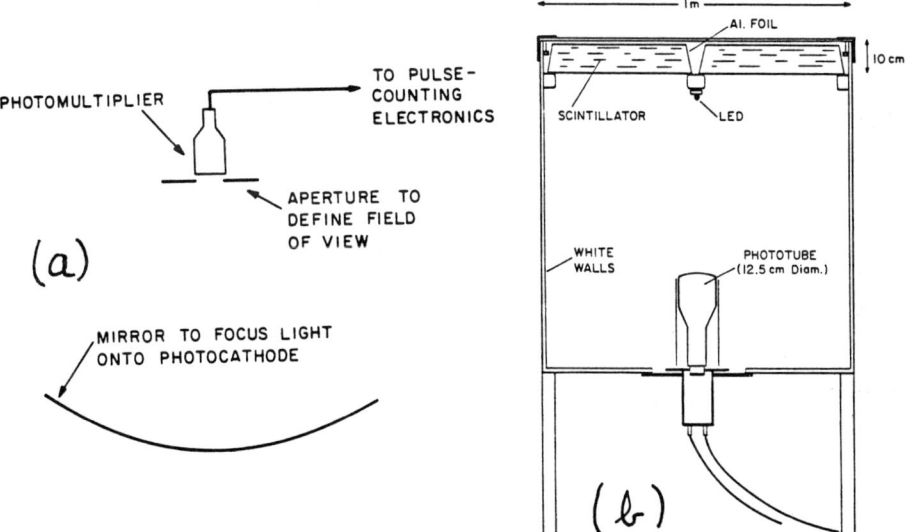

Figure 2. The basic telescope components in TeV and PeV gamma-ray telescopes: (a) the light detector (b) the air shower array element.

effective discriminant. The only technique that has been shown to be as effective as predicted is the Cherenkov imaging technique which rejects 98% of the background on the basis of the shower image parameters[2,3].

There are now more than a dozen each of TeV and PeV telescopes in operation around the globe. Initially there was a bias in the distribution towards the Northern Hemisphere; SN1987A in the Large Magellanic Cloud provided an impetus to develop more telescopes in the Southern Hemisphere. Because of the concentration of land masses there is a region of the southern sky that is only conveniently

Table 2

Gamma-ray Discrimination Characteristics.

Technique	Parameter	Gamma-ray Shower	Prediction
Atmospheric Cherenkov		Lateral distribution	Broader, more regular
"	"	Time structure	Faster, one component
"	"	Ultraviolet light	Less uv, more visible
"	"	Angular distribution	Narrower, directed
Air Shower Array		Muon-to-electron ratio	Less muons
"	" "	Shower age	No difference

reached from Antarctica. This is a region that is particularly rich in close binaries and radio pulsars[4].

The SPASE telescope[5], which has been in operation at the South Pole since December, 1987, is an example of a first generation air shower array; the decision to concentrate on a well-established technique was well justified by the speed with which this experiment was brought on-line and the efficiency with which it has been operated since then.

SOURCES.

The chief interest in VHE gamma ray astronomy is the detection of discrete sources of gamma rays; diffuse sources may exist but their detection will be an order of magnitude more difficult. We divide candidate sources into those which are expected (based on observations at other wavelengths and/or theoretical models) and sources which are unexpected (those for which no firm prediction of emission existed prior to observation). Examples of each are shown in Table 3.

Table 3

Source classification.

Type	Predicted	Candidates	Verification
Supernova	Yes	SN1987A	No
Supernova remnants	Yes	Crab	Yes
Radio Pulsars	Yes	PSR0531, 0833	Maybe
Radio galaxies	Yes	M87, Cyg A, Cen A	Maybe
Quasars	Yes	3C273	No
Close binaries	No	Her X-1, Vela X-1	Yes
Radio binaries	Yes	Cyg X-3	Yes
100 MeV sources	Yes	Geminga	No
Msec pulsars	No	PSR	Maybe

No attempt will be made to give a comprehensive source summary here. The observational status of various sources is somewhat fluid. For a detailed account of the current status of the field the reader is referred to several recent reviews[1,6,7]. But the field can be briefly summarized as follows:
(1) Only four sources can be regarded as well established; these are the Crab Nebula, Cygnus X-3, Hercules X-1 and Vela X-1.
(2) There are more detections at TeV energies than at PeV energies. There is one claim for the detection of a flux of EeV gamma rays[8] but this has been disputed[9].
(3) Most of the detections have been of transient periodic emission indicating that the VHE gamma-ray sky is quite unlike that at other wavelengths.
(4) Only at TeV energies has the signal from a source been definitely confirmed to be photons; at PeV energies what evidence there is indicates that the "source" showers are very similar to the background.
(5) The source spectrum of most sources appears to be very flat (integral power-law spectral index of -1).
(6) The power in cosmic ray particles must be very great indicating that these objects are leading candidates for the origin of cosmic rays with energies > 100 TeV.
(7) In addition to the sources listed in (1) there are more than 20 objects that have been claimed as TeV or PeV sources by one or two observations; the gamma-ray emission from these objects must be independently confirmed before they can be treated as detected.
(8) Even the existence of such a well-established source as Cygnus X-3 has been questioned[10]; nothing should be taken for granted.

Below are some notes on particular sources.

SN1987A.

Despite the detection of neutrinos from the initial supernova outburst there has been no reliable detection of gamma rays at any energy beyond 10 Mev. If cosmic rays are being produced it could be that the source will not become visible for a number of years as the gas shell thins. The upper limits reported to date[11] including one based on the first weeks of operation of SPASE at the South Pole[5] are shown in Figure 3.

Crab Nebula.

There are numerous models predicting TeV gamma-ray emission from the Crab Nebula which is known to be a reservoir of relativistic electrons; it is constantly being replenished by the 33 msec pulsar at the center of the nebula.
(1) At least three experiments have detected TeV gamma rays with a statistical significance greater than 5 sigma[2,12]; recently a detection at the 15 sigma level has been reported[3].
(2) There is no evidence for variability on any time scale; this is a standard candle for TeV gamma-ray astronomy.
(3) The energy spectrum appears to steepen beyond 1 TeV (Figure 4).
(4) The detected signal is consistent with gamma rays.
(5) There is weaker evidence for emission from the pulsar; if real it is apparently variable on time-scales of minutes to months.

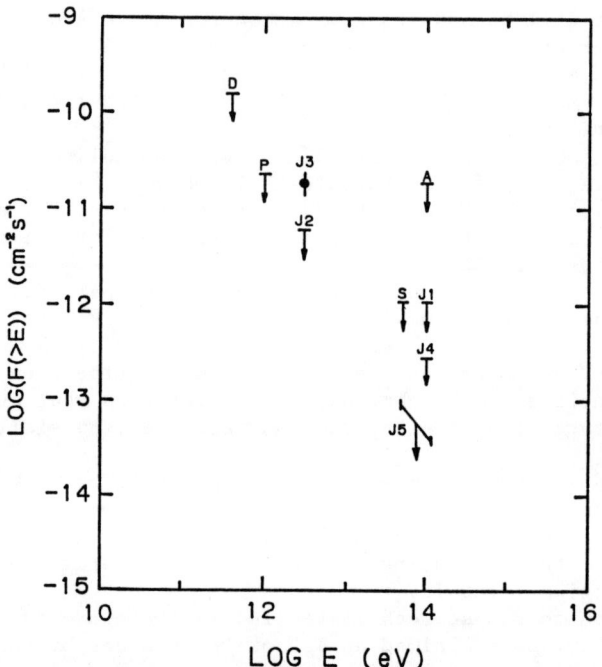

Figure 3. The upper limits (and one possible two day detection) of TeV and PeV gamma rays from SN1987A (from [11]).

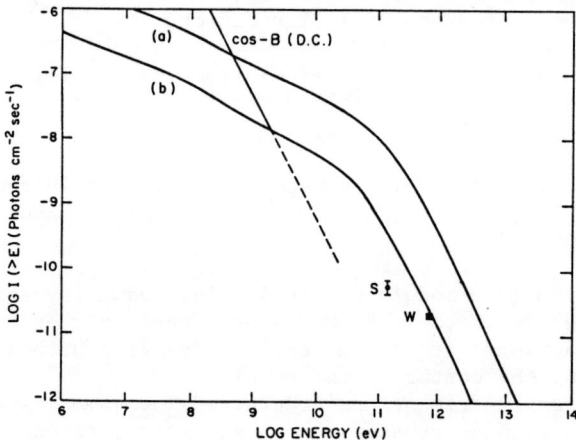

Figure 4. Spectrum of the Crab Nebula showing predicted inverse Compton spectrum and some of the detections (from [2]).

Geminga.
One of the strongest sources in the sky at 100 MeV energies this has not been detected at TeV energies (despite earlier claims

to the contrary which were based on the now discredited 59 sec periodicity).

3C273.
This nearby quasar is the only extragalactic source seen at 100 Mev energies. There have been no reports of detection at TeV energies; the TeV observations do severely restrain the extension of the spectrum from 100 MeV energies (Figure 5).

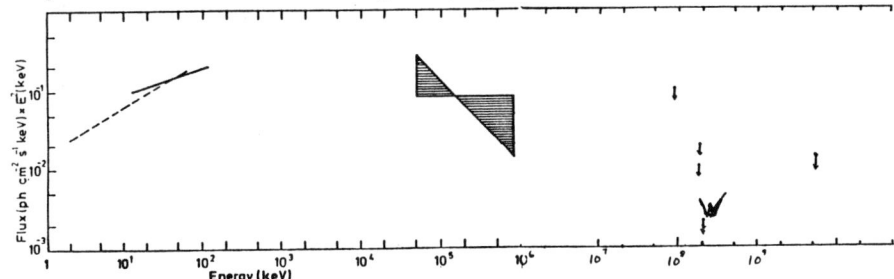

Figure 5. The power spectrum of 3C273 per decade of energy in x-rays and gamma rays showing that it peaks at 100 MeV energies; the upper limit marked W is from the Whipple Observatory[21].

Cygnus X-3.
This optically obscured source is unusual in that it is highly variable at all wavelengths; the emission of large radio outbursts indicates on-going emission of relativistic particles.
(1) Detections reported by more than 15 groups at energies ranging from 0.3 TeV to 0.5 EeV. Most detections saw the 4.8 hour modulation of the X-ray source.
(2) There is no evidence that the source is a standard candle; it appears that the observations are only consistent if there is variability on time-scales of days to months.
(3) The flat spectrum implies great energy at 1 to 10 PeV; this could be the most powerful source in the galaxy (Figure 6).
(4) There have been claims for the detection of short periodicities (at 9.22 and 12.59 msec) at TeV energies; these observations have not yet been confirmed.

Hercules X-1.
This is perhaps the most interesting binary detected because so much is known about the system from x-ray and optical measurements. Six groups have claimed to see Hercules in TeV and 0.1 PeV gamma rays. The strongest detections are:
(1) Discovery observation by the Durham group[13].
(2) Simultaneous observation of a 30 minute transient by the Whipple and Durham groups[14,15].
(3) Observation of transient emission in the spring/summer of 1986 by the Haleakala, Whipple and Los Alamos groups[16,17,18] (Figure 7) ; in each case the period of the emission was blueshifted from the x-ray period by 0.16%. The muon-to-electron ratio in the showers in the Los Alamos observations was not consistent with electromagnetic cascades.

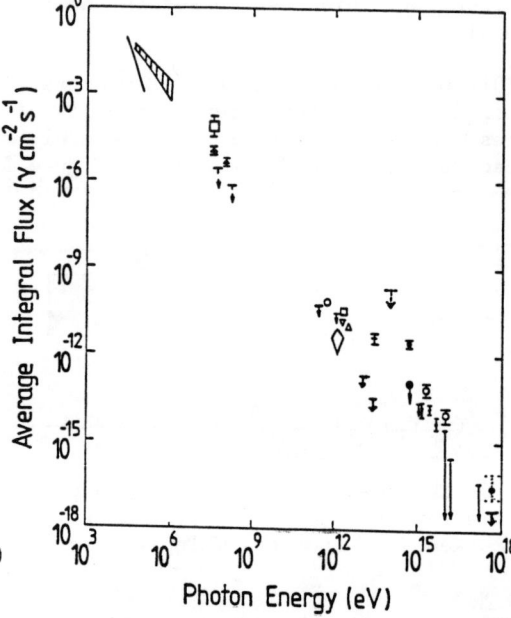

Figure 6. The energy spectrum of Cygnus X-3 showing the wide range of detections and upper limits.

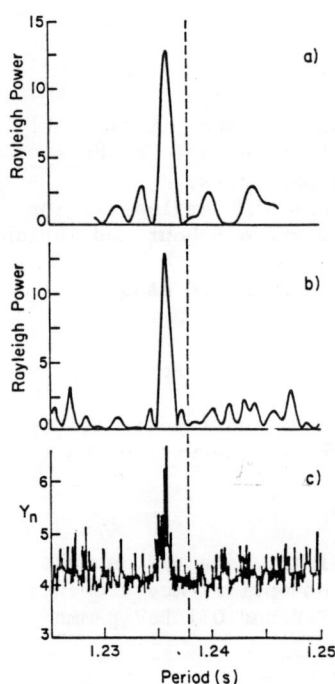

Figure 7. Periodograms from three observations of Hercules X-1 in 1986. Note that the observed period is offset from the x-ray period (dotted line) by the same amount.
(a) Haleakala observation: 86/05/13; 0.4 TeV; 15 min;
(b) Whipple observation: 86/06/11; 0.6 TeV; 25 min;
(c) Los Alamos observation: 86/07/24; 200 TeV; 30 min (x 2).

FUTURE REQUIREMENTS

For the discipline to make progress there are a few obvious requirements:
(1) Telescopes with greater sensitivity at all energies to provide deeper detections; at PeV energies this need will be addressed by the CASA project at Utah[19]; at TeV energies it could be satisfied by the GRANITE project[20];
(2) Telescopes that measure more shower parameters so that the nature of the primary can be ascertained; it is no longer satisfactory to measure just a directional or temporal anomaly. Again CASA (with associated muon detectors) and GRANITE (with high resolution imaging) will answer these needs;
(3) A greater number and distribution of telescopes to provide continuous coverage of variable sources such as the close binaries; there is a very definite need for a sensitive telescope at high negative latitudes to give good coverage of sources at high negative declinations;
(4) More conservatism amongst experimenters in claiming the detection of new sources; more scepticism amongst theorists in accepting such claims.

ACKNOWLEDGEMENTS

Research in VHE gamma-ray astronomy at the Whipple Observatory is supported by the U.S.Department of Energy and the Smithsonian Scholarly Studies Fund.

REFERENCES

[1] Weekes, T.C. 1987. Physics Reports. 160, 1.
[2] Weekes, T.C. et al. 1989. Ap. J. (in press).
[3] Lang, M.J. et al. 1989. Proc. Arkansas Workshop on Gamma-ray and Neutrino Astronomy, Little Rock, 10-13 May, 1989 (in press).
[6] Protheroe, R.J. in Proceedings of 20th International Cosmic Ray Conference, Moscow. August 2-15, 1987. 8, 21.
[7] Nagle, D.E. & T.K.Gaisser, R.J.Protheroe. 1988. Ann. Rev. Nuclear and Particle Sci. 38, 609.
[8] Cassiday, G.L. et al. 1989. Phys. Rev. Lett. (in press).
[9] Lawrence, et al. 1989. Univ. of Leeds preprint.
[10] Bonnet-bidaud,J.M. and Chardin,G.,1989. Phys. Rep.
[11] Bond, I.A. et al. 1989. Preprint.
[12] Akerlof, C.W. et al. 1989. Proc. Workshop on the Gamma Ray Observatory, Goddard, April 10-12, 1989 (in press).
[13] Dowthwaite, J.C. et al. 1984. Nature, 309, 691.
[14] Gorham, P.W. et al. 1986. Ap. J. 309, 114.
[15] Chadwick, P.M. et al. 1986. Proc. NATO Workshop on VHE gamma-ray Astronomy (Durham,1986), 121.
[16] Resvanis, L.K. et al. 1988. Astrophys. J. 328,L9.
[17] Lamb, R.C. et al. 1988. Astrophys. J. 328, L13.
[18] Dingus, B.L. et al. 1988. Phys. Rev. Lett. 61 (17):1906.
[19] Gibbs, K.G. et al. 1987. Nucl. Inst. and Meth. A264, 67
[20] Akerlof, C.W. et al. 1989. Proc. Arkansas Workshop on Gamma-ray and Neutrino Astronomy, Little Rock, 10-13 May, 1989 (in press).
[21] Vacanti, G. et al. 1989. Proc. Workshop on the Gamma Ray Observatory, Goddard, April 10-12, 1989 (in press).

THE SOUTH POLE AS A SITE FOR MONITORING 100 TeV COSMIC GAMMA RAYS BY MEANS OF AN AIR SHOWER ARRAY

A. M. Hillas
Physics Department,
University of Leeds, Leeds LS2 9JT, UK

ABSTRACT

The South Pole has special advantages as a site for monitoring spasmodic or weak periodic emission of ultra high energy gamma rays from many X-ray binaries, since those sources which are within view can be seen continuously with only slowly varying conditions of attenuation, and the very high altitude favours a high counting rate. It is well placed to view the large concentration of X-ray sources at southern declinations. It has proved possible to operate an air shower array efficiently in this hostile environment. Its view overlaps that from other Southern Hemisphere stations, but with higher possible exposure.

1. THE NEED FOR EXTENSIVE MONITORING OF POTENTIAL GAMMA-RAY SOURCES

The reported observation of 10^{16} eV gamma-radiation from Cygnus X-3 and some related objects is one of the most remarkable developments in high-energy astrophysics, as it implies the presence of a source of charged particles of energies up to 10^{17} eV, in such numbers as to be directly relevant to the origin of PeV cosmic radiation in the Galaxy. However, there appear to be a number of other weaker sources of TeV or PeV gamma-rays, suggesting that emission of such radiation is a common property of close binary systems containing a neutron star. Observations of other interacting neutron stars are needed to distinguish the accidental features (e.g. is the orbital phase of emission an accident of orientation or not?) from the essential ones, and to determine what features are necessary to produce acceleration (high or low magnetic field, fast or slow spin, accretion disc or accretion shock?).

The problem of seeing the pattern of emission is complicated by the fact that gamma rays are only emitted in short bursts, in some cases being restricted to certain orbital phases, but with further long-term factors causing considerable variation in output. Some of these variations are irregular, possibly related to episodic or unstable accretion.

We do not understand these sources, and they are sporadic: hence, in order to find the patterns underlying the emissions, there is a need for continuous monitoring of a large number of potential sources. Cerenkov techniques are not suitable for this purpose as the detectors have a small field of view and a limited duty ratio, so an air shower array is the appropriate detector: it might view all sources within about 45° of the zenith at the same time. The energy threshold should be reduced below 10^{14} eV to increase the counting rate and to bring more sources into view.

2. ADVANTAGES OF THE SOUTH POLE FOR MONITORING SHOWERS FROM X-RAY SOURCES

(i) *From the South Pole one can see sources* (in the far Southern sky) *for 24 hours per day, and at a constant altitude.* The latter condition means that, apart from modest barometric changes, there are no corrections for varying energy threshold due to changing atmospheric thickness, which should permit useful observations out to 50° from the zenith, bringing Vela X-1 into view. At most other sites, a typical source is

© 1989 American Institute of Physics

high enough for observation for about 6 hours per day for particle arrays, and Cerenkov detectors have a smaller duty ratio. The advantage of full-time observation will be displayed quantitatively below (Fig. 3).

(ii) *There is a concentration of potential gamma-ray sources - i.e. X-ray binaries - at high Southern declinations.* Figure 1 shows the declination distribution of identified compact X-ray sources. (This is not a critical compilation, but simply taken from the tabulation (1).) The marked "gamma sources" do not include sources mentioned recently in preprints from the Durham group. One is here only following the most plausible indicator in taking the X-ray binaries as a whole as tracers of the expected sky distribution of u.h.e. gamma-ray sources. SN1987a, pulsars, and unidentified and extragalactic sources are not shown.

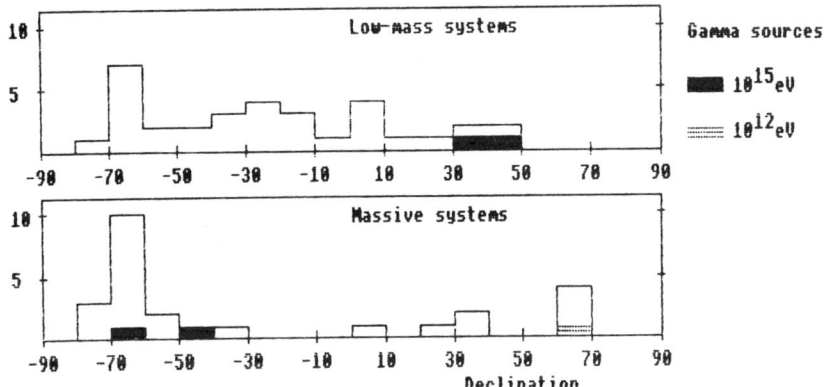

Fig. 1. Declination distribution of identified compact X-ray sources - tracers of the expected sky distribution of gamma-ray binaries.

If all X-ray binaries are regarded as potential sources, this diagram shows that their distribution on the sky peaks near 60° S, so an observing site far south has an advantage. In addition to the broad Galactic peak, well viewed from Australia, there is a narrow peak to which the Magellanic Clouds make a contribution. These are more distant sources, of course, but still intense in the X-ray band.

(iii) *Because of the very high altitude, one can readily observe showers to well below 100 TeV with a shower array.* Consequently, (a) one can have a high counting rate (necessary to detect short bursts of emission), and (b) one avoids the energy range where radiation from the Magellanic Clouds would be seriously attenuated.

Many of the X-ray binaries visible from the Southern sky, and also SN1987a, are in the Magellanic clouds, so any gamma rays they emit may be attenuated en route by interactions with microwave photons. Figure 2 shows the fraction of photons transmitted without such interaction over various path lengths. In 50 kpc there is serious attenuation above 300 TeV, but very little below 200 TeV. (Interactions with the 2.7° primeval radiation and with the far-IR flux (2) have been considered. This total ambient photon flux is well represented by the sum of a 2.35° Planck spectrum plus a dilute Planck spectrum of 4.0° and saturation 15%.) It would thus be desirable for the median energy to which the detector array responds to be less than 250 TeV, with an effective threshold well below 100 TeV.

The SPASE (South Pole Air Shower Experiment) array (3,4) has been operating at the Pole since the start of 1988, and the energy distribution of gamma rays that trigger it and coming from a distance of 50 kpc is estimated to be as sketched in the dashed curve (if only showers falling within the array are included). The median energy is about 90 TeV, and the response is mainly below the energy range where absorption is serious.

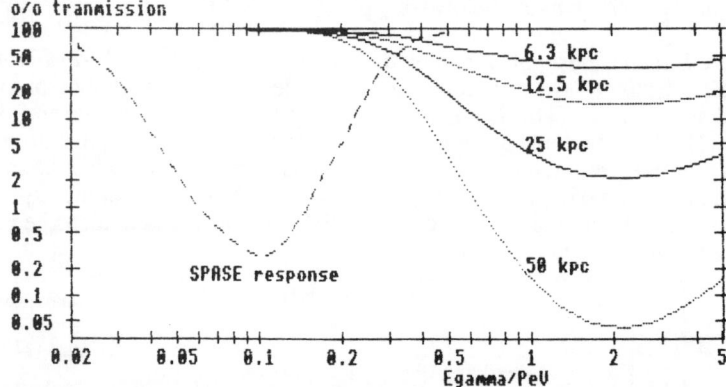

Fig 2. (a) Transmission of high-energy photons through a universal ambient photon field consisting of the primeval 2.7° photons together with a far IR flux as found by Matsumoto et al.. Path lengths 6.3 kpc to 50 kpc are shown. (b) Dashed line: estimated energy distribution of gamma rays from the LMC that would be recorded by the SPASE array (plotted downwards for ease of comparison, and on an arbitrary but linear vertical scale).

The array has 14 scintillators (each 1 m^2) spaced on a 30m hexagonal grid, and was designed as the simplest air shower array that could detect showers below 250 TeV at a high rate and measure their arrival directions accurately, so as to pick out point sources against an isotropic background of ordinary showers. Figure 3 represents an attempt to show the sensitivity of the SPASE array to sources in different declination ranges (curve labelled 90). The number of showers which trigger the SPASE array giving signals above about 1.5 particles in at least 6 detectors and having centroids within a

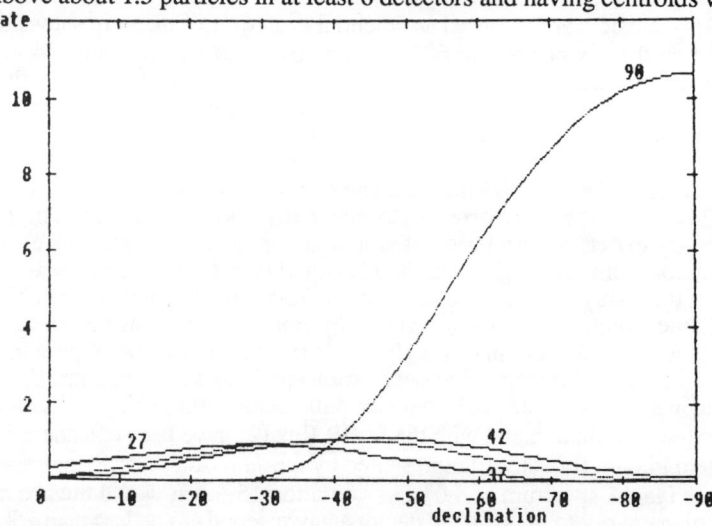

Fig. 3. Comparison of rate at which usable showers would be detected by the same array (as in SPASE) situated at the South Pole (S latitude 90), and at mountain altitudes (taken as 1800m) in New Zealand (S latitude 42), Australia(37) and South Africa (27). The rate of showers per m^{2}day per unit solid angle around a direction of interest is plotted versus declination of the source.

well-defined area inside the array has been counted as a function of declination of shower arrival direction, and the rate per unit ground area per day per unit solid angle is plotted. The zenith angle distribution will not be identical for showers due to gamma rays, but is not expected to differ very much from this. For comparison, one can calculate the number of showers per day that would be recorded from a particular area of sky by an identical array displaced to different geographical latitudes, assuming that the shower rate as a function of zenith angle remains the same if the altitude is the same. Non-polar observers generally see a given area of sky for a shorter length of time each day, before the source drops too low in the sky for appreciable counting rate - assumed to be where the counting rate falls below a quarter of the rate obtained on the meridian, though the results are very insensitive to this limit. But most practicable mountain sites are at a lower altitude than the Pole (where the atmospheric overlay is 695 g cm^{-2}), say 1800m rather than 2800m, so the effect of attenuation in an additional 1000m of air equivalent to 117 g cm^{-2} (vertically) has been incorporated, using a rate attenuation length of 230 g cm^{-2} (as is seen in the near-vertical SPASE trigger rate). This very roughly halves the rates. For the same number of detectors and area of array the polar site clearly has a great advantage for monitoring sources at a high southern declination, with a crossover point near a declination of -40°. Of course the size of the array can be altered; the shower rate will be nearly proportional to the number of detectors, for a given spacing.

As most of the difference arises from the 24 hour viewing per day at the Pole, as against perhaps 6 hours from the comparison site, the Pole accumulates more events overall per year; but during a burst the advantage will be smaller.

(iv) *The geographical position and continuous monitoring ensures that any source under observation in New Zealand, Australia or South Africa can be seen simultaneously at the Pole, for sources of declination from -40° southward.* Some sources can generate marginally detectable bursts of activity at a few hundred TeV lasting for less than an hour, according to reports from the Fly's Eye and CYGNUS experiments, and TeV bursts can be briefer. Since Australian and South African detectors view different longitudes of sky, and observing schedules differ, mutual confirmations will be rare. The SPASE array should thus play a very important role, especially as simultaneous observation at TeV and PeV energies is another key requirement in this field; thus it would be beneficial to increase the counting rate of the SPASE array to enhance the prospects for coincident detection in conjunction with the Cerenkov experiments of Adelaide and Potchefstroom.

3. ONE NON-OPTIMUM CHARACTERISTIC OF A POLAR ARRAY

For observing specific sources at particularly favourable declinations, it is possible that a much greater sensitivity may be obtained near 100 TeV by using the Cerenkov technique to observe the source when about 20° above the horizon, as suggested by Sommers and Elbert (5) and attempted at JANZOS (6). This does not give continuous monitoring, but may in suitable cases provide the greatest sensitivity to low fluxes.

REFERENCES

1. H V D Bradt & J E McClintock (1983) *Ann. Rev. Astron. Astrophys.* **21**: 13-66.
2. T Matsumoto et al. (1988) *Astrophys. J.* **329**: 567-71
3. N J T Smith et al. (1989) *Nucl. Inst. & Meth. in Physics Res.,* **276**: 622-7
4. J. C. Perrett; N. J. T. Smith; A. A. Watson: these proceedings.
5. P Sommers & J W Elbert (1985), *19th Int. Conf. on Cosmic Rays, La Jolla,* **3**: 457-60
6. I A Bond et al. (1989) JANZOS group preprint.

EXPERIMENTAL TEST OF THE ANOMALOUS DENSITY SPECTRUM OF COSMIC RAY SHOWERS, AS AN ADJUNCT TO THE SOUTH POLE AIR SHOWER EXPERIMENT?

A. M. Hillas
Physics Department,
University of Leeds, Leeds LS2 9JT, UK

ABSTRACT

Several very simple experiments on air showers, performed 20-25 years ago, gave indications of an unexpected behaviour of the shower core. If the particle density on a single small area was recorded, without regard to its position in the shower, its spectrum exhibited a sharp knee, apparently related, though not in a simple way, to the knee in the shower size spectrum. The position of this knee changed rapidly with altitude in a manner which implies that the primary particles have a very short interaction mean free path. To check the correctness of these indications that heavy nuclei may dominate the spectrum here and may generate unexpectedly dense and narrow showers at PeV energies, an experiment is being developed to measure this density spectrum with much better statistics than was done in the early work, at mountain altitudes, under various thicknesses of atmosphere. The location and nature of the SPASE array offers the possibility to test this phenomenon with a simple addition to the existing array.

1. THE ANOMALOUS DENSITY SPECTRUM OF AIR SHOWERS

It is well known that the spectrum of shower sizes shows a steepening or "knee" in the PeV region. It was not expected, though, that a measurement of the local particle density in a single small area would give much useful information about spectra, because occurrences of the same particle density can belong to showers of a very wide range of size, centred at different distances from the detector. One would expect to find any knee in the density spectrum very greatly blurred as a consequence. However, this is not what was found in several experiments in the 1950s and 1960s (1-5). There appeared to be a sudden change in the slope of the density spectrum, more pronounced than that seen in the shower size spectrum, and not a smoothed-out version of the latter. In experiments on Sulphur Mountain, Canada (5) two nearby cloud chambers were used, and the observations imply that in the case of events near the knee of the density spectrum, the detector used to trigger on a high density was typically within about 2m of the shower core: the second chamber showed a much smaller density than the one used for triggering. Hence a collection area and primary particle flux can be estimated, showing that the anomalous knee is indeed related to the knee in the shower size spectrum, though much more pronounced.

The knee occurred at much higher densities at the higher altitudes. Figure 1 shows an example of the variation of density spectrum with altitude found from experiments using proportional counters (2), together with arbitrary interpolated curves indicating the remarkable form of the spectra which will be discussed. The change in air density with altitude complicates matters: one has to make allowances for its effect on the particle densities and on the collection area for shower cores (7). The experiments by McCusker, McCaughan et al. using cloud chambers show very similar behaviour.

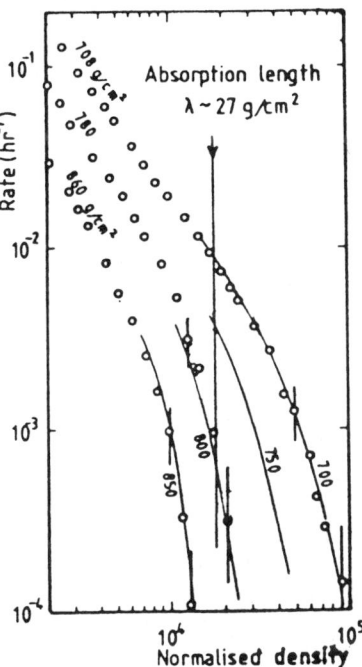

Fig. 1. Density spectra from proportional couters at various atmospheric depths. Densities are normalized to 780 g cm^{-2}. (Taken from 7)

Although not remarked on at the time, a striking feature of these spectra is the extraordinarily rapid fall in shower rate with increasing altitude (for fixed normalized density), at densities just above the knee. But the rate of showers of given magnitude cannot fall with a shorter absorption length than the mean free path of the primary particle in the atmosphere, since a particle interacting unusually deep can generate its shower displaced downwards in the atmosphere by a corresponding amount. Calculations had at that time indicated that, at a given shower size, proton showers would have much denser cores: heavy nuclei by contrast would generate a number of lesser sub-cores due to their individual nucleons: the spread of these superimposed sub-cores would lead to a less dense but broader shower core. Hence, if one selected showers by their particle density, one would expect to be picking out the proton primaries. So the evidence of a very short mean free path may perhaps indicate that there are extremely few protons in the PeV region (or else that protons fail to make the expected kind of interaction) (7), and that collisions of heavy nuclei produce denser particle jets than had been expected. Such an effect might arise from quark-gluon plasma formation, though it would be surprising if this happened in the majority of collisions - rather than in central impacts.

2. PROPOSED RE-INVESTIGATION OF THE DENSITY SPECTRUM AT HIGH ALTITUDES

The dual cloud chamber experiments analyzed particularly by McCaughan showed that the triggering detector became less close to the shower core at sea level, and the characteristics of the showers changed and were less striking. Hence it is necessary to make the observations at high mountain altitude.

If it is true that the high density events are generally within a few metres of the shower axis, it is possible to have detectors about 10m apart which will give almost independent measurements of the rate of events as a function of density, at least in the high-density range, and hence it is intended to gather much better statistics from a year's run than was obtainable in the pioneering experiments. By using thin Cerenkov counters one can avoid the possibility of mistaking the heavy ionization of a nuclear disintegration for a high particle density.

This experiment uses very small detectors and is easily portable. Where should it be operated? Initial plans include operation at different altitudes on Mt. Hopkins, in collaboration with the Whipple Observatory and Dublin groups, but because of the necessity to make corrections for differences in local air density when comparing rates seen under different air thicknesses, the interpretation will be clearest if one has the additional information of the angle of arrival of each shower, classifying individual showers by the slant air thickness. The local air density will then be the same for the different air thicknesses which are obtained simply from different zenith angles. If the detectors were to be operated within the boundary of the SPASE air shower array at the South Pole, showers near the knee of the spectrum would be certain to trigger the SPASE array also, and this would give shower directions accurate to better than 1°, and hence the slant air thicknesses, ranging from 695 g cm^{-2} vertically to 980 g cm^{-2} at 45°. The set of detectors used for density measurements would not give directions, nor would it locate shower cores by a density fitting process: it would only operate as an array in so far as it would check that each high local density was indeed accompanied by some kind of shower. (The detector areas have to be less than 0.04 m^2 in order to relate to the earlier observations - much too small to allow core location even if this complication were desirable.)

The typical low air temperatures of -80°C to -20°C, with consequent short Moliere unit for shower spread, should lead to much higher particle densities than would be found at the same altitude elsewhere. The comparison of spectra with those found at warmer sites could serve to check the altitude in the shower which determines the spread of the core. This will be very local if the hadronic backbone is very narrow and Coulomb scattering of electrons shapes the core, but would depend on temperatures somewhat higher up if one does have a bundle of separate nucleon showers.

3. SUMMARY: AN ADJUNCT TO SPASE?

If this auxiliary experiment were added to the SPASE array, with a data cross-link to correlate events, it would..
  i) involve only about 10 very small, light, detectors, not requiring fast timing,
  ii) not require a high data recording rate,
  iii) provide a check on accuracy of core location in the SPASE array,
  iv) give statistics much better than the previous experiments within three months,
  v) provide a slant-depth-attenuation measurement of the rate of occurrence of a fixed particle density made at fixed air density - much easier to interpret than the vertical displacement attenuation, and especially helpful if the latter is then determined later.

This is partly an "astrophysics" experiment, in so far as it tests the prominence of primary particles of very high cross-section in the cosmic radiation near the "knee", but it is also a pilot experiment in PeV physics, to show whether elaborate experiments to follow up the earlier results would be justified. (There is indeed evidence that shower cores in the PeV region have an unexpected structure, from the discharge experiments of A L Hodson's group, but with little indication of what underlies the unusual spreading that is observed.)

The experiment uses extremely simple apparatus, which is very powerful only in this context, where an array exists that gives shower arrival direction very accurately at this very high altitude, and operating in very stable conditions. The simplicity of the observation allows one to concentrate on a simple question of mean free path, undeterred by the complexity of shower cores. A one-year exposure is envisaged.

REFERENCES

1. R J Norman (1956) *Proc. Phys. Soc. A*, **69**: 804-20
2. R J O Reid et al. (1961) *Proc. Phys. Soc.* **78**: 103-12
3. D B Swinson and J R Prescott (1965) *9th Int. Conf. on Cosmic Rays, London*, **2**: 721-3
4. J B T McCaughan et al. (1965) *Proc. 9th Int. Conf. on Cosmic Rays, London*, **2**: 720.
5. J B T McCaughan (1973) Ph. D. Thesis, University of Sydney
6. J B T McCaughan (1982) *J. Phys. G, Nucl. Phys*, **8**: 413-32
7. A M Hillas (1981) *18th Int. Conf on Cosmic Rays, Paris*, **2**: 125-8.

DESIGN CONSIDERATIONS FOR A TEV TELESCOPE AT THE SOUTH POLE

K.Harris[1], M.A.Pomerantz[2], P.T.Reynolds[1], G.Vacanti[3], A.Walker[2], T.C.Weekes[1].*

1 Whipple Observatory, Box 97, Amado AZ 85645
2 Bartol Research Institute, University of Delaware, Newark. DE
3 Iowa State University, Ames, IA 50011

ABSTRACT

Some of the factors that should be considered in the design of atmospheric Cherenkov telescopes for use in gamma-ray astronomy at the South Pole are discussed.

INTRODUCTION

The South Pole offers an unusual challenge to the gamma-ray astronomer using the atmospheric Cherenkov technique. Any optical technique in this harsh environment faces problems. There are sufficient advantages to operating from this site that it is worth overcoming the various obstacles to establishing an operating telescope.

The gamma-ray astronomer working at PeV energies uses air shower arrays with enclosed detectors; the advantages of working at this site have been documented[1] and they include: (a) high elevation; (b) constant source elevation angle; (c) visibility of high negative declination sources. The TeV astronomer using an optical technique is normally restricted to limited observations under dark sky conditions; at the South Pole there is the added advantage of unusually clear skies and the possibility of uninterrupted observations over two week periods. This is especially important for overlapping observations with other southern hemisphere observatories and with planned space missions such as the Gamma Ray Observatory.

However the optical technique does face some problems not encountered by the higher energy air shower technique. Although manmade light pollution is minimal (in principle at least) there is a troublesome background of the aurora. In addition the optical detectors must be exposed in a harsh environment and must be suitably protected. For a telescope to be really efficient it must track the source and this presents the problem of operating a moving mount at one of the coldest sites in the world. In addition the problem of maintaining flexibility in coaxial cables at these temperatures must be solved.

PILOT EXPERIMENT

Since there is only limited information about the operation of optical telescopes during the Antarctic night at the South Pole and

----

* Paper presented by T.C.Weekes

phototubes if the current reaches catastrophic levels. Every hour a pulse derived from the SPASE clock attempts to reset the relays to the next highest level i.e. to restore the system to its normal operating voltage.

The phototube current from both channels is recorded on a two-pen Rustrak chart recorder which is operated continuously. An analog signal derived from the phototube current is also recorded every two minutes on the Apple. Under normal conditions the telescope is inclined from the zenith by 21°.

Results: at the time of the workshop the telescope was only coming into operation. Although redundancy had been planned for most component parts the High Voltage Supply was not duplicated and this was the first item to fail! Fortunately a spare supply was available from the SPASE experiment.

A problem that was less easily solved was that of blowing snow accumulating on the front glass window. The Winter Over operator (A.W.) devised a heated Plexiglass window which, with 200 watts of power distributed by a fan, kept the front window clear. It was noted that blowing snow adheres to glass surfaces much more readily than to plexiglass. The interior optics (lens, filter, phototube) which were at the ambient temperature did not show signs of frosting. With an aperture of 2° the shower counting rate was about 10 per minute with less than 10% random contribution. The ambient current in the unfiltered tubes was on average almost twice that recorded at the Whipple Observatory in Arizona.

Based on our limited experience with the operation of this system we offer the following maxims for atmospheric Cherenkov telescope design:
(1) optics must be protected from the environment both from radiative cooling and from direct exposure to blowing snow;
(2) the telescope design should be kept as simple as possible; the South Pole is not the place for trying new techniques;
(3) the number of component parts should be minimized with emphasis on redundancy;
(4) all optical detectors suffer from the disadvantage that the telescope cannot be tested in situ before the end of the summer and the W/O is left to face the inevitable first turn-on problems alone;
(5) although the high elevation and clear skies offer the promise of a low energy threshold this is offset by the window over the optics and the use of filters to avoid the principal aurora bands;
(6) the circumpolar paths of all the sources offers the principal advantage for TeV telescopes; to capitalize on this feature it would be best to have multiple small telescopes rather than one large one;
(7) the observing targets of choice are the many binary sources at high negative declinations.

TELESCOPE DESIGN

Below we offer a simple concept for a working multiple atmospheric Cherenkov telescope design. The basic concept is to have three simple telescopes on a single mount (Figure 2). Each telescope would consist of two 1.5 m aperture searchlight mirrors (rhodium on

no experience at all with the operation of an atmospheric Cherenkov telescope it was decided to operate a pilot experiment during the austral winter of 1989. The purpose of the P-SPACE (Pilot-South Pole Atmospheric Cherenkov Experiment) was fourfold:
1. To demonstrate the feasibility of operating an atmospheric Cherenkov telescope at the South Pole; this was to be accomplished by a few hours of successful operation in the pulse counting mode.
2. To amass some qualitative information on the auroral background at the South Pole as it pertains to a Cherenkov detector.
3. To monitor the supernova SN1987A at 10 TeV energies.
4. To verify the pointing of the SPASE array[2] by comparing events that trigger the two systems.

P-SPACE consists of two parallel detectors which are operated in coincidence (Figure 1). Each detector is a 5 cm diameter phototube at the focus of a simple plastic Fresnel lens (aperture 40 cm by 50 cm). The field of view is 3.5 degrees approximately. Above the Fresnel lens is a heated glass window. The analog signals are taken to the control hut where they are separated into their fast pulsed components and their slow d.c. component.

The fast signal is taken through amplifiers to a trigger. The outputs of the discriminators are taken to two coincidence units; one of these has a delay line so that the coincidence unit monitors Random coincidences. The resolving time of the coincidence units is approximately 15 ns; the threshold of the discriminator is adjusted to give triggering rates in each channel of about 500-1000Hz. Every two minutes (sidereal time) the coincidence counts and singles rates are recorded on floppy disk by an Apple computer; each disk has a capacity for 48 hours of data. Each data record consists of the Date, Universal time, Sidereal time, Singles Rates (2), Prompt and Random Coincidence Rates, Analog signals (2).

Fig. 1. P-SPACE shown with one glass window removed.

The slow d.c. signal is used to control the high voltage in the event of a brightening of the general night-sky background due to the aurora (or some other source (moon, man-made, meteor, dawn, etc.)): however the meter relay is set so that it is not triggered by starlight or the Magellanic Clouds. When this relay is triggered it causes a drop in the voltage to both tubes so that the current falls by about 60% and the pulse counting rate falls to about 50% of its original value. A second meter relay is triggered if a higher level of brightness is encountered even with this reduced voltage. The current and counting rate fall by a similar factor. Finally a third meter relay will turn off the system H.V. supply to the

brass) with 5 cm diameter phototubes at their foci. All three
telescopes would be on a common mount with azimuthal drive--
essentially a carousel. Three telescopes would be manually
adjustable in declination; they could be adjusted in right ascension
(azimuth) over a limited range. By selecting three sources that are
approximately eight hours apart in right ascension, it would be

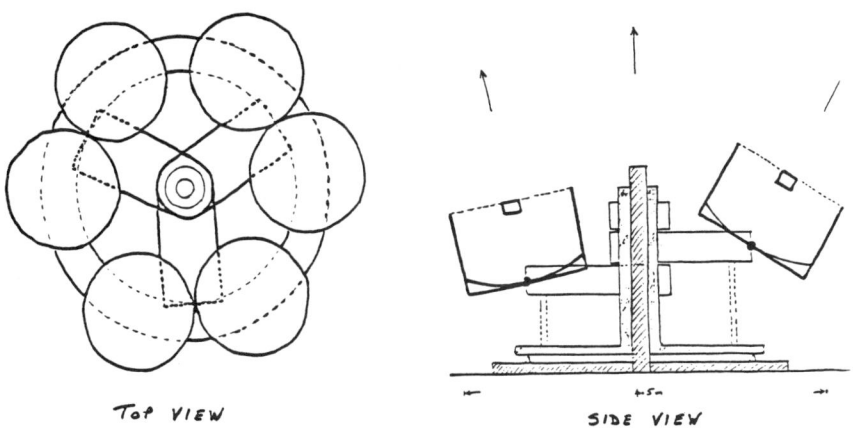

Figure 2. The South Pole carousel telescope.

possible to simultaneously monitor all three. Generally the same
three sources would be monitored for most of the observing season so
that mount adjustments would be minimal. The azimuth drive would be
at the sidereal rate and a simple star tracker would fine tune and
verify the pointing. Every 24 hours the telescope would be slewed
back through 360 degrees. Each telescope would be covered by a
heated front window of uv transmitting plastic and the entire optics
would be sealed and kept dry with a desiccant. The field of view
would be 2 degrees and the energy threshold would be about 2-3 TeV.
The counting rate would be 1 Hz. One central remote data-recording
system would be used. To minimize cable runs and provide easy access
to the mount control the carousel should be mounted on the roof of
the Control Building whose excess heat would also heat the mount
drives and the cable duct. The mount motors would be inside the
building and accessible from inside.

ACKNOWLEDGEMENTS

We are grateful to members of the SPASE collaboration for their
assistance in setting up the experiment and to the Leeds group
(Michael Hillas, John McMillan, Alan Watson) for the environmental
testing of optical components of P-SPACE. We acknowledge the
assistance of Lynn Gray, Teresa Lappin and Dave Martinez in the
construction of the telescope.

REFERENCES
1.   A.M.Hillas, in: Proc. Workshop on HE-UHE Behavior of Accreting
X-ray Sources (Vulcano,1986) in press.
2.   T.K.Gaisser et al., 1989, Phys. Rev. Lett. **62**, 1425.

A SOUTH POLE FACILITY TO OBSERVE VERY HIGH ENERGY GAMMA RAY SOURCES[*]

Bartol, Purdue, Smithsonian, Wisconsin Collaboration[**]

ABSTRACT

We plan to construct an atmospheric Cherenkov telescope (ACT) at the south pole to study pulsars and other possible sources of very high energy gamma rays. It is believed that a modest number of such sources may account for the bulk of the cosmic radiation. The ACT is the instrument of choice for gamma ray study in the very high-energy (VHE) range of ($10^{11} < E < 10^{14}$ eV). A prototype instrument is scheduled for installation at the pole during the 1989-90 austral summer. The south pole is an ideal location in several respects: (i) the long polar night permits almost continuous observation of a source, up to 400 hours per month, and up to 1700 hours/year; (ii) there is a particularly large concentration of interesting candidate sources at high southern declinations; including the remnant of supernova SN1987a, and (iii) the circumpolarity of the sky simplifies data analysis; (iv) the ACT would also operate in conjunction with the South Polar Air Shower Experiment (SPASE) to provide a comprehensive cosmic gamma ray facility.

INTRODUCTION

The discovery of 10^{16} eV gamma radiation from Cygnus X-3[1] implies the existence of mechanisms which can accelerate particles to energies of 10^{17} eV. Hillas[2] has proposed that a few such sources could supply all of the cosmic rays in the galaxy. Observations of very high energy gamma radiation have been reported from enough X-ray binary systems to suggest that this radiation is a major component of the binaries' emissions.[3] The discovery of 10^{12} eV gamma radiation with the characteristic period of well known X-ray binaries such as Hercules X-1,[4,5,6] and the measurement of the anomalous Hercules period,[7,8,9] now suggests ACT methods are a viable way to study these objects. Making precise pulsar period and energy spectrum measurements will allow a more complete understanding of their VHE emission mechanisms.

The sources of primary interest for gamma ray astronomers are pulsars and accreting X-ray binaries. The southern sky, especially at near-polar declinations is particularly rich in potential VHE gamma ray sources.[10] The galactic center, and a majority of known X-ray binaries and young pulsars, are southern objects, as is the extragalactic source Centaurus-A.[11,12] Several binaries, including some black hole candidates, are located in the Centaurus region and in the Magellanic Clouds.[13,14] We intend to monitor

---

[*] Summary of papers presented by R. Morse, University of Wisconsin and J. Gaidos, Purdue University at the Astrophysics in Antarctica Conference , Bartol Institute, June 8, 1989.
[**] Collaboration members are: M. Pomerantz, Bartol; J. Gaidos, F. Loeffler, G. Sembroski, C. Wilson, G. Zirnstein, Purdue Univ.; P. Slane, T.C. Weekes, Smithsonian; U. Camerini, M. Frankowski, W. F. Fry, F. Halzen, J. Jacobsen, M. Jaworski, A. Kenter, R. March, R. Morse, M. Skinner, Univ. of Wisconsin.

all of these objects. The southern sky also contains the remnant of SN1987a; this will probably be the object of our most intensive observation.

Supernova remnants are believed to be the principal source of cosmic rays. A typical remnant of age 10^3 years can only accelerate particles to energies up to 10^{14} eV. It is likely that energies above this level originate from younger remnants, beginning one to three years after formation.[15] A comparison of UHE and VHE observations will provide a crucial test of models of cosmic ray acceleration. The gamma ray luminosity in both regions of the spectrum is expected to change rapidly during the first few years of the remnants' existence, so it is important to have observations that are as continuous as possible. The two most southerly ACTs now operating are restricted to about 100 hours a year on SN1987a, however, no signal has yet been reported. A polar ACT would be able to observe this source for 1000 to 1500 hours per year.

SOUTH POLE SITE

The south pole is an ideal location for such an instrument in several respects: (i) the long polar nights permit almost continuous observation, up to 400 hours a month per source, limited only by moonrise; (ii) more total observation time each year, about 1700 hours or twice the viewing possible by ACT devices located at other latitudes; (iii) more hours per year on a specific source, about 1700 hours, or ten times the viewing time available at other latitudes since the source never passes out of sight; (iv) the large concentration of interesting candidate sources at high southern declinations; and (v) the constant zenith angle of a source observed from the south pole greatly simplifies the data analysis.

An ACT device located in the temperate or tropical zone can usually average about 80-90 hours a month or about 900-1000 hours of viewing a year. An ACT device is usually restricted to viewing a particular source between $\pm 45°$ of zenith. Thus, at non-polar sites one never gets more than about 6 hours of observation per evening, even on a completely moonless night. In the 2-3 months a year when the source is visible one can collect about 160 hours of data.

At the pole the sun is greater than $12°$ below the horizon from April 12th through August 20th, a period of 120 days. Night does not occur until May 12th when the sun drops to more than $18°$ below the horizon and continues for 81 days until August 1st. Our ACT experience at Haleakala indicates that we can operate one hour after sunset or at $15°$ solar depression. Using that $15°$ benchmark we can safely operate at the pole for 101 days, from May 2 until August 10.

At the pole the moon is below the horizon for about 14 days. When it appears on the horizon, it is in 1/4 phase or *half-illuminated*. It then moves in a circular fashion, higher and higher for 7 days until it attains its maximum height of $23.5\pm5°$ above the horizon. At this stage the moon is *full*. This height variation is caused by the moon's $\pm 5°$ orbital variations out of the ecliptic plane. The moon then starts to set and 7 days after a full moon the 3rd quarter half-illuminated moon drops below the horizon.

For 4 of those 14 days when the moon is above the horizon, it is below $12°$ and only about half-illuminated. At this low angle its light is considerably scattered so that its small angle spectrum is displaced toward the red. By using PMTs with bialkali photocathodes and filters with a bandpass in the blue near-UV region the ACT viewing can be extended to include these 4 days. This would bring the total number of contiguous viewing days per month to about 18 or about 400 hours.

The moon reduces our viewing efficiency to 60%. For the 100-120 days per year when the sun is far enough below the horizon to operate, the effective viewing time is reduced to 60-72 days or 1400-1700 hours.

VHE observations indicate that all known VHE sources are episodic in their emissions, producing short bursts of radiation of duration from 100 to 1000 seconds, with a duty cycle of no more than a few percent.[7,16,17] It is difficult to detect these outbursts in the short observation periods possible in the tropic and temperate latitudes. The almost continuous observation from the pole, with data runs measured in days rather than hours, will make it easier to detect such episodes.

Another advantage of the south pole is that all sources remain at constant elevation. The threshold and effective target area for an ACT vary with source elevation, an effect that can make it difficult to interpret changes in counting rate as the source moves across the sky. Below $40°$ from the zenith the gradient becomes steep enough to potentially compromise data analysis. With a constant elevation, however, it should be possible to work closer to the horizon.

It should also be noted that the effective altitude of 3200 meters (695 g/cm^2) is near that at which the product of signal strength and effective target area is a maximum. Winter viewing conditions at the pole are reported to be excellent. The average temperature is -70 C, with no moisture in the air and very little wind, snow, or cloudy weather.

The long night at a polar site, and the circumpolarity of the sky, would permit a particularly close coordination of an ACT with the SPASE air shower array[18] which is now operating at that site. They could operate as a common gamma-ray facility covering both the VHE and UHE portions of the gamma spectrum. An ACT could operate almost continuously for up to four months of the year, with as much of this time on one source as desired, while SPASE could provide continuous coverage of the sky within $50°$ of zenith.

If during all-sky observation the SPASE array detects new activity, the ACT could be immediately directed to this new source. The ACT, with its energy threshold of 5×10^{11} eV, nearly 100 times lower than SPASE, will provide higher-statistics data on the source of interest. Finally, by itself, a polar ACT would collect data in a region of the sky that is now not well studied.

INSTRUMENT DESIGN

A detailed discussion of the principles that underlie the design of the ACT for the Haleakala Gamma Observatory are given in the literature.[19] Here, we will summarize these points briefly, and then discuss those modifications that will be required for the polar ACT.

An ACT operates by detecting the Cherenkov light emitted by the relativistic electrons and positrons in the electromagnetic cascade initiated by a cosmic gamma ray incident on the atmosphere. The shower particles need not propagate to the ground, so gammas below the energy threshold for shower arrays are detectable.[19]

Light is collected by a multiple-mirror telescope pointed in the source direction, and detected by photomultipliers. The light signal is spatially diffuse, with a width of roughly a degree, so high-quality optics are not required. But it is extremely time-coherent, so if PMTs and electronics with nanosecond time resolution are employed, the PMTs may be operated at single-photoelectron threshold, using coincidence techniques to reject light from the night sky.

The PMTs will have bialkali photocathodes and UV transmitting glass windows such as Corning 9741, in order to guarantee about a 20% quantum efficiency in the bandpass of 285-380 nm. The blue end of this band is defined by a UV filter chosen to eliminate auroral light as well as reducing the normal night sky signal. In Fig. 1(a) we show the signal, bandpass, and quantum efficiency of the proposed PMT-filter combination.

Auroral researchers[20] have informed us that the shortest-wavelength nitrogen line in this bandpass region that is sufficiently bright to regularly disrupt our observations is at 391.6 nm. The ozone cutoff occurs at 285 nm. Thus a promising spectral window is from 285-380 nm in this region of minimal polar auroral activity. The blue aurora should prove troublesome only 15-20% of the time, most likely at magnetic dawn and dusk. The telescope's very small angular acceptance, about 10^{-4} steradians, and the use of a blue bandpass will further reduces this aurora-induced excess photon counting rate. In Fig. 1(b) we show the bandpass, and the regions of intense airglow and aurora activity.

The energy threshold for this device should be in the vicinity of 5×10^{11} eV. With an angular aperture of $.75^o$, it will view an effective target area of 1.5×10^4 m^2 at the altitude of shower maximum, 10 km above the instrument. It thus has an effective area of about an order of magnitude less than a typical shower array. It also operates below the threshold for strong gamma absorption by pair production on the 3° Kelvin cosmic background, and is thus more suitable for detecting extragalactic sources.[21]

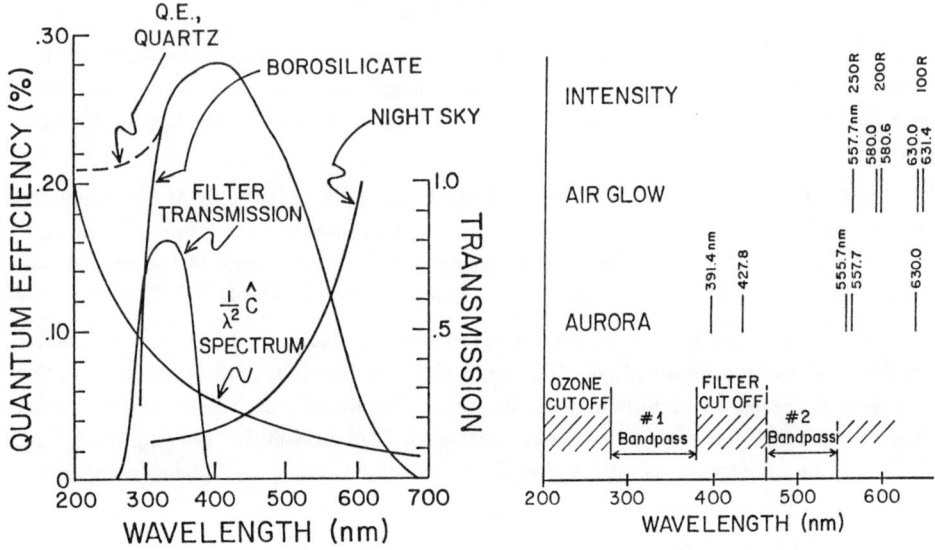

Fig. 1 (a) PMT quantum efficiency, filter transmissivity, Cherenkov spectrum, and night-sky spectrum as a function of wavelength, (b) Wavelengths and strengths of prominent airglow and aurora lines.

One feature unique to our design is the use of two independent, fully instrumented apertures; one aperture observes the *source* while the other monitors the *off-source*, a nearby area of the sky. The off-source aperture provides a concurrent background measurement to compare with the signal.

1. South Pole Telescope

Current plans call for the installation of the final device in austral summer 90-91. Given a reasonably successful prototype test during the austral winter of 1990, we would expect to install a steerable device with 18-22 F/1 mirrors of 0.8 meter diameter mounted on a common altitude-azimuth mount.

It was decided to modify the original 30 multi-mirror design to a smaller multiple mirror design with the possible extension to a second multi-mirror device. These two devices would allow the flexibility of observing multiple sources, or the training of both devices onto a single weak source, while keeping the electronic complexity at the level of the original design. All other aspects of the experiment would remain as proposed.

Light is collected by 0.5 m^2 mirrors and brought to focus on a PMT. Each mirror is fitted with two PMTs each behind its own aperture which has an angular acceptance of about $1°$. The PMTs are separated by about $3°$ and monitor simultaneously the source under study and the background from a nearby portion of the sky.

The entire mirror and PMT system is enclosed in an insulated cylindrical barrel with a front face of UVT plexiglass. The design is driven by the three major problems faced at the pole: snow collection on the mirrors, frost formation on the optical surfaces and stray light. The bottom surface of the mirrors are both heated and insulated to guarantee that heat flows uniformly through the mirror and up to the plexiglass covers. This will keep the mirror temperatures above the dew point, preventing frost and sublimating any snow build-up.[22] The schematic arrangement is shown in Fig. 2.

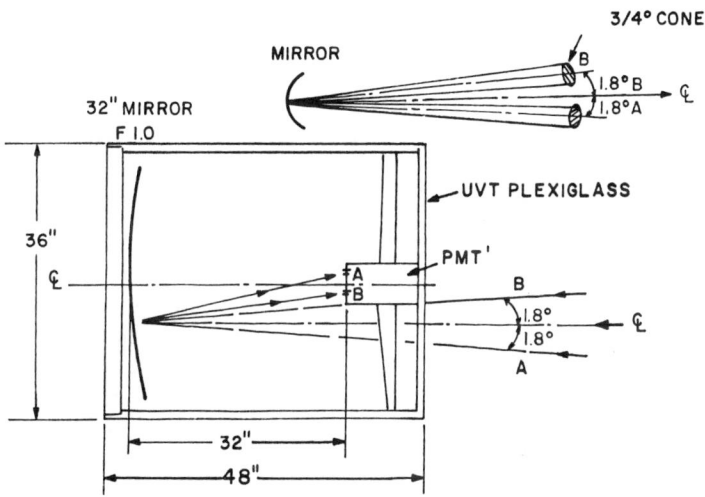

Fig. 2 Mirror and PMT arrangement. Optics designed to allow simultaneous viewing of two $3/4°$ segments of sky separated by $3.6°$.

The insulated barrel protects the mirror from off axis light and snow collection, and funnels the heat from the mirrors to the plexiglass entrance windows that seal the system. The heat from the mirrors will also heat the enclosed region and keep the plexiglass surface free of frost crystals.

The plexiglass windows also prevent the radiant cooling of the mirrors. Radiant cooling of the mirrors can reduce their temperature below the ambient air temperature and possibly the dew point. At temperatures of -70 C the maximum radiation occurs at 15 μm. This effect can be reduced by covering the mirrors with UVT plexiglass which is transparent in the visible and near UV but is opaque at wavelengths longer than 2.0 μm.

2. South Pole Prototype 89

A 10 mirror prototype is scheduled to be installed during the 89-90 austral summer, so that possible problems from aurora and specific polar conditions can be studied and corrected.

The device will be a non-tracking telescope consisting of 10 mirrors with their associated photomultipliers and electronics.[23] This will test the mirrors, PMTs, and electronics in the polar environment. The device, mounted on top of the electronics hut, will be about 15 feet in the air; this will keep the optics above blowing snow crystals and ground haze. The tests should expose a sufficient number of optical and electronic channels to polar conditions to allow drawing statistically meaningful conclusions. The schematic view is show in Fig. 3.

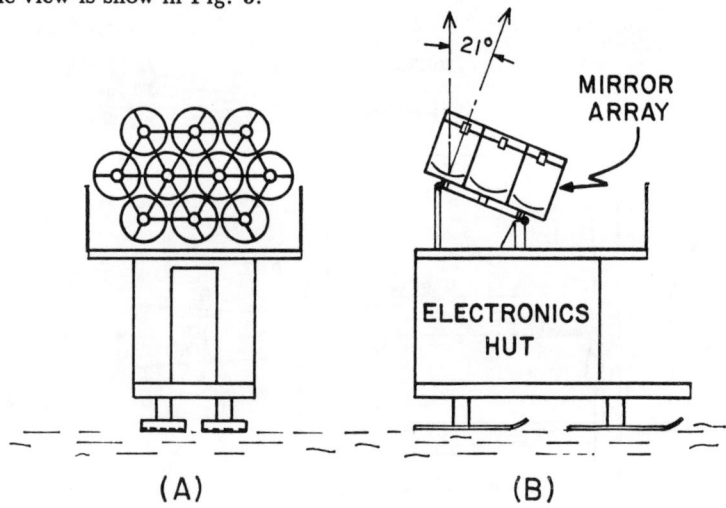

Fig. 3 View of the prototype 10 mirror array. (a) Front view showing mirrors mounted on roof of the electronics hut. (b) Side view showing the array pointing at 21 ° toward SN1987a.

The device would be pointed 21^0 from zenith to allow drift scans of SN1987a. It would be instrumented at the same level as the final instrument, only with fewer channels. This prototype will allow us to test and evaluate every aspect of the experiment under polar conditions, except the tracking capability.

The prototype telescope would be 11 feet wide, 6 feet high, and 7 feet deep, and would weigh about 800 lbs. It will be of modular design, to allow for the bolting together of sub-assemblies.

DATA PROCESSING, COMMUNICATIONS AND CLOCKS

During normal operation, the telescope would be operated remotely from computers located in the laboratory or dome area of the station. The experimental area will be visited by personnel only for purposes of maintenance. The communication link to the computer would consist of an Ethernet or coaxial link from the computer in the dome to the on-site computer. In addition, other readouts are planned, such as telephone-quality twisted pair links for low-speed communications.

The polar site can be reached only between November 1st and February 15th each year. This means the entire austral winter's data set, taken over a period from April through August, only becomes available for analysis in November. This is too long a period to *dead-reckon* an experiment without course corrections, so on-site analysis is planned.

We plan to install a VAX workstation at the polar base to allow a preliminary analysis of the data on-site. The VAX will be tied in with the main on-site computer via Ethernet. In this manner, partially-analyzed data will be continuously available for transmission by the satellite link to our home institutions for analysis.

It would be desirable to have a higher-speed link to the outside world than the present 9600 baud link now available. It is anticipated that future higher-speed satellite links between the pole and the U.S. will be available. In the interim, about 5% of the raw data can be sent back via the 9600 baud satellite link, so that monitoring of the device's operation can be made.

A time standard of absolute precision $10\mu s$ or better traceable to NBS will eventually be required. This is because many of the sources of interest are short-period pulsars. To study these pulsars fully it is essential to know the phase of emission relative to that of radiation in other parts of the spectrum.

It is possible to obtain this accuracy with the present rubidium clocks at the pole if the clock is set and checked daily. Time signals from the Global Positioning Service (GPS) satellites, which are visible from the pole many hours a day, could provide this accuracy. This service would provide precision of $1\mu s$ or better and would allow *after-the-fact* corrections to UTC.

POLAR AIRGLOW AND AURORA

Airglow is radiation emitted by the atmosphere not subject to direct sunlight. Airglow does not depend upon the observer's latitude, and the intensity of the radiation is about the same at all azimuths. There is a zenith angle dependence, the intensity being greater at larger zenith angles. At zenith the intensity is about 5 Rayleighs/nanometer (R/nm) in the blue at 400 nm, and rising to about 10 R/nm in the Green at 550 nm, and continuing to 20 R/nm in the red at 700 nm.

There are particularly bright emission lines in the green at 557.7 nm with an intensity of 250 R, in the yellow at 580.0 and 580.6 nm with variable intensity between 30-200 R, and in the red at 630.0 and 631.4 nm at about 100 R. Generally, there is a greenish glow to the earth's atmosphere. In the visible, the airglow corresponds to photon rates of 700 MHz/cm^2, but in the infrared (OH radiation) this rate can be as high as a 1400 GHz/cm^2. Generally the night sky brightness at Haleakala is of the order of 700-1000 Rs. This translates into effective photon rates of of about 70 MHz/cm^2/steradian.

The aurora is characterized by major emission lines at 630.0, 557.7, 427.8, and 391.4 nm. Their strengths vary from hundreds of Rs to several thousand Rs, so that in intensity, the aurora can be as bright to several times brighter than the night sky. By operating in the 285-380 nm region we avoid all of the red-green aurora and the N-lines at 391.4 and 427.8 nm, leaving only the weaker Vigard–Kaplan molecular N-bands to deal with.

The aurora cycle is such that it is overhead at the pole at magnetic dawn and dusk, and well away at midday and midnight. The probability of seeing the aurora reaches a maximum in an oval-shaped zone aligned along the Earth-Sun direction, with a radii from the geomagnetic pole of 12° in the solar direction and 22° in the anti-solar direction. The most intense aurora is seen in the solar midnight region of the oval. Aurora are usually oriented in the magnetic E–W direction.[24] We believe we are in the least populated aurora region, and expect the aurora only to be a problem about 20% of the time.[20]

Our experience at Haleakala is that blue and UV filters reduce the ambient night sky airglow rates by 70% and 90% respectively, while the shower rate reduction is less than 10%. This fact is qualitatively stated in Fig. 4, where we show the response of standard and UV sensitive PMTs to the night sky, and the $\frac{1}{\lambda^2}$ Cherenkov spectrum. We expect that the green-red aurora will be much like the airglow with which we have already successfully dealt by using filters. We are currently operating our Haleakala telescope in wavelength region 1 (285-380 nm) as shown in Fig. 1. However, we were less sure of the effectiveness of the filters on the nitrogen aurora at 427.8 and 391.4 nm.

In order to quantitatively estimate the effects of aurora on the ACT techniques, comparative measurements using identical instruments were carried out in Alaska, a region of active aurora, and in Hawaii, a region where only airglow is a factor. The Alaskan measurements were done at 65 N latitude, from 28 March to 14 April 1989 during a *new-moon* period, when the solar depression was quite shallow–about 15°. This location is one of the most intense regions on the aurora belt, and thus provides aurora conditions more intense than would be encountered at the south pole.

A long well-baffled cylinder with an aperture of 3 degrees with a PMT mounted at one end was used to collect the light at specific zenith angles. Three types of PMTs were tested: (i) a standard borosilicate glass PMT with bialkali photocathode having a mostly visible response, (ii) a bialkali PMT with a fused silica UV transmitting window (quartz), and (iii) a solar-blind PMT with response from 160-320 nm.

A medley of bandpass filters such as the Hoya B-370 and B-390 blue filters and Hoya U-340 UV filter were tested for transmission characteristics. In addition, a set of Melles Griot filters with bandpasses of 40 nm and transmission of 50% were utilized to measure the night sky intensity in intervals of 50 nm from 400 to 700 nm and at 340 nm.

Results from the Alaskan and Hawaiian tests displayed in Fig. 4. PMT-1 has a borosilicate window and PMT-2 has a quartz window and thus, extended sensitivity in the ultraviolet region. Both of these PMTs were operated at the single photoelectron level. Shown are the single photoelectron counting rates for a bandpass corrected for filter transmission. With no filter over the aperture, the R1450 PMT counted single photoelectrons at a rate of 230 kHz in Alaska compared to 125 kHz on Haleakala for the same zenith angle; rates of 500 kHz were not uncommon during visually bright aurora.

The auroral contribution to the night sky appears to be slightly more intense in the blue with very strong contributions at 427.8 nm and 555.7 nm, corresponding to well known spectral lines. Apart from periods of exceptionally intense auroral activity, particular filter combinations should provide for manageable data rates. These filters also allow data taking in the presence of a reasonable amount of moonlight, depending upon the zenith angle of the source and the angular distance from the moon.

To conclude, the Alaskan sky is about 2-3 times brighter than the Hawaiian sky at the blue end of the spectrum. This is undoubtably due to broad-band aurora and increased airglow activity caused by the decreased solar depression. The only aurora lines strong enough to be uniquely detected in the 50 nm wide scans were 427.8 and 555.7 nm and their contributions to the total rate were substantial. The filters were able to remove these line contributions in all but the brightest conditions. These studies also confirmed the existence of two ACT windows: a blue one between 285-380 nm and a green one between 440-540 nm. Both windows have the same width, but the blue one is preferred because that is where the PMTs' photocathode is most efficient and where more of the Cherenkov light is present.

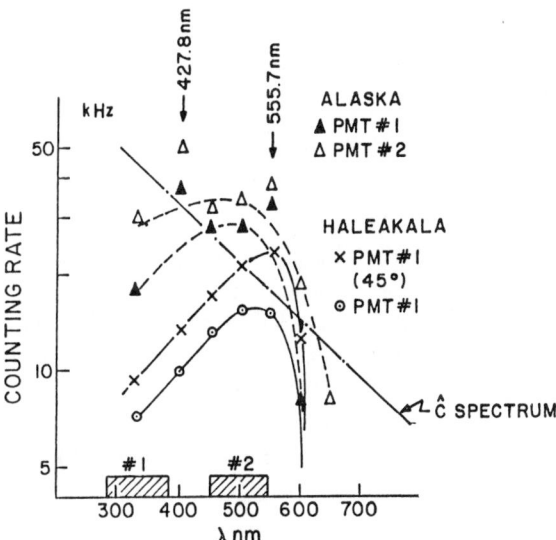

Fig. 4 The night-sky airglow and aurora spectra convoluted with PMTs' quantum efficiency for sites in Alaska and Hawaii.

SUMMARY

An ACT at the south pole would be a truly unique instrument. Its implementation faces many challenges–such as snow and frost accumulation on the optics, aurora interference, high-flying snow crystals, and high-ambient light levels. We believe that our construction techniques will allow us to solve the snow and frost problems. Our demand of high-multiplicity coincidence triggers will make us less prone to aurora and other phenomena that raise the ambient light levels, which make ACT work via conventional methods so difficult. The prototype device will be installed at the pole in austral summer 89-90, and data taking will start in the austral winter 1990.

The collaboration would like to thank Dr. John T. Lynch of the Polar Programs Division of the NSF for his continued help and encouragement in making this prototype test possible.

REFERENCES

1. M. Samorski, et al., Ap. J. Lett. **L17**, 268 (1983).
2. A.M. Hillas, Nature **312**, 50 (1984).
3. R.J. Protheroe, Proc. of 20th ICRC (Moscow, 1987).
4. J.C. Dowthwaite, et al., Nature **309**, 691 (1984).
5. R.M. Baltrusaitis, et al., Ap. J. Lett. **L69**, 293 (1985).
6. P.W. Gorham, et al. Ap. J. Lett. **L11**, 308 (1986).
7. L. Resvanis, et al., Ap. J. Lett. **L9**, 328 (1988).
8. R.C. Lamb, et al., Proc. of 20th ICRC (Moscow, 1987).
9. B.L. Dingus, et al., Proc 20th ICRC (Moscow, 1987).
10. H.V. Brandt, H. V., et al., Ann. Rev. Astron. and Astrophys. **21**, 13 (1983).
11. J.E. Grindlay, et al., Ap. J. Lett. **L9**, 197 (1975). **21**, 13 (1983).
12. R.J. Protheroe, et al., Ap. J. Lett. **L47**, 280 (1984).
13. A.R. North, et al., Nature **326**, 567 (1987).
14. B.C. Raubenheimer, et al., submitted to Astron. and Astrophys. (1988).
15. T.K. Gaisser, et al., preprint, U. Wis., MAD/PH/377 (October 1987).
16. L. Resvanis, et al., Very High Energy Gamma Ray Astronomy, ed. K. E. Turver (Dordrecht:Reidel, 1987), p. 131.
17. T.C. Weekes, Harvard-Smithsonian Observatory, private communications (1988).
18. M.A. Pomerantz, et al., South Pole Air Shower Experiment (SPASE), Proposal to NSF (1986), now active experiment (1988).
19. L. Resvanis, et al., Nucl. Inst. and Meth. **A269**, 297 (1988).
20. R. Eather, Boston College, private communication (1988).
21. R.J. Protheroe, Proc 19th ICRC, OG2.7-13 (San Diego 1985).
22. C. Stearns, Univ. of Wisconsin, private communication (1988).
23. M.A. Pomerantz, Bartol Institute, private communication (1988).
24. J. Romick, NSF,Washington D.C., private communication (1989).

PROSPECTS FOR GROUND-BASED GAMMA RAY ASTRONOMY AT THE SOUTH POLE
: WORKING GROUP REPORT :

A. A. Watson
Department of Physics, University of Leeds, Leeds LS2 9JT, UK

ABSTRACT

The discussions of the two working group sessions on ground-based gamma ray astronomy at the South Pole which were held during the Bartol Conference on 'Astrophysics in Antarctica' are reviewed. The sessions covered prospects at energies in the region of 1 TeV, where air Cherenkov telescopes (ACT) must be used, and in the region of 100 TeV, where telescopes comprising arrays of detectors sensitive to extensive air showers are the proven technique. It seems possible that the latter technique might be pushed down to 20 TeV and both will be particularly useful for providing overlap with the two week exposures of the EGRET instrument on the Gamma Ray Observatory during the 1990's.

INTRODUCTION

The motivations for studying the emission of astronomical objects in the very high energy (VHE) gamma ray band near 1 TeV and in the ultra high energy (UHE) gamma ray band near 100 TeV have been well described in the talk given by Weekes[1] at this meeting. The South Pole site, in addition to offering a unique view of the sky, provides the possibility of continuous monitoring of candidate objects at constant elevation. As there is an increasing body of evidence which suggests that most of the objects which emit in these wavebands do so only sporadically, the prospect of continuous monitoring gives the observer from the South Pole a considerable advantage over observers at other latitudes, provided, of course, the problems of operating equipment in the unusual conditions found there can be overcome. In the case of astronomy at 1 TeV, where air Cherenkov telescopes (ACT) must be used, a number of problems have to be solved before an operational system can be realised. For astronomy above 50 TeV the traditional air-shower technique in which an array of widely-spaced counters is used to detect the extensive air-showers generated by cosmic particles and cosmic gamma rays has already[2] been successfully implemented at the South Pole. Extension of this technique to even lower energies looks feasible.

In what follows I will try to summarise the discussion about the implementation of the two techniques and show how the existence of a South Pole Astrophysics Research Centre will benefit these activities.

OBSERVATIONS OF VHE AND UHE GAMMA RAYS AT THE SOUTH POLE

Cosmic rays of about 50 TeV and greater produce a sufficient number of particles at the altitude of the South Pole to allow detection of events at a useful rate (~ 1Hz) with a particle-array of modest (~ 6000 m^2) area[2]. The advantages of the South Pole site were first pointed out by Hillas[3]: in addition to its high altitude and the circumpolarity of the sources, an accident of position means that many low-mass and massive binary X-ray systems lie within the beam of the array which extends to ~ 50° from the zenith. In his talk during the workshop[4] Hillas reported an estimate of the rate at which usable showers could be detected by an array identical to the SPASE array located at a nominal altitude of 1800 m (rather than 2800 m at the Pole) and at latitudes corresponding to sites in New Zealand ($\lambda = 42°$ S), Australia (37° S) and South Africa (27° S). For sources with declinations less than -40° the SPASE array is always superior. For example, for a source with $\delta = -60°$ the SPASE array would record over six times as many events per cm^2 day per unit solid angle as an identical array at 41° S. The SPASE array has a median energy of detection of about 90 TeV for proton primaries.

Although the area of the SPASE array is about 6200 m^2 this is in fact somewhat less than the effective collecting area for gamma rays of about 1 TeV which can be detected using air Cherenkov telescopes. The latter technique, however, can only be used on dark, cloudless nights and, at the South Pole, there is clearly the possibility of achieving long and continuous exposures on individual sources during the three months of non-astronomical twilight in the depth of the winter. Implementation of this approach under these conditions is challenging. A first step was described by Weekes[5] who has developed a Fresnel lens/ photomultiplier system which is presently located near the centre of the SPASE array. The SPASE winter-over scientist, Alistair Walker, has reported (by E-mail!) on the extreme difficulty encountered in keeping the cover over the Fresnel lens clear of snow. It appears that snow adheres rather rapidly (within hours in windy conditions), probably aided by electrostatic effects, to plastic or glass covers. Attempts to keep the surface clear with heating wires have had only limited success; however use of a double cover through which a draught of warm, dry, air can be blown, may be more promising.

Morse[6] described the present thinking about a prototype mirror system which is to be deployed by the Purdue-Smithsonian-Wisconsin collaboration during the next austral summer. A heated 32" mirror will be viewed from above by two photomultipliers. The mirror and two photomultiplier tubes will be located inside an insulated box with a transparent end-plate. Heat convected up

from the mirror may be sufficient to keep the end-plate clear of
snow. Gaidos[7] reported on observations made at Poker Flats,
Alaska, which have shown that it is possible to counter the
effects of aurora to a considerable extent by observing in filter-
defined spectral bands. The filters will increase the threshold
of the planned detector to about 2 TeV.

Weekes[5] also described a carousel design having six mirrors
which would permit the simultaneous observations of several
sources: the constant elevation of an individual source makes this
approach feasible and indeed essential if the effort and expense
of a South Pole ACT is to be justified. Although there are
clearly severe difficulties to be overcome, the considerable
potential of the ACT method for long, uninterrupted, observations
lasting several days rather than several hours make the effort
highly desirable.

MULTI-WAVEBAND OBSERVATIONS OF PARTICULAR OBJECTS

The recent history of high energy astrophysics teaches that
multi-waveband observations of objects are particularly helpful to
attempts aimed at elucidating their properties. However such
observations are notoriously difficult to organise, particularly
when one or more satellite detectors are involved. The problem is
further compounded if the emission is sporadic as is believed to
be the case for many sources at high energies. For example Cygnus
X-3 was observed to flare in the radio region of the spectrum in
late September/early October of 1983, 1984 and 1985. A multi-
waveband observation of it was arranged for the fall of 1986 but
the object was found to be inactive! For this type of observation
the South Pole has particular advantages, especially for the SPASE
array which continuously monitors all objects with $\delta < -40°$, and
plans evolved during the workshop to extend the sensitivity of
SPASE so that the median energy is about 20 TeV. This could be
achieved by increasing the number of detectors within the present
array to give a 15 m rather than a 30 m separation. It would be
particularly useful to make this change in time to overlap with
the operating period (planned for eight years) of the Gamma Ray
Observatory (GRO) which is due for launch in 1990. The EGRET
instrument on this satellite will observe sources to 1 GeV for
periods of up to two weeks. Simultaneous observation at a useful
sensitivity could be made at about 20 TeV with an enhanced SPASE
array and at about 2 TeV with the planned air Cherenkov
telescope. Coverage at more northerly latitudes will be sparse
and broken because of weather, longitude, rising and setting of
sources etc. The problem is particularly acute above 100 TeV in
the Southern Hemisphere since the only arrays expected to be
operational, in addition to the SPASE array, are in Australia and
New Zealand and so are rather closely spaced in longitude.

USEFULNESS OF A SOUTH POLE ASTROPHYSICS CENTRE

The development of SPASE array with the rapidity which was achieved[2] would have been impossible without the help and advice of those in the Bartol Research Institute who had experience of deploying cosmic ray equipment at the South Pole. The experience of the SPASE collaborators is similarly benefitting the Purdue-Smithsonian-Wisconsin air Cherenkov telescope plans. Participants in the ground-based gamma ray workshop believed that there was a need to formalise this depository of expertise, as would be possible through the creation of a Centre. In addition there are sufficient similarities of technique in the particle-array and ACT work that there will be considerable scope for cost-effective sharing of computers, electronic spares, electronic test equipment, spares etc. Similarly there are common needs for power supplies and time references, the definition and supply of which should be one of the roles of the Centre. Additionally the sophistication of both types of gamma ray telescope, which require winter operation, requires the support of highly trained winter-over scientists capable of carrying out sophisticated data-analysis and interpretation at the Pole. Such people will need to be supported by computer and electronics experts who should be trained at the Centre where computing facilities and workshops duplicating those to be developed at the South Pole station should exist.

ACKNOWLEDGMENTS

I should like to thank the Bartol Research Institute for inviting me to participate in the AIA Conference and for their typically generous hospitality.

REFERENCES

1. T. C. Weekes, this conference: Overview paper on TeV and PeV gamma ray astronomy.
2. N. J. T. Smith et al, Nuclear Inst and Methods in Physics Res. **276**, 622-7, 1989.
3. A. M. Hillas, Proceedings Vulcano Workshop 1986, He-UHE Behaviour of Accreting X-Ray Sources (Editors F. Giovannelli and G. Maunocchi), Bologna (1986) p 331.
4. A. M. Hillas, 'The South Pole as a site for monitoring 100 TeV Cosmic Gamma Rays by means of an air shower array': this conference.
5. T. C. Weekes, this conference: Ground-based gamma ray astronomy workshop.
6. R. Morse, this conference: Ground-based gamma ray astronomy workshop.
7. J. Gaidos, this conference: Ground-based gamma ray astronomy workshop.

NEUTRINO ASTRONOMY

Francis Halzen
Department of Physics, University of Wisconsin, Madison, WI 53706

John Learned
Department of Physics and Astronomy, University of Hawaii, Honolulu, HI 96822

Todor Stanev
Bartol Research Institute, University of Delaware, Newark, DE 19716

ABSTRACT

We review the arguments supporting the claim that the observation of PeV ($E \gtrsim 10^{15}$ eV) γ-rays from cosmic sources at the flux-levels recently reported, guarantees the detection of neutrinos by detectors with an effective area of order 1 km^2. We emphasize the unique opportunities of neutrino astronomy as well as its multi-disciplinary facets touching astronomy, astrophysics, cosmology, and particle physics. At present we do not know of any cost-effective method to commission neutrino telescopes with $O(1\,\text{km}^2)$ effective area. We draw attention to the possibility of instrumenting Antarctic ice as a deep underground telescope detecting Cherenkov, radio or acoustic radiation from neutrino-induced electromagnetic showers.

NEUTRINO ASTRONOMY

Some important facts about neutrino astronomy:

- Like the photon and neutron, the neutrino is a neutral particle carrying directional information about its sources. It is not deflected by galactic or extra-galactic fields along its path.

- Because of their weak interaction with matter, neutrinos carry the only direct information on cosmic processes shielded from view by more than a few hundred grams of intervening matter.

- The discovery[1] of γ-rays with energies in excess of 10^{14} eV virtually guarantees the birth of neutrino astronomy provided we build telescopes with effective area of order 1 km^2. The hadronic processes which are the source of PeV γ-rays from $\pi^0 \to \gamma\gamma$ decay are also the source of neutrinos from $\pi^\pm \to \mu\nu$ decay. PeV γ-rays originate indeed in hadronic, not electromagnetic, processes. Electrons would inevitably be decelerated by synchrotron radiation in the same high B-fields required for acceleration to PeV-energies.

- We have witnessed 10 seconds of neutrino astronomy with SN1987A.

Enough generalities. Nothing illustrates the potential of neutrino astronomy better than a specific example.

CYGNUS X-3:
A STANDARD (?) CANDLE FOR NEUTRINO ASTRONOMY

Figure 1 shows a potential cosmic accelerator. A compact star, in orbit with a star that has not yet collapsed, accelerates protons, perhaps in the high fields of a pulsar or through conversion of gravitational energy from accretion of matter from the companion. Consider the configurations, shown in Fig. 1, where protons accelerated along the line-of-sight from Earth interact with the companion like a giant cosmic beam dump experiment. The beam initiates a hadronic cascade in the companion, the stable

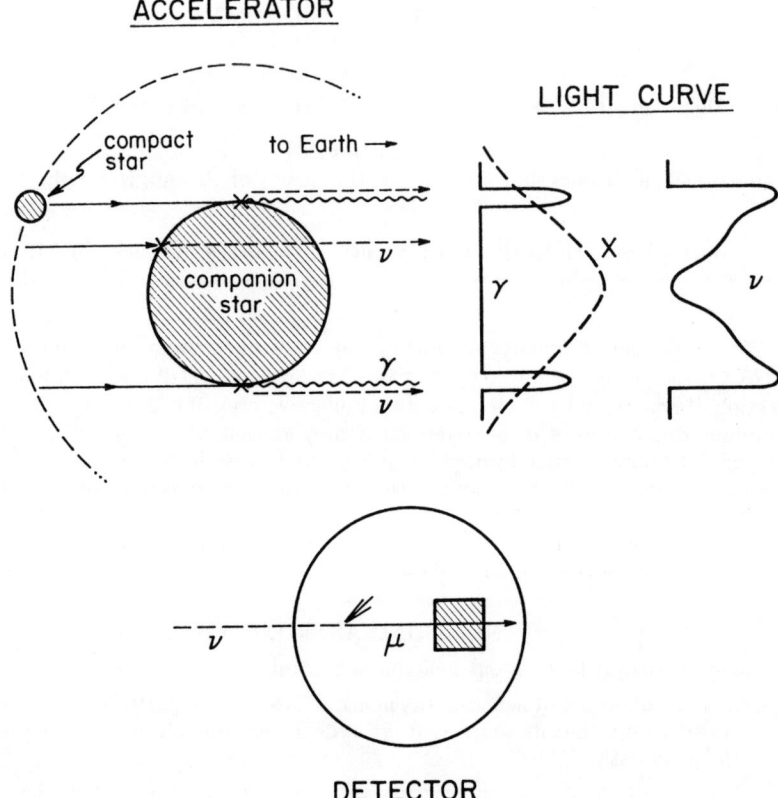

Fig. 1

end-products of which are photons, neutrinos, $e^{\pm}$ and $p, \bar{p}$. The charged particles are injected as cosmic rays into the galaxy. A fraction of the photons and neutrinos will escape from the companion and are detected at Earth. Photons will be absorbed in the target except for beams grazing the atmosphere of the companion star; see Fig. 1. Because of their weak interaction with matter, neutrinos, on the contrary, escape for all, or most, of the time the accelerator is eclipsed by the companion. The duty cycle for the delivery of a high energy ν-beam D_ν is larger than the one for delivering γ-rays. Typically,

$$\frac{D_\nu}{D_\gamma} \simeq \frac{\text{radius of the companion}}{\text{characteristic thickness of atmosphere}}.$$

We will apply this accelerator blueprint to Cygnus X-3, although this type of model cannot accommodate the detailed structure of this source. Vela X-1 is a more likely candidate for the accelerator shown in Fig. 1. Episodes of photon emission exactly coincident with the time the pulsar enters eclipse, have been reported. The TeV- and PeV-photon flux from Cygnus X-3, shown in Fig. 2, is well-described by an E^{-2} spec-

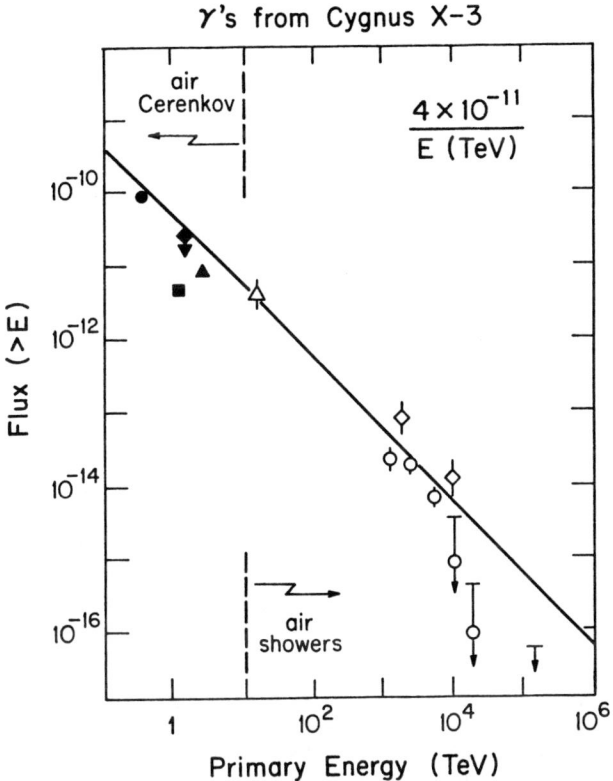

Fig. 2. Pre-1985 observations of Cygnus X-3 in the TeV and PeV energy bands.

trum, inspired by theoretical prejudice, with

$$\frac{dN_\gamma}{dE_\gamma} = \frac{4 \times 10^{-11}}{E_\gamma^2} \text{ cm}^{-2}\text{s}^{-1}. \quad (2)$$

Energy units will be TeV throughout. The dominant cascade in the beam dump is

$$
\begin{array}{ccccccc}
p + \text{matter} & \to & \pi^0 & + & \pi^+ & + & \pi^- & + & \cdots \\
& & \downarrow & & \downarrow & & \downarrow & & \\
& & \gamma\gamma & & \mu\nu_\mu & & \mu\nu_\mu & & \\
& & & & & & \downarrow & & \\
& & & & & & e\nu_e\nu_\mu & &
\end{array} \quad (3)
$$

and therefore we basically produce equal numbers of γ-rays and ν_μ. If the dump is such that muons decay, more neutrinos originate by $\mu \to e\nu_e\nu_\mu$ decay. We therefore predict equal fluxes

$$\frac{dN_\nu}{dE_\nu} = \left[1 - \frac{m_\mu^2}{m_\pi^2}\right] \frac{dN_\gamma}{dE_\gamma}, \quad (4)$$

up to a kinematic factor explicitly shown. At Earth, neutrino fluxes are further enhanced

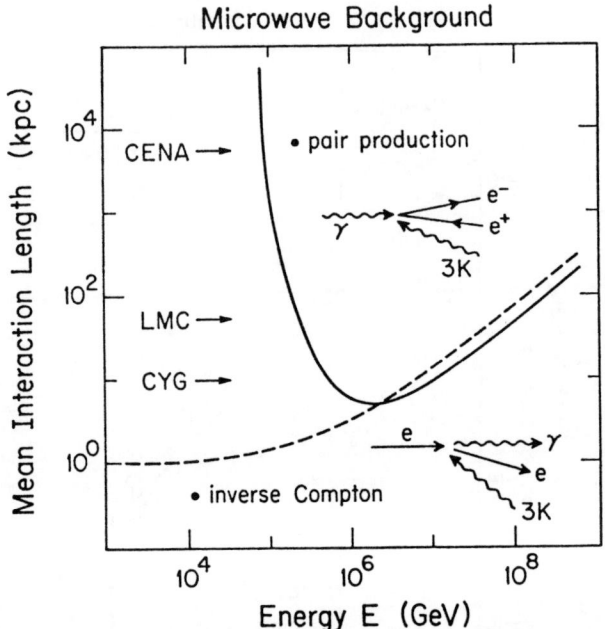

Fig. 3. Mean interaction length of a photon of energy E in the microwave background[3] due to the process $\gamma\gamma(3K) \to e^+e^-$. At higher energies the absorption is reduced due to the regeneration of photons by the secondary electrons $e\gamma(3K) \to \gamma e$. The arrows indicate the distance in kpc to Cygnus X-3 and some tentatively identified extra-galactic high energy γ-emitters LMC X-4 and Centaurus A.

by a factor $\frac{D_\nu}{D_\gamma}$ because γ-rays are preferentially absorbed in the target. There is a further enhancement factor f_{3K}^{-1} because a fraction f_{3K} of the photons are absorbed on $3°K$ cosmic photons along their travel to Earth. This fraction depends on the energy of the photons and their distance to Earth; see Fig. 3. For Cygnus X-3 $f_{3K} \simeq 1/3$. In summary, at Earth Eq. (4) becomes

$$\frac{dN_\nu}{dE_\nu} = \frac{1}{f_{3K}} \frac{D_\nu}{D_\gamma} \left(1 - \frac{m_\mu^2}{m_\pi^2}\right) f(\mu \to e\nu\bar{\nu}) \frac{dN_\gamma}{dE_\gamma}. \qquad (5)$$

The factor $f(\mu \to e\nu\bar{\nu}) \simeq 2$ represents the increased ν-flux from μ-decay in the dump.

Some observations of Cygnus suggest that $D_\gamma \simeq 0.02$. Therefore $D_\nu/D_\gamma = 50$ for $D_\nu = 1$. One can, however, imagine cosmic dumps with $D_\nu/D_\gamma \simeq 1$, e.g. a system where accelerator protons are confined magnetically for very long times in a region with little target matter from which photons, as well as neutrinos, freely escape. We will therefore investigate the range of values

$$1 < \frac{D_\nu}{D_\gamma} < 50. \qquad (6)$$

Although one can reasonably expect a greatly enhanced ν-flux, this is only part of the story as the detection of neutrinos is a non-trivial matter. The earth is the detector.

An underground experiment identifies muons produced by cosmic neutrinos interacting with the rock below the detector as shown in Fig. 1. The number of observed muons for a flux dN_ν/dE_ν is

$$N_\nu = \text{Area} \int dE_\nu \frac{dN_\nu}{dE_\nu} P(\nu \to \mu). \qquad (7)$$

Besides the area of the detector, the crucial quantity $P(\nu \to \mu)$ represents the probability that a neutrino actually interacts with the earth and deposits a muon in the detector

$$P(\nu \to \mu) = \rho(\text{rock})\,\sigma_\nu(E)[\text{Range}]_\mu. \qquad (8)$$

It depends on the density of the earth, the neutrino interaction cross section and the range of the produced muon. This is standard particle physics and for back-of-the envelope calculations we can approximate P by

$$\begin{aligned}P(\nu \to \mu) &\simeq 1.3 \times 10^{-6} E^{2.2} \quad \text{for} \quad E = 10^{-3} - 1 \text{ TeV}, \\ &\simeq 1.7 \times 10^{-6} E^{0.8} \quad \text{for} \quad E = 1 - 10^3 \text{ TeV}.\end{aligned} \qquad (9)$$

The two energy regimes directly reflect the energy dependence of the neutrino cross section σ_ν in Eq. (8) with $\sigma_\nu \sim E$ at low energy and a slower E-dependence at high energy because of the effect of the W-propagator.

Event rates in a future neutrino telescope can be readily calculated from Eqs. (2), (5), (7), and (9). We obtain[4]

$$30 < N_\mu < 1500 \, \frac{\text{events}}{\text{year}}, \qquad (10)$$

corresponding to the range of D_ν/D_γ values given by Eq. (6). The rates are for a 0.1 km^2 detector with a threshold of $E_\nu > 10$ GeV. The largest contribution to Eq. (10) is from neutrinos far above threshold with E_ν in the vicinity of 1 TeV where the interaction cross section becomes sufficiently large for the neutrinos to be trapped by the earth. For yet higher energies the flux (5) is too small to produce a signal inside a 0.1 km^2 area.

Can we elevate the result of Eq. (10) to the role of "standard candle" of neutrino astronomy? These event rates are conservative from the computational point of view. We assumed a $E^{-\alpha}$ γ-ray flux with $\alpha = 2$. For a "steeper" energy spectrum, say $E^{-2.7}$ as for cosmic rays, the relative flux of neutrinos is greatly enhanced with[5]

$$N_\mu \simeq e^{8(\alpha-1)}, \qquad \text{for} \quad 2 < \alpha < 3. \qquad (11)$$

From the point of view of the input γ-ray flux, our estimate is, on the contrary, likely to be very optimistic. A source like Cygnus X-3 is highly variable. All data in Fig. 2 precede the summer of 1985.[2] Since then no steady flux has been detected, only relatively short bursts. We also remind the reader that a conversion of the observed flux in Eq. (2) to luminosity at the source yields a result exceeding the Eddington limit by a factor of ten. Following Hillas,[6]

$$\mathcal{L} = \epsilon \frac{1}{f_{3K}} \frac{1}{D_\gamma} 4\pi d^2 \, N_\gamma(> E) \simeq 10^{39} \, \frac{\text{ergs}}{\text{sec}}, \qquad (12)$$

assuming $N_\gamma(> E) = 4 \times 10^{-11} \text{cm}^{-2}\text{sec}^{-1}/E(\text{TeV})$ following (3), $1/f_{3K} = 3$, $D_\gamma = 0.02$, and $d = 10 \text{ kpc} = 3 \times 10^{22}$ cm. $\epsilon \simeq 10^{-1}$ is the efficiency with which luminosity

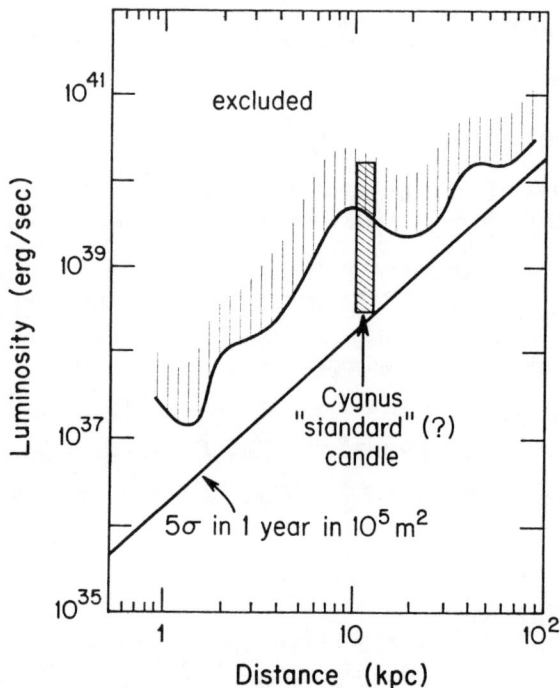

Fig. 4. Compilation[7] of limits on neutrino fluxes from various candidate sources. Our estimate for a Cygnus X-3 type source is plotted at 12 kpc. Also shown is the flux-sensitivity of a detector with $10^5 m^2$ effective area.

is converted into produced particles. This luminosity is adequate to supply the galaxy with all its cosmic rays of energy 100 TeV and above.

It is therefore not unreasonable to assume that Eq. (10) represents a realistic goal within very large ambiguities. Clearly detection of such a flux is not guaranteed by present detectors which have an effective area of a few times 10^{-4} km^2, but it is within reach of proposed detectors such as Dumand and Grande which are not much smaller than the 0.1 km^2 detector previously considered. The situation has been summarized in Fig. 4 where it can be seen that present detectors exclude the most optimistic values in Eq. (10), whereas a 0.1 km^2 detector guarantees observation of our standard candle with a 5 standard deviation signal in one year. The 5σ observation is above a background of upcoming neutrinos produced in cosmic ray air showers at the other side of the earth; see also Fig. 5.

CYGNUS X-3 AND A LOT MORE:
INTERDISCIPLINARY SCIENCE WITH A NEUTRINO TELESCOPE

If neutrino astronomy sounds by now a bit routine, let us try again. In Fig. 5 we have tabulated[8] the flux of contained neutrino events, *i.e.* events where the neutrino interacts to produce a muon or electron inside the detector, from various sources. Sandwiched between the large flux of MeV neutrinos from SN 1987A and the small flux of TeV neutrinos from a Cygnus-type source, we tabulated the event rates from neutrinos produced in cosmic ray-induced air showers. The figure inadequately illustrates the

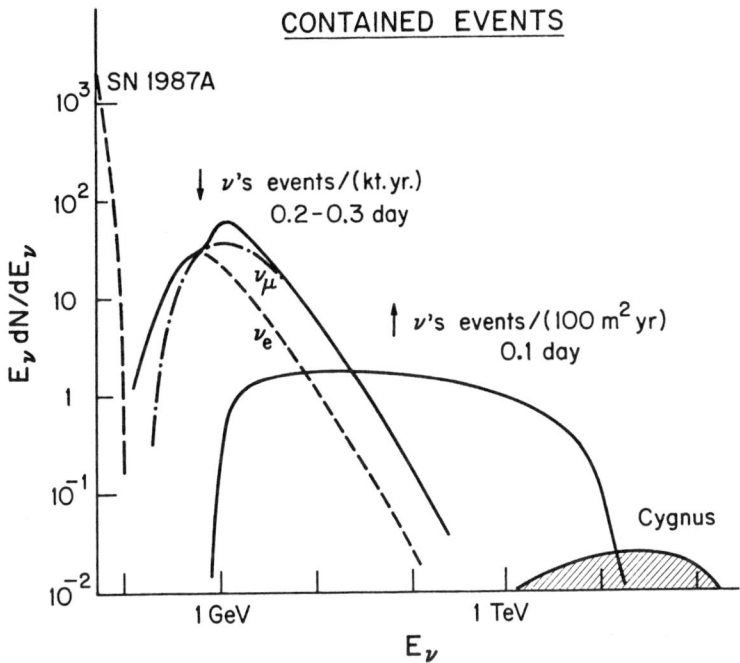

Fig. 5. Compilation of neutrino fluxes: neutrinos from Supernova 1987A, down- and upgoing neutrinos produced in cosmic ray air showers and neutrino beam from Cygnus X-3 type source.

wide scope of physics within range of a neutrino detector. This was most dramatically illustrated by the observation of neutrinos from SN 1987A. We present an incomplete list of the goals of neutrino astronomy:

- Astrophysics:
  - Observation of shrouded sources, sources opaque to γ-rays, or sources shielded from our view by the 3°K cosmic background.
  - Besides binaries such as Cygnus X-3, Hercules X-1, and LMC X-4 supernovas and active galactic nuclei can be the sites of hadronic processes. Neutrons escaping from AGN's could be responsible for the production of high energy γ's and the associated neutrinos.[9]
- γ-ray Astronomy:
  - The discussion of the accelerator in Fig. 1 clearly illustrates the power of the complementary γ-ray and neutrino detection for deciphering the structure of sources.
  - If high energy γ-rays produce muons in γ-initiated air showers in excess of conventional expectations, neutrino telescopes detecting muons become powerful PeV γ-ray detectors, perhaps competitive with the largest air shower arrays.[10]
- Particle physics:
  - The case has been dramatically illustrated by SN1987A where ten seconds of ob-

servation matched particle physics experiments on neutrino mass and improved on limits on the neutrino lifetime and magnetic moment.
- Neutrino oscillations can be studied over Earth-size distances of $1\text{-}10^3$km, thus probing mass differences of order $\delta m^2 \sim 10^{-3}$ MeV^2

• Cosmology:
- Dark matter particles (WIMPS, photinos, ...) can produce neutrinos by annihilation.
- There is also a standard candle of neutrinos of cosmological origin. The neutrinos are produced by extragalactic cosmic rays interacting with the 3°K photon background. The rates are small, but could be dramatically underestimated if cosmic strings exist or, perhaps more conservatively, if there were epochs in the history of our universe when cosmic ray activity was significantly enhanced.

We feel this list presents a strong case for the development of neutrino detectors with effective area in excess of the 0.1 km^2 discussed. Present techniques are unlikely to be cost-effective on such a scale. We discuss next how Antarctic ice might present us with extraordinary opportunities.

NEUTRINO DETECTION IN SOUTH POLE ICE

The experimental challenge for neutrino astronomy is to conceive detectors of large area in an environment shielded from the large continuous cosmic ray backgrounds of particles at the surface of the earth. Our basic idea[11] is to exploit the clear optical characteristics of deep polar ice in order to use it as a medium in which to sense the Cherenkov radiation of muons produced by neutrinos (and possibly γ-rays). TeV neutrino interactions will produce muons traveling hundreds of meters upwards through the ice; see Fig. 6. The muon trajectory, identified by photomultipliers sensing the Cherenkov light, determines the parent neutrino arrival direction to a precision of order 1°.

Independently of the cost effectiveness, which we will discuss further on, doing neutrino astronomy at the South pole has compelling physics advantages:

(i) a cosmic source can be observed continuously and remains at a fixed zenith angle relative to the detector, making background subtraction extremely simple;

(ii) the best studied cosmic accelerators are located in the northern hemisphere; their neutrino radiation, upcoming through the earth, will be in the detector's field of view 24 hours a day;

(iii) finally, instrumenting polar ice with phototubes allows for the experiment to follow science: the detector can grow, be gradually increased in resolution or (alternatively) expanded in coverage over the essentially unlimited expanse of the stable icefields.

The importance of the unique opportunity of uninterrupted observation of a source should be emphasized because point sources appear to be impulsive and episodic emitters, with variations on perhaps as many as four time scales.

Any proposal or detailed design of this type of detector is held hostage to the fact that optical attenuation of deep ice in the mid UV range (~ 0.45 μm), characteristic of the Cherenkov light emitted by high energy muons, has not been directly measured. We proceed on the hope that a simple test will confirm the belief that it is similar to the observed 25 m attenuation length for blue to mid UV light in clear water in deep ocean basins. It is of course essential to place the PMT's below the firn layer. It is known that ice becomes bubble free below typically 1 km, but there are areas where bubble-free ice

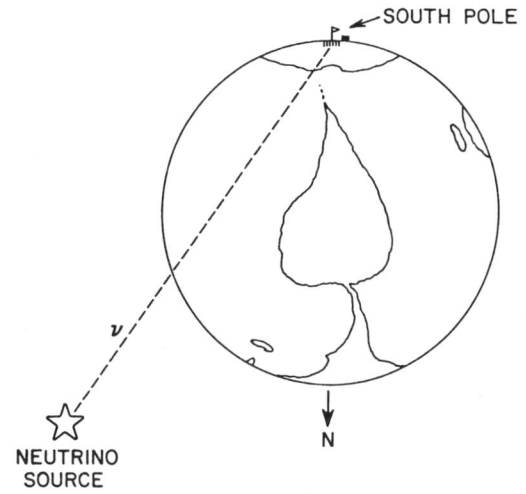

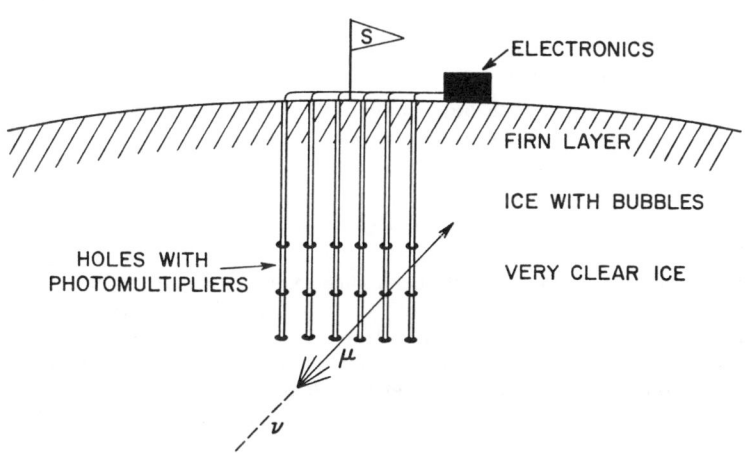

Fig. 6. Artist concept of Antarctic neutrino telescope.

can be found at depths of order 150 m. This includes domes at 3000 m altitude near the pole and floating shelves which can be uniform over a range of 100 km or more. These areas are free of shear forces and although some layers contain volcanic ash or cracks, they can be identified and avoided. One can envisage a detector with three layers of detection, say at 500 m, 1000 m, and 1500 m, although these depths depend on detailed test measurements of attenuation of light and on the science targeted. This geometry of detector, see Fig. 6, can be achieved by lowering photomultipliers into separate holes, or lowering instrumented strings of three phototubes into holes which can be (perhaps) frozen in. Alternatively, the holes can be backfilled with liquid that doesn't freeze (a standard technique) and has reasonable optical quality. It can possibly be supplemented with wavelength shifter. What makes the idea attractive is that coring equipment is readily available and drilling holes without retrieving cores is relatively easy and fast. The size of the holes is set by the PMT's needed for detection of muons at ranges of a few times ten meters (40 $\sim$ 50 cm for PMT's developed for DUMAND or KAMIOKANDE). Alternatives employing light collection might allow for substantially smaller hole diameters.

Although the ultimate spacing and depth of the phototubes will depend on the tests, we can envisage the development of a detector in the following stages. One can use existing equipment to lower a string of PMT's in an existing coring hole and measure the cosmic ray rate as a function of depth. One could further drill two (or more) holes 10 m apart in order to detect a source in one hole with a phototube placed at various depths in the other hole. It appears that the *in situ* optical properties (the diffuse attenuation coefficient being the relevant parameter for this application) have not been measured. The results would have stand-alone scientific value.

Given encouraging results of these exploratory measurements one could install a few electronically instrumented strings and operate them for several months.

A significant physics experiment could be done with a grid of 36 holes (108 PMT's) spaced by 40 m, covering an area of $40,000 \, \text{m}^2$. An investment of \$1M in hardware may be sufficient for a 36 hole prototype with real neutrino physics potential as it approximately matches the $6.25 \times 10^4 \, \text{m}^2$ area of the biggest detectors proposed elsewhere. Note that we do not include the cost of hole preparation, transportation, or laboratory maintenance in this estimate.

This experiment could then be expanded in a (up to that time) flexible fashion in order to achieve $O(1 \, \text{km}^2)$ coverage after several years. For the sake of comparison with other proposals we define the final telescope as a grid of 25×25 holes ($3 \times 25 \times 25 = 1875$ PMT's) covering $1 \, \text{km}^2$. This is a number of channels similar to the existing IMB water Cherenkov detector, although the PMT's are likely to be larger here. A guess at the ultimate cost is $O(\$10\text{M})$.

The electronics and data collection for such experiments would be straightforward, employing standard HEP techniques with the addition of fiber optic data transmission from phototubes to an electronics van. Clearly handling a neutrino data rate of order 1 event every 10 minutes is not a problem for our standard candle. Data transmission (and even software development) by satellite is feasible.

Another aspect of the potential for physics of an array of light detectors in the deep ice has to do with the observation of downgoing clusters of muons associated with UHE cosmic rays incident upon the top of the atmosphere. Several projects are proposed (GRANDE and several others) or in the building stage (*e.g.*, the Chicago-Michigan array in Utah) to study point source UHE radiation from celestial objects. The radiation already observed by several EAS arrays is presumably of gamma ray origin, but yet the two experiments that have made significant accompanying muon measurements of the showers associated with the point sources find that the showers are not deficient in muon content, as previously expected. While this news may be bad for the EAS arrays that are hoping for muon-poor-shower tagging to increase their signal-to-noise ratio, it may be good news for a shallow under-ice array that could do point source searches with the muon detection alone. An experiment in conjunction with a surface scintillation array, and/or an air Cherenkov detector would also have attractive characteristics, which we shall not further dwell upon herein.

In conclusion, we suggest that field investigations begin as soon as possible to examine the relevant optical properties of deep Antarctic ice. It is amusing to envisage the Antarctic ice sheet as a giant neutrino telescope, with the whole earth as its rotating neutrino bandpass filter.

NEUTRINO DETECTION BY RADIO ANTENNAS

We also note that there is a program already in active feasibility demonstration stage at the Soviet Vostok Station to detect UHE neutrino interactions in deep cold ice via the microwave emissions from interaction generated cascades.

The radio emission comes from the excess negative charge that develops toward the tail of neutrino-induced particle cascades in the ice[12] due to the charge asymmetry of the low energy processes—Compton effect, δ-electrons and positron annihilation in flight. The emission in the radio band follows the well-known energy spectrum of the Cherenkov radiation and the signal in a frequency band $\Delta\omega$ is

$$W = \frac{q^2}{c^2}\omega\Delta\omega l(1 - 1/\beta^2 n^2),$$

where q is the charge excess ($\simeq 10^{-10}E_0$ (eV) for 20% charge excess), l is the length of the cascade and n is the refraction index of the medium. For frequencies below 10^9 Hz the emission is coherent and the radio emission is proportional to the square of the cascade energy.

The suggestion was made to arrange radio antennas on the surface of the ice in Antarctica and exploit the very low absorption of radio waves in cold ice[13]. The radio signal will be distributed around the Cherenkov cone and the timing of the signal in different antennas will give enough information to determine the neutrino direction. The estimates in Ref. 13 show that such an arrangement will be sensitive to 10^{14} eV cascades to a depth of 100 m of ice. The depth will be 1 km for cascade energy of 10^{15} eV. Detection is possible if the ambient noise level at the detection site is less than 1000 K. The measurements at Vostok[14] show only half of that noise level.

The high detection threshold, however, makes the effectiveness of this technique questionable for detection of point neutrino sources. A simple calculation using a neutrino threshold of 100 TeV, the cascade depth estimates of Ref. 13 and the standard neutrino candle gives only 1/10-5 neutrino events per 1 km^3 per year. This is actually an upper limit of the rate as it does not account for the absorption of the neutrino flux in the Earth. A useful event rate can be achieved if the detection threshold is decreased by at least one order of magnitude or if the detector area is drastically increased by a suitable arrangement of the antennas[15].

HEARING NEUTRINO INTERACTIONS

High detection threshold may also be the main shortcoming of the idea of acoustic detection of neutrinos in ice, originally suggested by Bowen[16]. Acoustic waves are produced by the heating of the medium by the ionization loss of the cascade particles. Shower direction can also be determined by the timing and amplitude of the signal in different detectors[17] (the beam pattern is coherent, and reinforcing, in a plane perpendicular to the cascade direction).

We envisage a detector consisting of pressure transducers placed in an array in the deep ice. The optimal depth is not as yet clear, but it would have to be deep enough to avoid the problem of acoustic wave scattering due to bubbles in the ice, and perhaps also to escape high frequency surface noise. The firn layer will certainly provide a good acoustical shield for downgoing noise of surface origin, and the great depth of ice will provide a shield for upcoming noise of rock-ice interface origin. One may expect the inherent noise in the polar ice at a depth of several hundred meters to be close to the thermal limit. It seems that high frequency (by which we mean about 10kHz) noise has not been measured in the deep antarctic ice.

The acoustic losses set the size scale of the detector. We were also not able to find measuements of acoustic attenuation in the high frequency range in antarctic ice. One report[18], however, does present temperature and content dependant data collected in the laboratory which indicates that pure ice has a sound velocity (of compressional waves) of $v = 3200$ m/s and a Q of perhaps 1000 at low temperatures (South Pole ice near the surface is below $-50°C$). With an attenuation length of Qv/pf M 100m, one can imagine detectors in a lattice of roughly 100 m spacing. Using Bowen's[16] figure of merit, we estimate a signal-to-noise ratio of unity in a single detector at 100 m distance from a 6 PeV cascade, with the signal-to-noise ratio scaling as the square of energy/distance. The frequency maximum is about 20kHz.

Despite the dauntingly high detection threshold in comparison to the optical, and perhaps the radio technique, acoustic detection remains interesting enough for further investigation for several reasons. First, acoustic sensors are compact, pressure tolerant, inexpensive (compared to photomultipliers), and could be installed relatively easily (a several cm diameter string of piezoelectric sensors might be melted down to depth, without any need for drilling). It does not seem outrageous to imagine a $30x30$lattice of strings extending downwards for several kilometers (thus covering a volume of the order of 30 billion tons!). Second, while the production of acoustic pulses by particles in liquids, water in particular, has been explored experimentally, nothing has been done about ice, particularly at low temperatures and under high pressure. Moreover, in contrast to liquids, ice transmits shear waves. Pure ice has a shear wave velocity of about 1900 m/s, and a Q value similar to that of the compressional wave[18]. Thus depending upon the particle cascade coupling to the shear wave, one may potentially detect a delayed shear pulse, which pulse contains range information in the time delay. Furthermore there is, of course, also the possibility of unsuspected coupling mechanisms as well which could lower the threshold detection energy.

Thus we suggest that experimental investigation of the acoustic properties of deep antarctic ice are well worth carrying out.

SUMMARY

We have pointed out the importance of neutrino astronomy in several disciplines, and we emphasize the importance of aiming for square km sized detectors. We have superficially explored the use optical, radio, and acoustic techniques taking advantage of the unique properties of antarctic polar ice. The optical technique is well understood, but we do not know the ice clarity versus depth. The radio technique is not well understood yet, nor are the radio noise backgrounds. The acoustic approach also suffers from an untested acoustic pulse generation mechanism in ice, and the acoustic properties of the deep antarctic ice are not measured. We conclude that while there are attractions to all three methods, we do not have sufficient information, particularly from *in situ* physical measurements, to make decisions about the best method, if any, to pursue. We recommend a program to accumulate this data.

ACKNOWLEDGEMENTS

This research was supported in part by the University of Wisconsin Research Committee with funds granted by the Wisconsin Alumni Research Foundation, and in part by the U. S. Department of Energy under contract DE-AC02-76ER00881. The research of T. S. is supported in part by the U. S. National Science Foundation.

REFERENCES

1. T. C. Weekes, Physics Reports 160, 1 (1988); J. M. Bonnet-Bidaud and G. Chardin, Physics Reports 170, 325 (1988); R. Protheroe, in Proceedings of the 20th International Cosmic Ray Conference, Moscow, 1987, eds. V. A. Kozyarivsky et al. (Nauka, Moscow, 1987), Vol. 8; D. E. Nagel, T. K. Gaisser, and R. J. Protheroe, Ann. Rev. Nucl. Part. Sci. 38, 609 (1988).
2. F. Halzen, Proceedings of the International Europhysics Conference on High Energy Physics, Bari, 1985, edited by L. Nitti and G. Preparata.
3. R. J. Protheroe, Mon. Not. R. Astro. Soc. 221, 769 (1986).
4. This is consistent with previous estimates, see T. K. Gaisser and T. Stanev, Phys. Rev. Lett. 54, 2265 (1985); E. W. Kolb, M. S. Turner, and T. P. Walker, Phys. Rev. D32, 1145 (1985); 33, 859(E) (1986); T. P. Walker, E. W. Kolb, and M. S. Turner, Proceedings of New Particles '85, Madison, Wisconsin, edited by V. Barger, D. Cline, and F. Halzen (World Scientific, Singapore, 1986); F. W. Stecker, A. K. Harding and J. J. Barnard, Nature 316, 418 (1985); A. Dar, Phys. Lett. 159B, 205 (1985); V. S. Berezinsky, C. Castagnoli, and P. Galeotti, Nuovo Cimento 8C, 185 (1985).
5. T. K. Gaisser, Proceedings of the 1988 Snowmass Summer Workshop.
6. A. M. Hillas, Nature 312, 50 (1984).
7. R. Svoboda, Proceedings of the Workshop on Physics and Experimental Techniques of High Energy Neutrinos and VHE and UHE Gamma-Ray Particle Astrophysics, Arkansas, USA (1989), edited by G. B. Yodh and D. C. Wold.
8. T. Stanev, Proceedings of the Workshop on Physics and Experimental Techniques of High Energy Neutrinos and VHE and UHE Gamma-Ray Particle Astrophysics, Arkansas, USA (1989), edited by G. B. Yodh and D. C. Wold.
9. A. Mastichiadis and R. J. Protheroe, Mon. Not. R. Astro. Soc., to be published.
10. F. Halzen, Proceedings of the Workshop on Physics and Experimental Techniques of High Energy Neutrinos and VHE and UHE Gamma-Ray Particle Astrophysics, Arkansas, USA (1989), edited by G. B. Yodh and D. C. Wold.
11. F. Halzen and J. Learned, Proceedings of the International Symposium on Very High Energy Cosmic Ray Interactions, The University of Lodz Publishers, edited by M. Giler (1989).
12. G. A. Askaryan, Zh. Eksp. Theor. Fiz. 25, 276 (1961); Sov. Phys. JETP 48, 988 (1965).
13. M. A. Markov and I. M. Zheleznykh, Nucl. Inst. Meth. A248, 242 (1986).
14. I. N. Boldyrev et al., Proceedings of the 20th International Cosmic Ray Conference, Moscow (1987), vol. 6, p. 472.
15. J. P. Ralston and D. M. McKay, this volume.
16. T. Bowen, Proceedings 16ICRC, Kyoto 1979, 11, 184, T6-1 (1979).
17. J. G. Learned, Phys. Rev. D19, 3293 (1979).
18. H. Spetzler and D. L. Anderson, Journal of Geophysical Res. 73, 6051 (1968).

ICEMANd: MICROWAVE DETECTION OF ULTRA-HIGH ENERGY NEUTRINOS IN ICE

John P. Ralston and Douglas W. McKay

Department of Physics and Astronomy

University of Kansas, Lawrence, Kansas 66045

ABSTACT

A muon from an ultra-high energy neutrino interaction produces an electromagnetic shower of considerable length. Coherent Cerenkov emission at microwave frequencies from the electric charge imbalance developing in such a shower serves as an efficient signal of the event. We discuss detecting upward going UHE neutrinos ($E_\nu \gtrsim 10^{15} eV$) in the antarctic ice by detecting this microwave signal with comparatively cheap and simple antennas located on the ice surface. We conclude that a pilot experiment to measure UHE neutrinos from point sources such as Cygnus $X - 3$ is feasible.

Detecting Neutrinos

A great deal of human effort has gone into detecting neutrinos. From the early discovery experiments at reactors to solar neutrino experiments at the Homestake mine and neutrino detection at underground proton–decay facilities, the effort has paid off. As a bonus, almost every new detection creates a new mystery: witness the generation puzzle, discovered in part through separate neutrino lepton number conservation; recall the solar neutrino puzzle, and its time–dependent reincarnation; recall the amazing detection of supernova 1987A by neutrinos and its stunning implications for astrophysical models. All of these experiments have been with neutrinos in the low energy, MeV range where new physics is comparatively difficult to discover.

© 1989 American Institute of Physics

The Energy Region

Here we discuss a promising method for detecting ultra–high energy ($E \gtrsim 10^{15} eV$) cosmic ray neutrinos. The approach is exciting because it is both practical, efficient and cheap. The energy threshold of $10^{15} eV$ is set by considerations of microwave coherence for emission from an electromagnetic shower, as we discuss below. Coincidentally, the energy is one where new physics might be discovered. In the standard model, the reaction $\bar{\nu}_e e^- \to W^-$ is resonant for neutrino energy $E_\nu \sim 7 \times 10^{15} eV$ hitting an electron at rest. The resonance energy for producing a new particle X that could, e.g., produce generation-changing leptonic interaction via $\nu e \to X$ is $E_\nu = 0.98 TeV \, (m_X \, GeV)^2$. Thus the energy region is a characteristic scale of fundamental interest. In addition, the standard model cross section $\sigma(E)$ for ν–nucleon interaction rises to roughly $10^{-33} cm^2$ for $E_\nu = 10^{15} eV$ and continues to rise thereafter due to a substantial QCD enhancement[1,2] as shown in Fig. 1.

Finally, the energy region of $10^{13} - 10^{20} eV$ is arguably the most interesting energy region ever studied in cosmic ray physics. It seems that nearly all of the cosmic rays in this region emanate from point sources, apparently neutron–star binaries. Cygnus $X - 3$ is the most famous example.

The data from Cygnus $X - 3$ and other point sources present some baffling mysteries.[3] In every model studied a sizable neutrino flux at UHE is predicted which must be measured to make progress in solving these mysteries and achieving true understanding of the sources and the physics of the particles they are producing.

Corerent Microwave Cerenkov Mission

We propose measuring upward–going neutrino interactions through coherent microwave Cerenkov emission. Microwave antennas are easy to mass produce and are well understood. The issue of coherent microwave

Cerenkov emission is not so well understood and requires a discussion, as well as further study. The idea is as follows.

A charged–current muon–neutrino interaction will produce an electromagnetic (em) shower with a large number of e^+e^- pairs. To a rough approximation, the pairs in a fully developed shower have energy distribution dN/dE going like E_0/E^2, where E_0 is the primary energy. In the classic estimates of Rossi,[4] the integral number of electrons with energy greater than $E_>$, $N(E_>) = \int_{E_>} dE' dN(E')/dE'$, is of order $E_0/E_>$, i.e. $N(E_>) \sim 10^7 (E_0/10^{15} eV)$ for $E_> = 100 MeV$: there are many pairs in an UHE shower.

So long as the shower is neutral it is not efficient at Cerenkov emission. An imbalance between negative and positive charge can develop because the cross sections for electrons and positrons are different. The interaction lengths are significantly different in the region of energies less than about 100 MeV: a substantial fraction of positrons annihilate. Letting $\Delta N(E_>)$ be the electron minus positron integral number, the effective charge of the shower is $-e\Delta N(E_>)$.

The shower's effective charge can emit Cerenkov radiation. The radiation is coherent so long as the wavelength λ of the Cerenkov radiation is much larger than the longitudinal shower thickness. The Cerenkov power radiated for such wavelengths goes like the net charge *squared* and becomes sizeable when the effective charge is large. This is the case for an ultra–high energy event, as we discuss below. Details of the coherence, angular distribution and dispersion of the Cerenkov radiation are important but do not affect the order of magnitude discussion we present here. They are straightforward and have been discussed elsewhere, but the details of the electromagnetic shower merit further study.

The idea of detecting an electromagnetic shower by microwave coherence is not new. Askaryan suggested the idea in the early 1960's.[5] By the late 1960's signals in simple microwave antennas had been obtained in coincidence with ground–based extensive air shower (EAS) arrays.[6,7]

The effect is real. The experiments, difficult due to atmospheric natural and man–made microwave noise, were used mainly to complement the EAS measurements and were eventually abandoned.

For the Cerenkov emission we use the usual formula[8] for the energy dW coherently emitted at angular frequency ω by a charge qe in distance dx,

$$\frac{dW}{dx} = \frac{q^2 e^2 \omega d\omega}{2c^2}(1 - 1/\beta^2 n^2) = 10^{-40} q^2 \frac{ergs}{cm} s^2 \ \omega d\omega \ .$$

For an estimate we replace the integral over $d\omega$ by $\omega \Delta \omega \sim 10^{18} s^{-2}$. This corresponds to coherent microwave emission at wavelength $\lambda = \frac{2\pi c}{\omega} \cong 2m$ over a frequency interval $\Delta \omega \cong \omega$. We believe this is a conservative value, satisfying $\lambda \gg \ell$, with ℓ the effective shower thickness of the ultra–relativistic moving shower.

Shower Charge Imbalance

For electrons with energies in the 100 MeV region of an em shower Askaryan estimated $\Delta N(E_>) \approx 10\% N(E_>)$ on the basis of qualitative arguments. A more complete treatment requires study of the evolution equations for the distributions. We emphasise that we only report on a preliminary, "first–look", at this very detailed problem.

Let $N_\pm(E_>, x)$ be the integral number of $e^\pm$ in the shower at depth x, $N_\pm = \int_{E_>} dE' n_\pm(E', x)$, and let the effective positron–electron inverse interaction length difference be $(\mu/E)\ell n(E/m_e)$. [The difference vanishes at high energy, where the electron and positron bremmstrahlung and pair production cross section become equal.] Then a shower imbalance evolution equation can be written for $\Delta N(E_>, x) = N_+(E_>, x) - N_-(E_>, x)$:

$$\frac{\partial \Delta N}{\partial x} = \frac{1}{2} \int_{E_>}^{E_0} dE' \, n(E', x) \frac{\mu}{E'} \ell n(E'/m_e) \ .$$

The solutions to this can be obtained for the conventional ansatz $n_\pm(E, x) = F(x) E^{-s-1}$, i.e.

$$\frac{\partial \Delta N}{\partial x} = \frac{\mu \ell n(E_>/m_e)}{2 E_>}(1 - \frac{1}{s+1}) N \ ,$$

to leading log accuracy. For s close to 1, N is nearly independent of x, the case of a fully developed stationary shower. Then we obtain

$$\frac{\Delta N}{N} \simeq \frac{\mu ln(E_>/m_e)}{4E_>} \Delta x \ .$$

The effective Δx is set by the optimum shower depth, which is essentially equal to $ln(E_0/E_>)$ for $E_0/E_> \gg 1$. Then for a crude estimate we use $\Delta x = ln(E_0/E_>)$, $E_> = 100 MeV$ and $\frac{\mu}{4E_>} ln(E_>/m_e) \cong 0.1$, obtaining $\frac{\Delta N}{N} \cong 0.1 ln(E_0/10^8 eV)$. This differs from Askaryan's estimate (which ignores shower evolution) by the logarithmic factor $ln(E_0/E_>)$.

Logarithmic factors are not insignificant. Since the radiated microwave power goes like $(\Delta N)^2$, a factor of $ln(E_0/10^8 eV)$ in ΔN gives an increase of $16^2 = 256$ in the signal to the antenna at $E_0 = 10^{15} eV$. However the ansatz $n_\pm = F(x) E^{-s-1}$ is only accurate ignoring logarithmic factors and does not exactly represent a fully developed shower. A more accurate analytic estimate for $N(E_>)$ is $N(E_>) \simeq E_0 / \left(E_> \sqrt{ln(E_0/E_>)} \right)$, implying that ΔN may be only of order $\sqrt{ln(E_0/10^8 eV)}$ larger than one would estimate using Askaryan's method.

This analysis is presented to show that the shower imbalance requires further study. More complicated analysis persistently gives theoretical uncertainty of order $\sqrt{ln(E_0/E_>)}$, or an order of magnitude in the microwave power. The problem justifies a serious combination of analytic and numerical work, which is in progress. Fortunately the Askaryan value seems to be an underestimate, which we use in a conservative spirit in the following event rate calculation.

Event Rates in a Large Antenna Array on Ice

Consider a large array of n microwave antennas arranged to look downward to see upward going events (Fig. 2). The depth at which a shower's microwave emission can be detected is limited by the intrinsic microwave attenuation length. Ice at $^-50 - {}^-60°C$ is an ideal medium, having a long attenuation length, $\gtrsim km$ for frequencies in the $10^9 Hz$ region. Thus Markov and Zhelezhnykh[9] suggest using the antarctic ice pack as a neutrino detector. The antarctic ice is an ideal medium of high homogeneity and sufficiently well known chemical composition. There is also a lot of it.

The event detection rate R is estimated as follows. For neutrinos with energies $E_0 > 10^{15} eV$, the cross section[1,2] is $\sigma(E_0) \gtrsim 10^{-33} cm^2$, the integral flux from point sources[10] such as Cygnus X-3 is $\phi(E_0) \cong 2 \times 10^{-13} cm^{-2} s^{-1}$, and the target (rock and ice) has a number density of protons $N \gtrsim 6 \times 10^{23} cm^{-3}$. The rate $R = N\phi(E_0)\sigma(E_0)V\mathcal{E}$, where V is a target volume and $\mathcal{E}$ is a detector threshold factor.

The target volume V depends on how the antennas are distributed. Optimizing this, we find that the separation distance d between antennas goes roughly like the primary energy, so long as $d \lesssim 1.0 km$. For $10 m^2$ antennas to receive a signal roughly 3 times a $300°K$ noise, at threshold, with 10–16 antennas within the Cerenkov cone at the surface constituting an event, we use $d = 400m$. Then $V \cong 4 \times 10^{14} n cm^3$ is the target volume directly covered by n antennas using a (conservative) range of 2.5 km for the upward going, multi-TeV muon. Events occurring at the edges of the covered volume increase the effective target volume by a factor of roughly two. The threshold factor $\mathcal{E}$ has been set to zero for energies below $10^{15} eV$, or unity given 10–16 struck antennas with $E_0 \gtrsim 10^{15} eV$. Then the rate $R \cong 10^{-7} n s^{-1}$. Another way of putting this is a rate of 3 events/antenna–year, or 150 events/year for 50 antennas.

The primary uncertainties in these figures are from the neutrino flux, conservatively estimated on the basis of the high–energy photon flux, and the shower charge imbalance, conservatively estimated by dropping logarithmic enhancements.

An array of the size 50 antennas would be sufficient for a pilot experiment. This would give a few hundred events/year, sufficient to verify the signal, as well as do limited timing, directionality, and physics studies. If the events are seen then further investment is justified. If the events are not seen, then a bound on the flux can be established (itself a useful piece of scientific information) and there would be no justification for going to a larger array.

Concluding Remarks

What would one learn from detecting UHE neutrinos? The physics of UHE neutrino interactions has never been directly checked. It seems reasonable that at least a flux times cross section could be extracted reliably. Such information could teach us much about what is coming to us from Cygnus $X - 3$. Conceivably there could be particle physics surprises.

On the astrophysical side, the accelleration mechanisms are not understood. There is a prejudice that neutrinos at UHE should be proportional to the flux of photons. Is this true? There are calculations showing an upper limit of order $10^{16} eV$ from a cosmic accellerator, contradicted by data[3] upwards of $10^{18} eV$. What is going on? It is fair to say that the astrophysical impact of detecting UHE neutrinos would be comparable to the detection of Supernova 1987A by neutrinos. Thus the stakes are high and the project is challenging, which is as it should be.

Acknowledgements

We thank our colleagues on this project, R.K. Moore and E. Zeller, for useful discussions. This research was supported in part under DOE Grant No 85–ER40214.A005.

References

1. McKay, D., and Ralston, J., *Phys. Lett.* **167B**, 103 (1986).

2. Quigg, C., Reno, M., and Walker, T., *Phys. Rev. Lett.* **57**, 774 (1986).

3. Dingus, B.L., *et al.*, *Phys. Rev. Lett.* **61**, 1906 (1988); Dingus, B.L., *et al.*, *Phys. Rev. Lett.* **60**, 1785 (1988); Markov, M., *et al.*, *Phys. Rev. Lett.* **55**, 1965 (1985); Hillas, A., *Nature* (London) **312**, 50 (1984); Weekes, T., *Phys. Rep.* **160**, 3 (1987); Baltrusaitis, R., *et al.*, *Ap.J.* **293**, L69 (1985); Cassiday, G., *et. al.*, *Phys. Rev. Lett.* **62**, 383 (1989).

4. Rossi, B., *High Energy Particles*, Chap. 5, Prentice Hall, New York (1952).

5. Askaryan, G., *Zh. Eksp. Teor. Fiz.* **41**, 616 (1961). [*Soviet Physics JETP* **14**, 441 (1962)]

6. Mandolesi, N., Morigi, G., and Palumbo, G., *J. Phys. A,* **9**, 815 (1976); Allen, H., *et al.*, *Proc. 14th Int. Conf. on Cosmic Rays*, Munich Vol. 8 pp. 3077–81, (1975).

7. Hough, J., *J. Phys. A* **6**, 892 (1973); Allen, H., *Progress in Elementary Particles and Cosmic Ray Physics*, **10**, eds. J. Wilson and S. Wonthuysen (North Holland, 1971).

8. Jelley, J.V., *Cerenkov Radiation and Its Applications*, Chap. 3, Pergamon Press (1958).

9. Markov, M. and Zheleznykh, I., *Nuclear Instruments and Methods in Physics Research* **A248**, 242 (1986).

10. Weekes, T., *Phys. Rep.* **160**, 3 (1987).

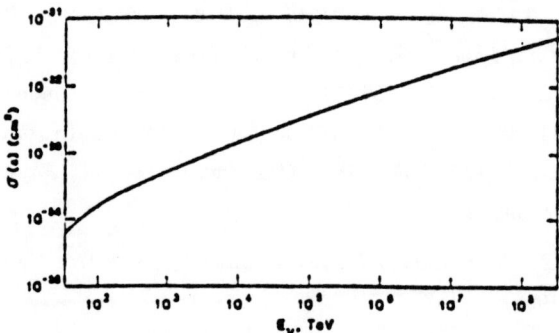

Fig. 1) The UHE νp total cross section showing the QCD enhancement from the growth of small-x quark distributions (Ref. 1). The solid curve is the sea-quark contribution, an underestimate for energies below $10^{15} eV$.

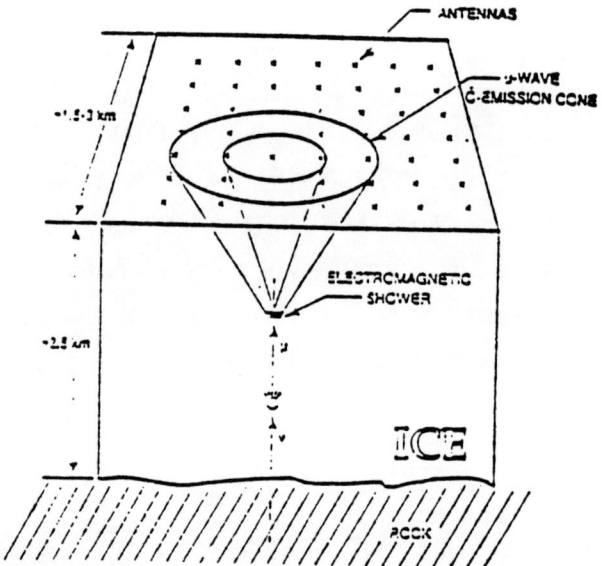

Fig. 2) Schematic layout of microwave detection of upward going neutrinos in ICEMANd, the ice muon and neutrino detector.

New Ideas in South Pole Experiments

D. Seckel

Bartol Research Institute, University of Delaware, Newark DE 19716

Many of the experiments discussed at this meeting could well have been called 'new ideas' a short time ago, but are now recognized as fitting into an established plan for South Pole experimental astrophysics. This rapidly changing state of affairs is due to the fact that the astrophysics community is still becoming familiar with the unique characteristics of the South Pole environment which make some experiments possible that may be done nowhere else on Earth. As these characteristics are more widely understood, it is to be expected that more 'new ideas' will continue to be generated by the community. Although Antarctic astrophysics has matured it has a long and interesting life ahead of it.

With these remarks, it is appropriate to introduce a session on new ideas with a review of those peculiar attributes of South Pole that make it worthwhile to invest relatively large sums of money and time in order to successfully complete an experiment there. The characteristics in the following list are not all independent, but where one feature follows from another I have tried to order them in a logical manner. Some of the qualities attributed to the pole are still under debate. These are included, but with caveates.

Latitude - South Pole is, after all, at the geographic south pole ...

> **24 hr observations** - Due to its polar location, astronomical sources in the southern sky remain above the horizon all the time. Similarly, during the austral summer long period solar observations are possible.
>
> **Constant zenith angle** - Another consequence of the pole is that astrophysical sources remain at constant zenith angle, which simplifies the analysis of backgrounds in high sensitivity experiments.

Elevation - The elevation at South Pole is 2835 m, but the low ambient temperature leads to an equivalent pressure altitude of about 3520 m. This fact makes the pole a good location for studies of high energy cosmic rays.

Climate - There are several aspects of the polar climate which contribute to experimental opportunities; specifically, it is cold, dry, not very windy, and is presumably fairly stable.

> **Temperature** - As mentioned above, the low temperature helps make an equivalent pressure altitude of 3520 m.
>
> **Humidity** - The low humidity, in combination with the low pressure, makes for a small column density of water vapor, which makes South

Pole the best site on Earth for infra-red and sub-millimeter observations.

Stability and Wind conditions - Meteorological conditions were a major topic at the AIA meeting. Although the surface conditions are calm on average, that does not necessarily assure good astronomical seeing. Issues such as average cloud cover, local turbulence, and auroral intensity at selected wavelengths all go into determining the quality of an astronomical site. It is possible that South Pole allows unique seeing conditions (in addition to the low column depth of water vapor), but this is yet to be demonstrated. Another advantage of benign meteorological conditions is the possibility of long duration balloon flights, but again, survey work and test flights are needed before this can be listed as a confirmed attribute of South Pole.

Geomagnetic latitude - South Pole is at geomagnetic latitude of 79°. For surface cosmic ray studies South Pole is not a unique site (although it is a good one), but in combination with favorable meteorological conditions it could provide a unique opportunity for cosmic ray ballooning.

Ice - South Pole is located on top of a glacier more than 2 kilometers thick. The possibility of using Antarctic ice to experimental advantage is an idea which was discussed at some length in the Neutrino and 'New Ideas' workshops.

Most of the discussions in the New Ideas workshop centered around the role of Antarctic ice. For high enery cosmic ray experiments there are Antarctic ice analogs to water detectors proposed at other sites in warmer climates. Two obvious possibilities are icy versions of DUMAND[1] and GRANDE[2]. In the first, a number of bore holes would be drilled to great depth in the ice. Phototubes would be lowered into the ice creating an array capable of detecting high energy neutrino cosmic rays. The ice plays the role of shielding atmospheric cosmic ray showers, target material for the neutrinos, and detector by converting some of the shower energy to the Čerenkov light that is seen in the phototubes. The second experiment would be a combination of extended air shower and neutrino experiments, located at or near the surface. Phototubes embedded in ice, but looking upwards could detect air showers that reach the surface. A second, deeper layer, also looking upwards, could be deployed to detect muons in coincidence with the air showers. Finally, one or more planes of phototubes looking downwards could detect upward going muons from deep underice neutrino interactions. All of this is in direct analog to the GRANDE proposal except that the medium would be ice instead of water. Both these experiments require clear ice. The transmissivity of ice at depth is yet to be determined. Surface ice is not very clear, so an icy GRANDE would require construction and melting of a large quantity of ice.

Another discussion relevant to high energy cosmic rays concerned the possibility of observing Čerenkov emission at radio frequencies from cosmic ray cascades developing in ice (see John Ralston's contribution). Again the ice would play the role of shield, target, and detector.

Mark Ander discussed the use of Antarctic ice as a thick, laterally homogenous platform for sensitive gravitational experiments. There has been much interest recently in gravitational anomalies, or 'fifth force' experiments. The anomolies could be manifested either as an isotope dependence to the gravitational force, or as a deviation from a $1/r^2$ law (presumably in the form of a Yukawa potential), or both. At the moment, the experimental constraints on Yukawa potentials are weakest for length scales of order a kilometer. One way to test for Yukawa type forces is to look for anomalies in the Earth's gravitational field. The problem with such experiments lies in developing sufficiently accurate models of the Earth's density profile near the experiment. There are sampling problems with geological surveys in rock which seem to preclude definitive conclusions, but if the gravity meter can be placed in a laterally homogenous medium, such as ice or water then these problems may be simplified. Deep ice has the advantage of being relatively stable over the lifetime of the experiment.

Experiments measuring gravitational gradients in ice have been performed in Greenland[3]. The results are marginally inconsistent with Newtonian gravity, suggesting the existence of extra forces. However, the length scale of the inferred Yukawa potential is uncomfortably close to the depth of the ice, and so a correction must be performed for possible density anomalies in the rock below the ice. Needless to say these corrections are a matter of some dispute. It is at this point that the South Pole becomes interesting. First, the ice is nearly twice as deep as that of the Greenland experiment. Second, a judicious location of the Antarctic experiment will allow for the necessary corrections due to the underlying rock to be made more reliably than at Greenland. An Antarctic gravitational anomaly experiment would be costly (a few million dollars) but would be the definitive experiment of its type. If the hints and rumors that exist in the current experimental results are to be borne out, South Pole is a logical place to do it.

References

[1] P. Bosetti, et al., *U. Hawaii DUMAND Center Preprint* HDC-2-88 (1988)

[2] R. Ellsworth, et al., *Irvine Preprint* UCI Neut No. 87-38 (1987)

[3] M.E. Ander, et al., *Phys. Rev. Lett.* **62**, 985 (1989)

Millimeter, Sub-millimeter and Infrared Astronomy

TILTING AND VIBRATION OF THE "SKYLAB" BUILDING AT THE SOUTH POLE

Antony A. Stark
AT&T Bell Laboratories; Holmdel, NJ 07733

John Gress
Amundsen-Scott South Pole Station; Antarctica

ABSTRACT

The angular stability of a large building at the South Pole was measured using a tiltmeter for a period of 11 days. Both the top and bottom floors wobble by approximately 20 seconds of arc on a diurnal cycle, and show long-term drifts of up to 60 seconds of arc. The top floor also showed vibrations of up to 30 seconds of arc that were excited by people moving about in the building. These vibrations were not present in the ground floor. This building is not a suitable mounting platform for telescopes with stringent pointing requirements; its foundation does not vibrate, however, and indicates that such a platform could be built.

Ground-based telescopes usually have heavy foundations connected to solid earth. Space and aircraft-borne telescopes require elaborate and expensive active mounts. One problem connected with the siting of large telescopes on the Antarctic plateau is that the the foundation for the mount necessarily rests on ice, and this may not provide a sufficiently stable platform for a conventional passive mount. As a preliminary step in investigating this problem, we have measured the stability of the top and bottom floors of the "Skylab" building at the National Science Foundation Amundsen-Scott South Pole Base. This is a four-story wooden building, where the floorplan of each story is a square approximately 8 meters on a side. The building is supported by large wooden columns at the four corners. These columns rest on heavy timbers frozen into the snow. On the top floor are a number of experiments which observe the sky.

© 1989 American Institute of Physics

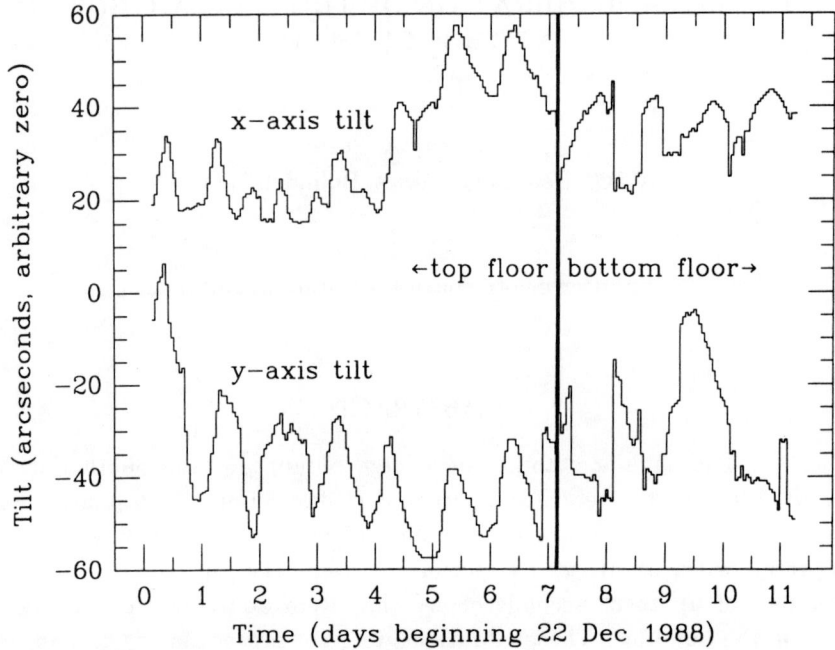

Fig. 1. Relative angular tilt of the top and bottom (after day 7) floors of the Skylab building at the South Pole. Samples were taken every 66 minutes. Time 0 is 0^h UT on 22 December 1988.

We used a biaxial tiltmeter manufactured by the Autonetics Division of Rockwell International (model SE541A). Each of the two orthogonal axes of this instrument consists of a bubble level electrode, an electronic bridge measuring circuit, and an amplifier, which we connected to a strip-chart recorder. A bubble level electrode is a device much like a carpenter's bubble level, where the movement of mercury liquid in a tube causes a change in capacitance which is sensed by the bridge circuit. The two electrodes are mounted in a heavy steel holder with leveling screws. These were adjusted to give a reading near the center of the range; the instrument gives an indication of the change in tilt over time, but does not measure absolute horizontiality. We calibrated the amplitude on the chart recorder using the display panel built into the tilt meter electronics box, which was pre-calibrated at the factory and reads out in arcseconds. The tilt

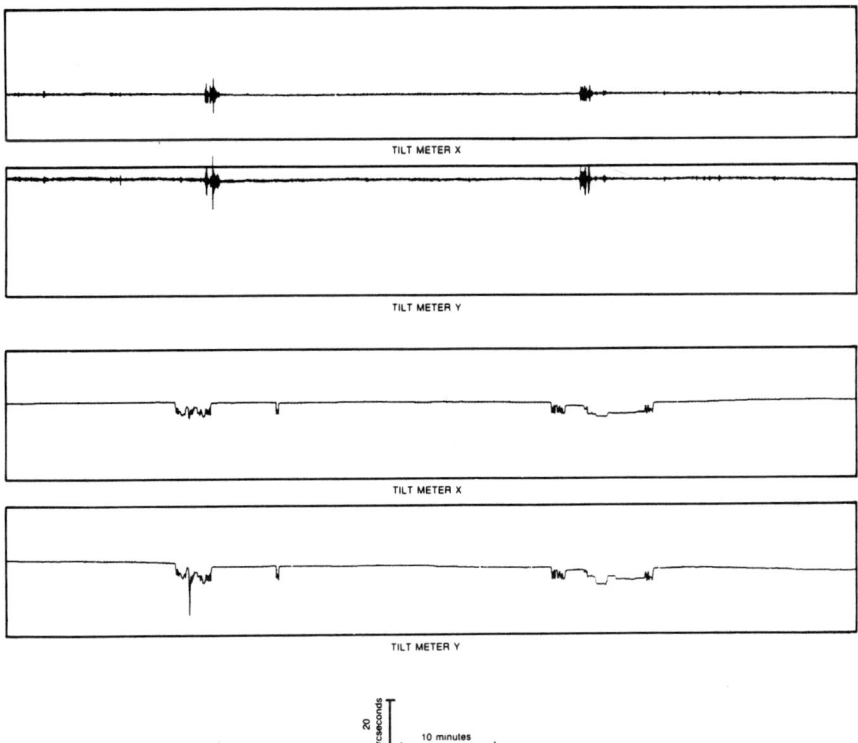

Fig. 2. Short timescale vibrations of the Skylab tiltmeter. This is a reproduction of the chart recorder trace for two 85 minute periods. The upper two traces are of the top floor showing vibrations; the lower two traces are of the bottom floor showing a 7 arcsecond rocking of the floor, but significantly smaller vibrations.

meter - chart recorder combination had a frequency response from about 1 Hz to DC, so we were not sensitive to high-frequency vibrations or soundwaves.

On 22 December 1988, we placed the tiltmeter on the wooden surface of the top floor of the Skylab building, and left it undisturbed for seven days. We then moved it to the ground floor, and made measurements for another four days. The long-term results are shown in Figure 1, where the actual value of the tilt is sampled every 66 minutes. No averaging has been done to arrive at the numbers in this plot. There are long-term drifts of about 60 arcseconds amplitude, and a diurnal periodicity with about 20 arcseconds amplitude. Both effects are present on both the top and bottom floors. In addition to these drifts with timescales of hours or more, the tiltmeter showed vibrations resulting from people walking about in the building and closing doors. These had different characters on the top and bottom floors, as shown in Figure 2. The top floor oscillates very rapidly by as much as 30 seconds of arc. This "earthquake-like" jitter is seen quite often, whenever someone moved in the building. The bottom floor is very much more stable, and never showed the jitter seen on the top floor; it did, however, rock by 7 arcseconds and then rock back a few minutes later in a manner suggestive of mechanical hysteresis.

The mechanical behavior of the Skylab building is inadequate for the mounting of large telescopes with pointing requirements of a few arcseconds. It would require a complex and expensive active mount to remove the short-timescale jitter of the top floor, which has an amplitude an order-of-magnitude larger than is acceptable for most millimeter or submillimeter telescopes, and two orders-of-magnitude larger than is acceptable for infrared and optical telescopes.

The measurements of the ground floor, however, are cause for optimism: it is possible to build a foundation at the South Pole which does not vibrate very much. The 7 arcsecond tilts are probably the result of inherent non-rigidity of the wooden floor, rather than a symptom of vibration. The long-timescale and diurnal tilts, presumably the results of wind and differential solar heating, can be removed by attaching a tiltmeter to the telescope and feeding the measured tilts as corrections to the telescope's pointing computer.

We conclude that while the Skylab building itself is not a suitable site for a large telescope, it is possible to make a passive telescope mount at the South Pole which will not vibrate excessively.

LOW-FREQUENCY MEASUREMENTS OF THE CMB SPECTRUM

A. Kogut, M. Bensadoun, G. De Amici, S. Levin, M. Limon, and G. Smoot
U.C. Berkeley, Lawrence Berkeley Laboratory, and Space Sciences Laboratory

G. Sironi
Physics Department — University of Milano

M. Bersanelli and G. Bonelli
IFCTR/CNR — Milano

ABSTRACT

As part of an extended program to characterize the spectrum of the cosmic microwave background (CMB) at low frequencies, we have performed multiple measurements from a high-altitude site in California. On average, these measurements suggest a CMB temperature slightly lower than measurements at higher frequencies. Atmospheric conditions and the encroachment of civilization are now significant limitations from our present observing site. In November 1989, we will make new measurements from the South Pole Amundsen-Scott Station at frequencies 0.82, 1.5, 2.5, 3.8, 7.5, and 90 GHz. We discuss recent measurements and indicate improvements possible from a polar observing site.

INTRODUCTION

The spectrum of the cosmic microwave background (CMB) preserves a record of the interactions between the evolving matter and radiation fields in the early universe. The Rayleigh-Jeans portion of the spectrum is particularly sensitive to energy releases occurring at redshifts between approximately 2×10^6 and 4×10^4, where inverse Compton scattering and bremsstrahlung can produce spectral features[1,2]. The Rayleigh-Jeans spectrum can also distinguish between such models of the apparent sub-mm CMB distortion as dust emission and Compton scattering, which predict different spectral behaviour below 90 GHz[3]. Starting in 1982, an international collaboration from Italy and the U.S. has measured the Rayleigh-Jeans portion of the CMB spectrum[4,5]. Our present results indicate a slightly lower CMB temperature below 30 GHz than do measurements at higher frequencies. In an effort to reduce the magnitude and variability of non-cosmological foregrounds, we plan new measurements in November 1989 from the South Pole.

MEASUREMENT TECHNIQUES

The measurement at each frequency compares the output voltage of a radiometer as it alternately views a cryogenic reference target and the zenith sky. The 0.82, 2.5, and 90 GHz radiometers are Dicke-switched, while the others are total-power. The antenna temperature of the zenith is

$$T_{A,\text{zenith}} = T_{A,\text{load}} + G(V_{\text{zenith}} - V_{\text{load}}), \qquad (1)$$

© 1989 American Institute of Physics

where $T_{A,zenith}$ and $T_{A,load}$ are the antenna temperatures of the zenith sky and the reference target, G is the calibration coefficient of the radiometer, and V_{zenith} and V_{load} are the output voltages viewing the zenith or the reference load. We then measure and subtract all non-cosmological foregrounds (principally atmospheric and galactic emission) to arrive at the CMB antenna temperature:

$$T_{A,CMB} = T_{A,load} + G(V_{zenith} - V_{load}) - T_{A,Atm} - T_{A,Galaxy} - T_{A,ground} - T_{A,RFI} - \Delta T_{Offset}. \quad (2)$$

Here $T_{A,Atm}$, $T_{A,Galaxy}$, $T_{A,ground}$, and $T_{A,RFI}$ are the antenna temperatures of the atmosphere, galaxy, the earth seen in the antenna sidelobes, and man-made radio interference. ΔT_{Offset} refers to possible changes in radiometer performance correlated with radiometer position.

The measurements depend critically upon the cryogenic reference target used. The target used above 1 GHz from 1982 through 1987 has been described elsewhere[5]. To allow calibration down to 1 GHz, and to reduce potential systematic uncertainties, we built a new target before the measurements in 1988 (Figure 1).

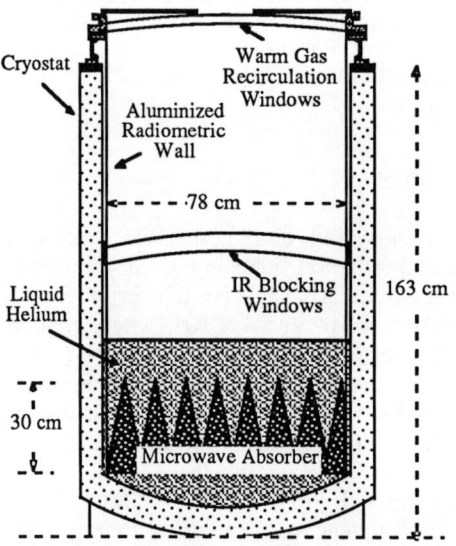

Figure 1: The cold load used in 1988.

The new cold load consisted of a microwave absorber immersed in liquid helium (LHe) within a large, open-mouthed cryostat. An aluminum-coated fiberglass cylinder surrounded the absorber and acted as an oversized multi-mode waveguide, preventing excessive radiation from warm portions of the cryostat from reaching the antenna aperture. Thin (25 μm) polyethylene windows at the top of the cryostat prevented air from condensing on the radiometric surfaces. There were two principal changes in the new cold load. Two IR-blocking windows replaced the old shutter system, which eliminated gaps in the radiometric wall and allowed easier operation with a much lower heat leak to the LHe. In addition, the absorber in the new cold load was more than 50% thicker to ensure low reflection at low frequencies. The antenna temperature of the target could be calculated from the temperature of the absorber (equal to the boiling point of helium at ambient pressure), with small contributions from emission or reflection from the windows and walls. The Teflon-impregnated glass cloth windows (76 and 150 μm thick) introduced new sources of reflection and emission, which we measured carefully. Total corrections were small (<50 mK between 1.5 and 10 GHz). The 0.82 GHz radiometer will use a separate LHe-temperature coaxial cold load[6].

Atmospheric and galactic emission define a natural observing window for ground-based CMB measurements between 1 and 20 GHz. We have measured atmospheric emission above 2.5 GHz by tipping the radiometers and comparing the signals from the zenith and the scan angle (typically 30° or 40°), and correlating the signal with the airmass in the beam. We estimated galactic synchrotron and HII emission by scaling maps at lower frequencies[7]. We have checked the galactic model by performing differential drift scans, comparing the signal difference between two points separated by ~30° in right ascension as the Earth's rotation moves the galaxy through the beams. The limiting uncertainty in the CMB determination typically has been the atmosphere for measurements above 2.5 GHz, and the galaxy below 2.5 GHz.

CURRENT STATUS

The current status of CMB spectrum measurements is summarized in Figure 2. There is an apparent spectral distortion at high frequencies[8] and a possible disagreement between the low-frequency ground-based measurements and results at higher frequencies. Measurements in this field are subject to a range of systematic errors. To assess these uncertainties, is highly desirable to measure the CMB spectrum under a variety of different conditions using different experimental techniques.

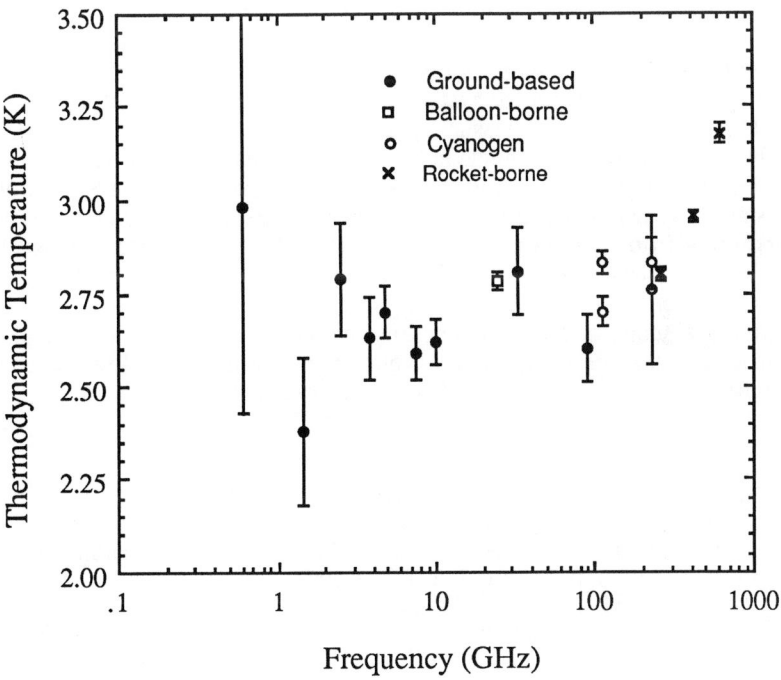

Figure 2: Recent precise CMB measurements

Future Work

Systematics are the bane of precise CMB measurements. Since 1982, we have changed many observing parameters as a cross check on the measurements. Such changes include changes in radiometer design, measuring technique for atmospheric and galactic scans, and an entirely new reference target. In November 1989 we plan to change the observing site by making measurements at 0.82, 1.5, 2.5, 3.8, 7.5, and 90 GHz from a site near the South Pole. The polar site allows a reduction in the magnitude and variability of various foreground signals.

Atmospheric emission of order 1 K is a foreground common to all of our measurements, and the only one without large frequency dependence; as such, it is a candidate for systematics in the ground-based set of measurements. A small error in the tip scans (e.g., larger than suspected ground pickup) could result in an overestimate of $T_{A,Atm}$ and an underestimate of $T_{A,CMB}$. We have tested extensively for such an effect; it is unlikely for our estimates of $T_{A,ground}$ to be in error by the same amount over more than a decade of frequency and a variety of radiometer designs. The importance of the low-frequency results, however, demands an additional check.

The measured $T_{A,Atm}$ tends to be both larger and more variable than predicted by models based on the oxygen and water vapor content of the atmosphere[9,10,11]. Although the data are probably better understood than the models, a cross-check is desirable. The low water vapor content of the polar atmosphere will reduce the magnitude of the atmospheric emission and eliminate essentially all variability, providing a convenient test for systematics in the determination of $T_{A,Atm}$. We will also attempt an atmospheric measurement at 1.5 GHz.

Interference from man-made radio transmitters has been a significant problem at several of our observing frequencies. We anticipate that RFI at the South Pole will be both smaller (perhaps negligible) and better understood. Nearly simultaneous observations from the same site at 6 frequencies will reduce the possibility of undetected systematics. The planned measurements will provide a precise determination of the low-frequency CMB spectrum from an ideal site. The reduction in foreground signals, and consequent improvement in the understanding of related systematics, will significantly improve our understanding of the low-frequency CMB spectrum.

References

1. L. Danese and G. De Zotti, Astron. Astroph. **107**, 39 (1982).
2. —, Astron. Astroph. **84**, 364 (1980).
3. F.C. Adams, K. Freese, J. Levin, and J.C. McDowell, Ap. J. (submitted 1989).
4. Smoot, G., *et al.*, Ap. J. Lett. **291**, L23 (1985).
5. —, Ap. J. Lett. **317**, L45 (1987).
6. M. Limon, C. Marchioni, and G. Sironi, J. Phys. E: Sci. Inst., in press (1989).
7. C.G.T. Haslam, C.J. Salter, H. Stoffel, and W.E. Wilson, Astron. Astroph. Suppl. Ser. **47**, 1 (1982).
8. T. Matsumoto *et al.*, Ap. J. **329**, 567 (1988).
9. G. De Amici *et al.*, Ap. J. **329**, 556 (1988).
10. A. Kogut *et al.*, Ap. J. **325**, 1 (1988).
11. —, Ap. J. (submitted 1989).

High Altitude Measurements of Fluctuations in the CMB

R.D. Davies
Nuffield Radio Astronomy Laboratories
Jodrell Bank
Macclesfield
Cheshire
UK

Abstract

The detection of fluctuations in the primordial CMB emission requires long integrations on limited areas of the sky; such extensive observations are best made from the ground on (high) dry sites. A current programme of measurements is described covering the frequency range 5 to 32 GHz using equipment at Teide Observatory, Tenerife, and in Antarctica. A bolometer system is under development for observations in the range 100 to 300 GHz to be made on Mauna Kea.

1. Introduction

The search for fluctuations in the Cosmic Microwave Background is one of the key observational programmes in modern cosmology. Such observations place fundamental constraints on scenarios of galaxy formation in the early Universe.

Limits to $\Delta T/T$ of less than 10^{-4} have been set over angular scales from arcminutes to tens of degrees. The lower angular scales refer to clusters of galaxies while the largest refer to structures at the upper end of those detected in the local Universe. At an upper limit of $\Delta T/T \sim 10^{-4}$, the low density adiabatic picture of galaxy formation is ruled out. However more sensitive observations approaching $\Delta T/T \sim 1-3 \times 10^{-6}$ are required to confront the wide range of models currently in vogue. For example Cold Dark Matter (CDM) scenarios require much lower fluctuation levels ($\Delta T/T \leq 10^{-5}$) than adiabatic scenarios (Bond & Efstathiou 1984, Vittorio & Silk 1984).

The detection of fluctuations at $\Delta T/T \sim 3.7 \times 10^{-5}$ on $8°$ scales by Davies et al. (1987) on the one hand suggests that fluctuations in the CMB may have been detected but on the other hand raises the possibility that at the observing frequency of 10.4 GHz foreground emission from the Galaxy or from extragalactic sources may be contaminating the results. The likelihood that the signals were some spurious response of the observing system has been ruled out with subsequent observations made on an independent system operating on an angular scale of $5°$ which detected excess emission in the same part of the sky. Sensitive CMB measurements are required at a number of frequencies in the astronomical window between 10 and 300 GHz. At the lower end of the window galactic thermal and synchrotron emission produces confusion, while at the upper end IRAS cirrus presents a problem.

2. A Programme of Observations

Long integrations are required to achieve the sensitivities (3-10 micro Kelvins) necessary for the detection and mapping of CMB fluctuations. Depending on the bandwidth available, observation times necessary per beamwidth are in the range hours to days with currently available system sensitivities. Since hundreds of beam areas are required for these programmes, long observing campaigns are essential. Ground-based facilities are necessary for these campaigns. At the higher frequencies atmospheric water vapour causes a serious deterioration of system performance, and as a consequence dry environments are necessary, such as mountain or polar sites.

I will describe below the programme of observations of the CMB to which we are committed at Jodrell Bank with our collaborators at the Instituto de Astrofisica de Canarias Tenerife, Mullard Radio Astronomy Observatory Cambridge, British Antarctic Survey Cambridge, Queen Mary College London and the Joint Astronomy Centre Hawaii.

2.1 5 GHz Interferometry

At Jodrell Bank, essentially a sea level site in the context of the present article, we are operating a 2-element interferometer on a baseline of 11 wavelengths. The response gives a lobe separation of 5° within an 8° beam. The system is receiver noise limited with a total system noise of 30K and a bandwidth of 400 MHz. The only obvious atmospheric effects are those produced during times of rain or heavy cumulus clouds. We are making further investigations of the effects of lesser humidities by comparison with continuously recorded meteorological data.

2.2 10.4 GHz beamswitching

A description has been given of preliminary results (Davies 1987) of the 8° beamswitching observations at Teide Observatory, Tenerife, located at a height of 2300m. Beamswitching on an 8° scale is more vulnerable to atmospheric effects than the interferometer described above. Nevertheless, our experience at Teide Observatory is that only 20 percent of the observations are affected by atmospheric noise. Such periods of increased noise are identified with the times when mist and cloud rise to the elevation of the observatory.

A major improvement in this system resulted from the incorporation triple-beamswitching which rejects long-term drifts in atmospheric emission. All the beamswitching systems described below are of the triple beam type.

2.3 15 GHz beamswitching

A system noise of 60K has been achieved by cooling the circulator switch between the 5° horns and the receiver. The bandwidth is 1000 MHz. As with the beamswitching system described above two independent channels record the in-phase and out-of-phase astronomical signals to provide a further $\sqrt{2}$ improvement in sensitivity. This system is running at Jodrell Bank and will be installed on Tenerife in the autumn.

2.4 32 GHz beamswitching

A triple-beam dual-channel system which is a scaled version of the 15 GHz system is under construction at Jodrell Bank. It is planned to take it to Tenerife in the coming winter. At the end of 1990 it will be installed at the British Antarctic Survey Halley Bay base, Antarctica. This site has low precipitable water vapour levels during the southern winter. Over substantial periods the water vapour content lies between 1-2 mm pwv and is often less.

2.5 Millimetre wave beamswitching

A consortium from NRAL, QMC, MRAO and JAC is now actively designing a beamswitching system with a 2° resolution based on bolometers cooled to 50-100 mK. The design has benefitted from the development of the Submillimetre Common User Bolometer Array (SCUBA) at the Royal Observatory Edinburgh for UKIRT on Mauna Kea. Again this is to be a triple-beam system; it will scan around a circle of radius $\sim 10°$. The estimated sensitivity in one hour of integration on a point is 2 μK. The aim of this experiment is to measure the CMB structure at 3 mm and to use 2 mm and 1.1 mm channels to measure and correct for atmospheric and galactic cirrus emission. Since CMB fluctuations can be up to 10 percent polarized (Bond & Efstathiou 1987), we plan to measure polarization simultaneously.

The fast scanning will allow the astronomy to be separated from the weather since they are on different timescales. Variations in noise power due to the atmosphere can be quantified by observations at 3 wavelengths and making use of their characteristic spectral form (Meyer, Jefferies & Weiss 1983). The data-gathering for atmospheric subtraction will be carried out by distributed transputers and the results fed back to the central computer where a real-time algorithm will provide a corrected sample value at each frequency and polarization.

The galactic cirrus makes a significant contribution at millimetric wavelengths. For example an HI cloud with a line integral of 10^{20} atom cm^{-2} corresponding to a 100μ brightness of ∼1 MJy sr^{-1} gives a 1 mm temperature of 120 μK, a 2 mm value of 20 μK and a 3 mm value of 10 μK. Thus at the potential sensitivity of this experiment, corrections will need to be made for the effects of interstellar cirrus even at high galactic latitudes. If there is a low temperature (∼10K) component of the cirrus as indicated by the observations of Matsumoto et al. (1988) the corrections at the longer wavelengths will be larger than given above.

References

Bond, J.R. & Efstathiou, G., 1984. Astrophys.J.Letters, 285, L45.
Bond, J.R. & Efstathiou, G., 1987. Mon.Not.R.astr.Soc., 226, 655.
Davies, R.D., Lasenby, A.N., Watson, R.A., Daintree, E.J., Hopkins, J., Beckman, J., Sanchez-Albeida, J. & Rebolo, R., 1987. Nature 326, 462.
Matsumoto, T., Hayakawa, S., Matsuo, H., Murakami, H., Sato, S., Lange, A.E. & Richards, P.L., 1988. Astrophys.J., 329, 567.
Meyer, S.S., Jefferies, A.D. & Weiss, R., 1983. Astrophys.J., 271, L1.
Vittorio, N. & Silk, J., 1984. Astrophys.J.Lett., 285, L39.

CBR ANISOTROPY AND GALACTIC EMISSION OBSERVATIONS FROM ANTARCTICA

P. Andreani[1], G. Dall'Oglio[1], M. Di Bari[1],
L. Martinis[3], L. Piccirillo[1,2], L. Pizzo[1], C. Santillo[1]

(1) University of Rome, Physics Dept, P.le A. Moro 2, 00185 Rome (Italy)
(2) Istituto Superiore P.T., V.le Europa 190, 00144 Rome (Italy)
(3) ENEA T.I.B., Frascati (Italy)

INTRODUCTION

Ground-based experiments devoted to the search for Cosmic Background Radiation anisotropies must be designed and realized taking into account two major sources of disturbance:
- Atmospheric attenuation and noise
- Contamination due to galactic emission.

Therefore, it seems to be better to choose a wavelength in the millimetric range for which the CBR is near its peak of emission, the atmosphere is transparent and the galactic emission is expected to be negligible.

In this paper we report the results of an experiment on the CBR anisotropy at 2 mm wavelength and at 2.5° angular scale, carried out during the antarctic summer 1987-88. Atmospheric noise is partly reduced by monitoring the sky with an additional detector channel at 350 μm or at 1 mm wavelength.

We find evidence for statistical fluctuations at a level of $\Delta T/T = 2\ 10^{-4}$, most of which are to be ascribed both to atmospheric residual noise and/or patchy galactic emission.

A portion of the observing time was devoted to the study of galactic emission in the region $b \approx -8° \div -16°$ and $l \approx 290° \div 310°$. A few southern clouds (the Chamaeleon region and a cirrus cloud) have been detected[1].

In addition, a run of observations has been devoted to the study of the Magellanic Clouds.

All of these structures show an excess at millimetric wavelength in their spectra suggesting the presence of a cold dust component[2] coexisting with the IRAS warm dust.

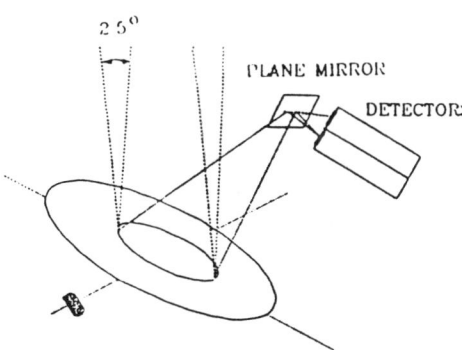

WOBBLING OFF-AXIS PARABOLOID

fig. 1

EXPERIMENTAL SETUP

a) <u>Instrumentation:</u> the experimental setup is shown in fig. 1. The optics consist of a f/2 off-axis paraboloid, 1 meter diameter, which defines a field of view (FOV) of about 1°, and a secondary plane mirror near its focus.

The matching between the optics and the detectors is provided by two f/4.5 Winston concentrators[3], cooled at the same temperature of the bolometers. Because of the large f-number, the area on the primary mirror viewed by the Winston cone is a circle of about half the diameter of the flux collector, in order to reduce diffraction.

An ^{3}He composite Ge-bolometer is coupled to a bandpass mesh filter cooled to 4K, which defines a bandwidth of 500 μm, around 2150 μm central wavelength, matching the atmospheric window between $4 \div 6$ cm^{-1}. The working temperature of the detector is about 0.35K.

The second detector is a ^{4}He monolithic Si-bolometer cooled to 1K and coupled either to a lowpass filter with a cut-on around 1 mm wavelength or to a bandpass filter centered at 350 μm with a bandwidth of 100 μm. All these filters are cooled[4] at 4K.

The 2 mm channel is devoted to the detection of radiation coming from the sky. The measured atmospheric transmission is about 90%.

b) <u>Calibration</u> is performed in two steps: first we measure the responsivity recording the DC output from the detectors when:
- the window is open looking at the zenith
- the window is open looking at the laboratory ($T \approx 273$ K)
- the window is covered with a metal sheet at room temperature polished to have high reflectivity.

In the last configuration the output signal is the smallest one suggesting that the equivalent radiative temperature of the sky is higher than the equivalent radiative temperature of the metal sheet. The former may be obtained by the product ηT of the emissivity η and the thermodynamic temperature T, which gives $T \approx 23$ K. Therefore we can safely affirm that the metal sheet has an equivalent radiative temperature lower than 23 K. By using experimental data an equivalent radiative temperature of $4 \div 8$ K is obtained.

With this procedure we obtain:

- at 2 mm: $R = (9.5 \pm 1.8) 10^{-6}$ V/K ≡ $(2.1 \pm 0.4) 10^6$ V/W
- at 1 mm: $R = (53 \pm 15) 10^{-6}$ V/K ≡ $(1.4 \pm 0.4) 10^6$ V/W
- at 350 μm: $R = (12 \pm 3) 10^{-6}$ V/K ≡ $(9.0 \pm 2.0) 10^3$ V/W

including all the experimental errors.

As for the second step we perform the following test: we record the DC output when the instruments are looking directly at the sky and when are looking at the sky through the telescope. Since the two signals are equal, within the errors, we may assume that the radiation coming from the sky fills entirely the detector FOV in both cases.

c) <u>Observations</u> are carried out as follows. The optics are aligned in order to modulate in the direction E-W and to point to the zenith, which means to observe at a declination of -74.7°. Signals are integrated for 30 s and sampled every 10 s. Observations have been carried out with the drift scans technique, i.e., by keeping the telescope fixed. This configuration eliminates the effects of different diffraction patterns surrounding the observing beam and provides a scanning of the sky following the Earth's rotation.

DATA ANALYSIS

<u>CBR anisotropy</u>: in order to define independent sky regions and to avoid overlap between two consecutive measurements, we average all the points corresponding to one FOV in the sky and we obtain, for each sky position, a value X_i with associated a statistical error ΔX_i. By using the 350 µm channel when this signal is well correlated with the 2 mm, we obtain a reduction of atmospheric noise of a factor of about 3. The resulting signals turn out to be well correlated with those obtained from the analysis of the IRAS maps at 100 µm wavelength[1,2]. As a consequence we believe that our data are partly contaminated by thermal emission of a very cold dust component well mixed with the one detected by IRAS. In order to estimate a value for the sky fluctuations we extract only those data corresponding to "quiet" sky regions, i.e., those regions for which IRAS 100 µm gradients at the same angular scale are the smallest. We find statistical evidence for fluctuations[5] at a level of $\Delta T/T = 2 \cdot 10^{-4}$. Since this result does not agree with more stringent upper limits (Melchiorri et al.[6] and Davies et al.[7]) at similar angular scales, we conclude that residual atmospheric noise and/or patchy galactic emission could be responsible of most of the fluctuations observed. Nonetheless our measurements are not in contrast with the upper limits given by Mandolesi et al.[8].

<u>Magellanic clouds and galactic emission</u>: during our scanning we observed the Large and Small Magellanic Clouds and a region at b=-8°÷-16°, l=290°÷310°, where two cirrus clouds (G299-16, Chamaeleon cloud) have been detected by IRAS.

The data, once reduced, show strong signals whose non-atmospheric origin is demonstrated by the following argument: a) there is a clear rise above the atmospheric noise in sky regions where the IRAS maps show the presence of infrared structures; b) the same signal has been recorded during a second and third scan.

We find a good correlation between our data and the IRAS ones for all the structures we detected (fig. 2). Our results show a millimetric excess with respect to the extrapolated emission from IRAS data, for both the Magellanic Clouds and the other sources. This could be explained by the presence of a cold dust component well mixed with the warm one inferred from IRAS. As far as the Galactic cirrus clouds are concerned we can make a grey-body fit with T≈22 K and spectral index $\alpha \approx 2$ for IRAS data, while T≈10 K and $\alpha \approx 1$ for our data (fig. 3).

In the case of the Magellanic Clouds, the millimetric excess is too high to be explained by the extrapolated radio spectra[9]. Therefore the presence of a cold thermal component seems, also in this case, a natural way to explain the spectra. In order to determine whether our observations provide evidence for such conclusion or not, we adopted Chini's model [10] and found a good agreement with the assumption T=(25÷35) K for the warm dust and T=(14÷16) K for the cold one.

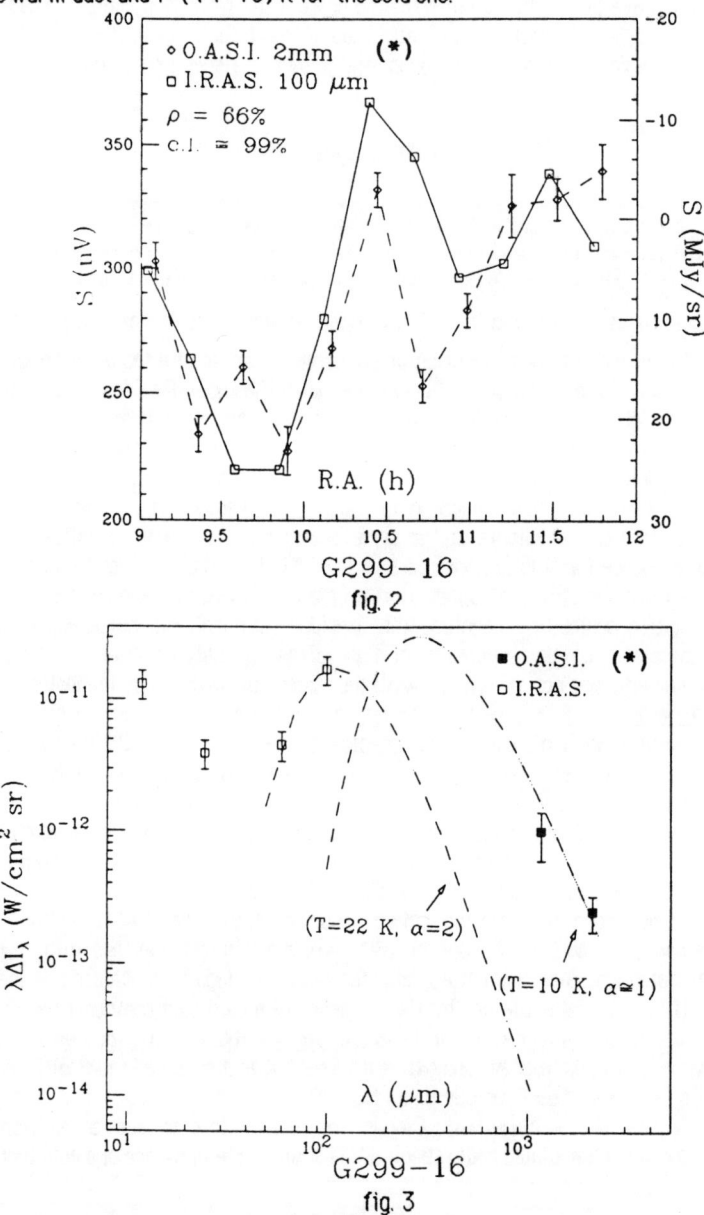

fig. 2

fig. 3

(*) this work.

REFERENCES

1. Andreani P., Dall'Oglio G., Martinis L., Piccirillo L., Pizzo G., Rossi L., Venturino C., submitted to Ap. J., (1988)
2. Andreani P., Ceccarelli C., Dall'Oglio G., Martinis L., Piccirillo L., Pizzo G., Rossi L., Venturino C., submitted to Ap. J. (1988)
3. Winston R., Journ. of Opt. Soc. Am., 60, 245 (1970)
4. Mason C., Ceccarelli C., Masi S., Dall'Oglio G., Ferri G., Radford S.J.E., Infrared Phys., 26, 273 (1986)
5. Andreani P., Dall'Oglio G., Martinis L., Piccirillo L., Pizzo G., Rossi L., Venturino C., submitted to Astron. Astrophys. (1989)
6. Melchiorri F., Melchiorri B., Ceccarelli C., Pietranera L., Ap. J., 250, L1 (1981)
7. Davies R.D., Lasenby A.N., Watson R.A., Daintree E.J., Hopkins J., Beckman J., Sanchez-Almeida J., Rebolo R., Nature, 326, 462 (1987)
8. Mandolesi N., Calzolari P., Cortiglioni S., Delfino F., Sironi G., Inzani I., De Amici G., Solheim J.E., Berger E., Partridge R.B., Martenis P.L., Sangree C.H., Harvey R.C., Nature, 319, 751 (1986)
9. Alvarez H., Aparici J., May J., Astron. Astrophys., 176, 25 (1987)
10. Chini R., Kreysa E., Krugel E., Mezger P.G., Astron. Astrophys., 166, L8 (1986)

COSMIC BACKGROUND ANISOTROPY STUDIES AT 10° ANGULAR SCALES WITH A HEMT RADIOMETER

Todd Gaier

Jeff Schuster †

Philip Lubin

University of California, Santa Barbara, CA 93106

ABSTRACT

An expedition to the Amundsen-Scott South Pole Station was recently mounted to measure medium to large angular scale fluctuations in the cosmic background radiation (CBR) at 15 and 25 GHz. Preliminary results are reported in this paper. No fluctuations have been detected as yet and data analysis is proceeding using likelihood ratio tests to set upper limits of $\Delta T/T$ for models which may be constrained by this experiment.

INTRODUCTION

The cosmic background radiation (CBR) is presently believed to be highly red-shifted photons from the era of matter-radiation decoupling when the universe was at a temperature $\approx 4000K$. The visible universe is highly non-uniform with density fluctuations in the form of galaxies and galactic clusters, and the CBR should contain information about density perturbations in the early universe which led to the inhomogeneities observed in the present day matter universe. CBR photons have been travelling unimpeded since the time of decoupling and so serve as an excellent measure of conditions of the universe at the time of decoupling.

To date measurements of angular fluctuations of the CBR have shown that the background is highly uniform. The only definite exception to this homogeneity is a dipole anisotropy of amplitude 3 mK which is believed to be the result of the peculiar motion of our galaxy and not intrinsic to the CBR. Recently two groups have reported controversial detections of anisotropy in the CBR. With the exception of these results, the CBR has been measured to have a temperature 2.75 K which is uniform to $\Delta T/T < 10^{-4}$. Our experiment is most sensitive to fluctuations at angular scales of 4-20 deg and is capable of measuring fluctuations of $\Delta T/T < 10^{-5}$.

OBSERVING STRATEGY AND THE INSTRUMENT

Each system in the radiometer uses low noise cryogenic HEMT amplifiers and a cryogenic latching Dicke-switch to achieve a total system noise of $\approx 3-4mK/\sqrt{Hz}$. Cryogenic requirements for the instrument were satisfied with a closed-cycle helium gas refrigerator, which only needs electricity to operate, allowing extended runs without liquid cryogens. Figure 1 is the schematic of the instrument.

---

† also Physics Department, UC Berkeley

When making ground based measurements, the atmosphere can be a formidable problem to deal with adding unwanted noise and signal. For this reason we chose to make our measurement from the Amundsen-Scott South Pole Station. By making zenith scans atmospheric temperatures (radiometric) of $2.78 \pm .35K$ at 15 GHz and $5.5 \pm .25K$ at 25 GHz were measured. Figure 2 shows these scans and the appropriate fit. In addition to choice of site, observing techniques can reduce the effects of atmospherics. We employ in our system, a slow second chop or "wobble" which measures differences on shorter time scales making the system less prone to long term instabilities which include the atmosphere. Figure 3 shows the FFT of a data scan of a single sky position. The 1/f knee occurs at about 400 sec. Subtraction of these chop states also makes the experiment less sensitive to atmospheric gradients.

We chose to observe using drift scans at a few elevations. This involves double-chopping at a given azimuth and letting the earth's rotation change the field of view with time. Because the entire sky is not free of contaminating signals (the galaxy and sun), the instrument was periodically slewed to a different azimuth to increase the amount of integration time at sky positions free from galactic emission and far from the sun. Unfortunately this technique had the effect of causing baseline offsets of $\approx < 1mK$ at different azimuths. These offsets are stable for a given azimuth set position and can be attributed to elevation deviations as a function of azimuth position. In analyzing the data, we only consider drift scans which pass through the galactic plane allowing us to determine offsets with great accuracy. This galaxy information is also useful for frequency scaling of galactic emissions between our two frequencies, which will hopefully increase our effective sky coverage by allowing us to subtract out known galactic sources.

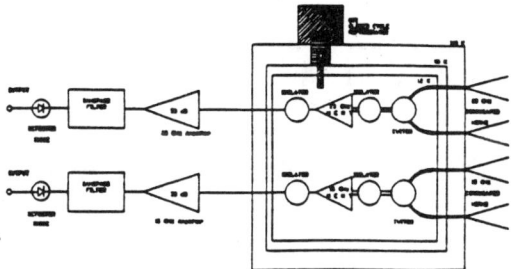

Fig. 1

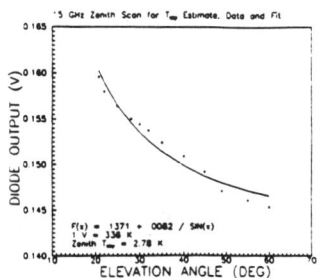

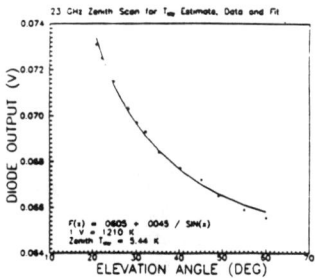

Fig. 2

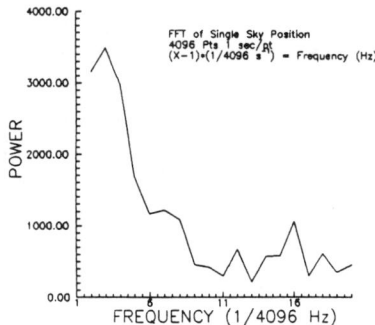

Fig. 3

RESULTS

In **Figure 4.** data from scans at $\delta = -60$ are displayed. The data have been binned, double-subtracted, and filtered for display purposes. The only data fitting performed is subtraction of scan offsets. The expected galactic contribution at 25 GHz is also displayed. This galactic contribution is modeled from a 408 MHz map with scaling indices of $\nu^{-2.7}$ and $\nu^{-2.1}$ for galactic synchrotron and HII emissions respectively. These values are not expected to be accurate over such a large range of frequencies and for positions off of the plane. We have not yet analyzed our data at 15 GHz to determine optimum scaling parameters.

Simple statistical tests have been performed on the 25 GHz data. Figure 5 shows a segment of the data with galaxy and dipole components as well as a small residual linear trend removed. The data are grouped into 1/3 beam width bins and slightly oversampled. The reduced χ^2 of this set is about .85 for 14 degrees of freedom or about 60 percent consistent with random noise with this σ. These results are preliminary.

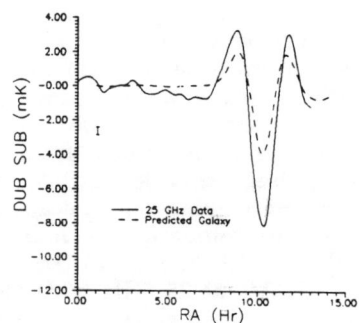

Fig. 4

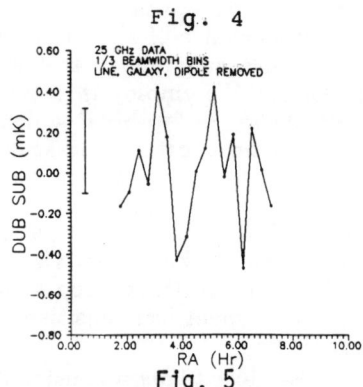

Fig. 5

CONCLUSION

We have described an instrument designed to make sensitive measurements of the CBR on 10° angular scales. The instrument was operated at the South Pole during the austral summer of 1988-89. Preliminary results reveal no obvious intrinsic structure. The sensitivity was limited by observation time not sytematics.

ACKNOWLEDGEMENTS

This work was supported by the National Science Foundation, the California Space Institute, and the University of California. This work would not have been possible without the support and encouragement of John Lynch. We thank Mike Balister, Sandy Weinreb, and Marian Pospieszalski of NRAO for their 23 GHz HEMT amplifier. Finally, we would like to thank Amundsen-Scott station manager Bill Coughran, L.G.N., and the whole 1988-89 ANS support staff for their constant support at the Pole.

REFERENCES

1. F. Melchiorri, B. O. Melchiorri, C. Ceccarelli, and L. Pietranera, *Ap. J.*, **250**, L1 (1981).

2. R. D. Davies, R. Watson, E. J. Daintree, J. Hopkins, A. N. Lasenby, J. Beckman, J. Sanchez-Almeida, and R. Rebolo, *Nature*, **326**, 6112 (1987).

3. C. G. T. Haslam, C. J. Salter, H. Stoffel, and W. E. Wilson, *Astron. Astrophys. Suppl.*, **47**, 1 (1982).

SOUTH POLE STUDIES OF THE ANISOTROPY OF THE COSMIC BACKGROUND RADIATION AT ONE DEGREE

Peter R. Meinhold †

Philip M. Lubin

Alfredo O. Chingcuanco ‡

Jeff A. Schuster †

Michael Seiffert

University of California, Santa Barbara, CA 93106

ABSTRACT

We have developed a system for making measurements of spatial fluctuations in the Cosmic Microwave Background Radiation at 3 mm wavelength, on an angular scale of .5 to 5 degrees. The system includes a telescope with a Gaussian beam with a full width at half max (FWHM) of 20 to 50 arc-minutes, an SIS (Superconductor-Insulator-Superconductor) coherent receiver operating around 90 GHz, and for balloon flights, a pointing system capable of 1 arc-minute RMS stabilization. We report on results from ground based measurements made from the South Pole station during December, 1988.

INTRODUCTION

Searches for structure in the spatial distribution of the Cosmic Background Radiation (CBR) are one of the few experimental tests of cosmological models.

Currently no definitive detections of anisotropy have been made except for the dipole term, and limits of 20 to 200 parts per million have been established from 10 arc seconds to 90 degrees angular scale (see Figure 1). In the region from 1 to 10 degrees few experiments have been done with sufficient sensitivity to seriously constrain cosmological models, galaxy formation scenarios in particular. Recent reports of detection in this region are suggestive but may suffer from systematic problems.

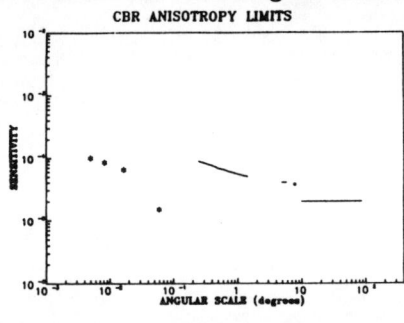

Figure 1

For theoretical and experimental reasons, interest in experiments in the .5 to 10 degree range has risen in the past few years. The two primary systematic difficulties with doing sensitive experiments in this angular range are the atmosphere, which has time varying structure, and galactic dust contamination, which must be modelled and possibly subtracted.

---

†also, Physics Dept., UC Berkeley
‡also, M.E. Dept., UC Berkeley

OUR EXPERIMENT

We have chosen to work at 3 mm, where emission from the galaxy is low. While this choice of frequency reduces the problem of galactic contamination, problems with atmospheric emission are increased. Figure 2 shows the antenna temperature due to the atmosphere as a function of frequency at sea level, South Pole, and balloon altitudes. The plot is based on an atmospheric model with a standard temperature and pressure versus altitude profile, using water, oxygen and ozone absorption lines. It is evident that in order to work at 90 GHz, one requires either a very stable atmosphere or a high enough altitude that the emission lines are not saturated and the measurement can be done between molecular transitions. For example, at sea level, the atmospheric emission is more than 6 orders of magnitude higher than a desired sensitivity of $\frac{\Delta T}{T} = 10^{-5}$. We have built a system to make measurements on .5 to 5 degree scales, and have carried out experiments at balloon altitude and at the South Pole Station. We chose the South Pole as a ground observation site because of the low water content and previously reported high stability of the atmosphere there. For example, Figure 3 shows precipitable water for the time we were observing. Following is a brief description of the experiment and some of the results from the South Pole expedition.

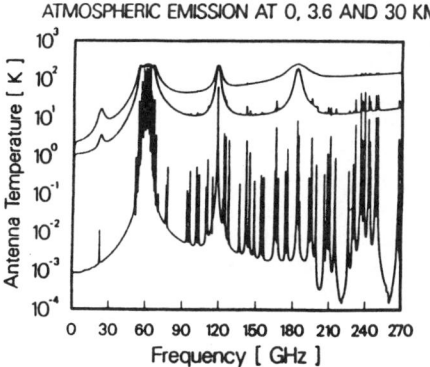

Figure 2

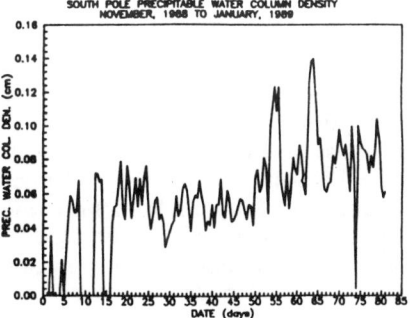

Figure 3

OPTICAL SYSTEM

Our optical system is an off axis Gregorian telescope, consisting of a 6.5 degree (FWHM) corrugated scalar feed, a 1 meter diameter, 1 meter focal length primary, with a confocal elliptical secondary mirror. The resulting beam can have a FWHM of 20 to 50 arc-minutes, depending on the secondary mirror used (our results are for a FWHM of 36 arc-minutes). Rotation of the secondary about the axis of the feed horn throws the beam horizontally on the sky. We chop the beam by a physical angle of 1 degree on the sky at 10 Hz to make a first difference measurement of temperature fluctuations. Our primary reason for using this configuration is the very low sidelobe response of such an antenna. For the central lobe, the beam is well approximated by a Gaussian of $\sigma = 15$ arc-minutes. $P(\Omega) = e^{-\theta^2/2\sigma^2}$ With a FWHM of 36 arc-minutes, the ratio of solid angle available for contamination to that in the beam puts stringent limits on the allowable sidelobe response. We measured our sidelobes down to -85 dB, without ground shields. In addition a ground shield was attached during data taking both during the balloon flight and at the South Pole.

SIS RECEIVER

A schematic of our detection system is shown in Figure 4. We use a Niobium SIS (Superconductor-Insulator-Superconductor) based coherent radiometer, operating at 90 GHz. Our mixer, HEMT IF amplifier (spot noise about 1 K), and cooled RF section enable us to achieve a system spot noise of about 33 Kelvin at a mixer physical temperature of 3.5 Kelvin. During data taking at the South Pole, our full band (0.6 GHz) noise was approximately 40 K, providing a theoretical system sensitivity (before chopping) of $\Delta T = 1.6 \frac{mK}{\sqrt{Hz}}$.

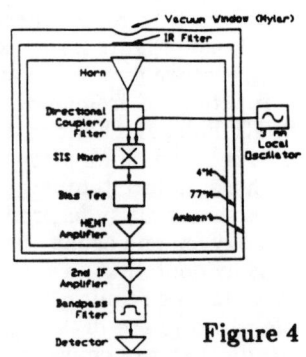

Figure 4

SOUTH POLE RESULTS

From late November, 1988 to early January, 1989, we made measurements from the South Pole station of CBR fluctuations and galactic emission.

Since galactic dust emission is a probable cause of error, we tried to get as much information for scaling known data to our frequency as possible. By comparing data from two of our balloon experiments at 90 GHz to the IRAS data at 100 microns, we obtain a cross calibration of approximately 10 $\frac{\mu K}{MJy/Sr}$ for the ratio of 3 mm emission to IRAS 100 micron emission. Using this number, along with a galaxy scan taken at the South Pole, we can estimate the contribution of dust emission to our data. We chose to measure in a region around RA=21.5, DEC=-73, where the IRAS 100 micron map shows a total intensity minimum of about 4-10 MJy/Sr, and first differences only of order 1-2 MJy/Sr. Using the galaxy data described above, this would be about 10-20 microKelvins in our data, which is small though not completely negligible compared to our errors (about 40 to 60 microKelvin per data point). We are currently at the point in sensitivity where even in the best parts of the sky, dust emission at 3 mm wavelength is near our detection limit. To do an order of magnitude more sensitive measurement will almost certainly require galactic subtraction preferably by multiple wavelength measurements.

CBR DATA

We observed 9 points with 1 degree physical chop angle on the sky, in a strip, spaced so that one beam from each point coincided with one beam from the next point. Several strips were measured to different sensitivities. This gives us a powerful test for systematic errors, as well as providing information on a variety of angular scales, from the beam sigma of 15 arc-minutes up to approximately 5 degrees. After time lost due to setting up, equipment problems and bad weather, we obtained about 80 hours of data, which reduced to about 70 hours after editing out radio interference and bad sky data. Our scan system gave us an efficiency (time spent on the measurement points) of only 60 percent, reducing the real data further to about 43 hours.

With a calculated statistical system sensitivity (on the sky) of 3 $\frac{mK}{\sqrt{Hz}}$, or 4

$\frac{mK}{\sqrt{Hz}}$ with sky shot noise included, we measured approximately 6 $\frac{mK}{\sqrt{Hz}}$ (RMS) on the sky for short time scales. Several runs were made of just atmospheric noise and are being investigated to help understand the nature of the sky noise.

RAW DATA FITTING

In order to work with the data, we have found it necessary to remove slow drifts in offset, which can be attributed to long term sky variations, changing electrical offsets, and temperature gradients on the primary. Our observing technique allows a natural way to remove such non- intrinsic shifts. Since we scan from one side of the strip to the other and then back in a period of about 30 minutes, linear variations on time scales long compared to 30 minutes can be removed without removing CBR structure. The results plotted in Figure 5 are the summed data for each point, with statistical error bars, where the raw data have been edited and piecewise linear fit in time, over times of approximately 3 hours. The results for a truncated Fourier fit subtraction, constructed to fit only structure longer than 3 scans, as well as a Legendre polynomial fit, are consistent with the linear fit presented. The error bars on this data set are consistent with the short term RMS fluctuations.

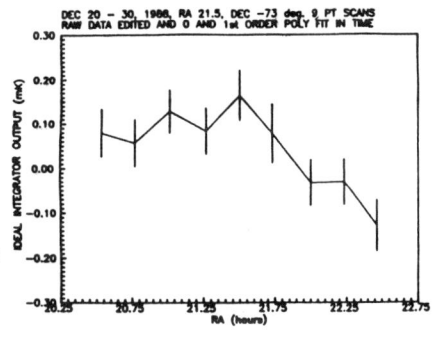

Figure 5

DATA ANALYSIS AND RESULTS

Looking at the data set in figure 5, a linear trend is evident across the points. Although this could be taken as an indication of intrinsic structure in the background radiation, we are unwilling to rule out some systematic effect to produce this. As an example, the sun was at RA of about 18 hours during our data taking, and contributions from this on the 100 μK level are not out of the question. We choose to remove the linear component from the data and consider the result to be our final set, which is shown in Figure 6. This set with error bars shown has a reduced chisquare of 1.53, corresponding to approximately 20 percent probability of being consistent with the null hypothesis. We are currently analyzing the data to test for various cosmological models, such as the cold dark matter galaxy formation model, scale invariant Gaussian fluctuations, etc.. These calculations will be presented in a forthcoming paper.

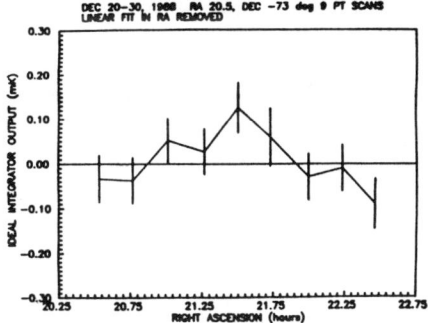

Figure 6

ACKNOWLEDGEMENTS

This work was supported by the National Aeronautics and Space Administration, National Science Foundation, the California Space Institute, the University of California, and the US Army. This work would not have been possible without the support and encouragement of Nancy Boggess, Buford Price, and John Lynch. Special thanks to Robert Wilson, Anthony Stark, Joe Stack, and Paul Moyer at Bell Labs for assistance in machining the primary and secondary mirrors. We gratefully acknowledge the support of Donald Morris. We wish to especially thank Anthony Kerr and S.-K. Pan of NRAO for supplying the exceptional 90 GHz SIS mixer. We would like to thank station manager Bill Coughran and all of the 1988-89 ANS staff at the Pole for their support.

REFERENCES

1. R. D. Davies, A. N. Lasenby, R. A. Watson, E. J. Daintree, J. Hopkins, J. Beckman, J. Sanches-Almeida, and R. Rebolo, *Nature*, **326**, 462 (1987).

2. E. B. Fomalont, K. I. Kellerman, M. C. Anderson, D. Weistrop, J. V. Wall, R. A. Windhorst, and J. A. Kristian, *Ap. J.*, submitted (1988).

3. A. N. Lasenby, Ph.D. Thesis (1981).

4. F. Melchiorri, B. O. Melchiorri, C. Ceccarelli, and L. Pietanera, *Ap. J.*, **250**, L1 (1981).

5. A. C. S. Readhead, C. R. Lawrence, S. T. Myers, W. L. W. Sargent, H. E. Hardebeck, and A. T. Moffet, *Ap. J.*, submitted (1989).

6. R. Sachs and A. Wolfe, *Ap. J.*, **147**, 73 (1967).

7. I. A. Strukov, D. P. Skulachev, and A. A. Klypin, *Large Scale Structure of the Universe*, J. Audouze et al. (eds.), IAU No. 130.

FIRST SUBMILLIMETER OBSERVATIONS FROM THE SOUTH POLE : THE INTEGRATED GALACTIC EMISSION

F. Pajot, R. Gispert, J.M. Lamarre, J.L. Puget
Institut d'Astrophysique Spatiale, BP 10, F-91371 Verrières le Buisson, France

M.A. Pomerantz
Bartol Research Institute, University of Delaware, Newark DE19716, U.S.A.

R. Peyturaux
Institut d'Astrophysique, 98 bis, boulevard Arago, F-75014 Paris, France

ABSTRACT

The EMILIE experiment was designed for ground based photometry of the galactic diffuse emission in the submillimeter range. The regions of the Galaxy visible from the south hemisphere were observed from the geographic South Pole during the Austral summer 1984-1985.

The characteristics of the site for submillimeter astronomy are deduced from these observations and specific measurements made with this instrument. The total precipitable water obtained this way is in good agreement with meteorological soundings carried from the station. Comparing this site with Mauna Kea, Hawaii, where the EMILIE experiment was first operated in 1982, the excellent quality of South Pole for submillimeter astronomy is proved (total precipitable water, atmospheric emission and transmission, fluctuations of this emission). This experiment led to new measurements of the galactic dust emission at 900 μm between the longitudes 330° and 6°.

1. INTRODUCTION

The EMILIE experiment has been designed for ground based observations of the galactic dust emission in the few atmospheric windows in the submillimeter range. The first photometric measurements of the diffuse galactic emission at 900 μm were carried from Hawaii in 1982. They validated the instrumental concept and led to the first evaluation of the emission at galactic longitudes between 0° and 30° and to a partial longitude profile and an averaged latitude profile of the Galaxy[1]. Nevertheless, the observations were limited by the characteristics of the atmosphere in Hawaii (fluctuations of the atmospheric emission, total precipitable water). The potential of South Pole for infrared and submillimetre astronomy has already been pointed out by several groups in the past[2]. After a few minor modifications, the EMILIE experiment was operated during a campaign at the Amundsen-Scott South Pole Station during the Austral summer 1984-85 in collaboration with the Bartol Research Institute.

2. INSTRUMENT CHARACTERISTICS AND OBSERVATIONAL METHOD

The configuration of the EMILIE experiment is shown in Fig.1. As in all ground-based submillimeter experiments, the emission of the sky is much higher than the astronomical signal (typically 10^4 times) and the choice of the modulation is crucial. A first modulation of the signal is performed at 35 Hz between the 0.55° main beam pointed to the source and a 7° wide reference beam aimed at the sky. The elevation of the reference beam is adjusted to balance the two signals and the 0.55° beam is scanned through the source at constant elevation at a typical rate of $6°s^{-1}$. The reference beam can be sent on an extended blackbody to determine the absolute sky emission. The observation beam

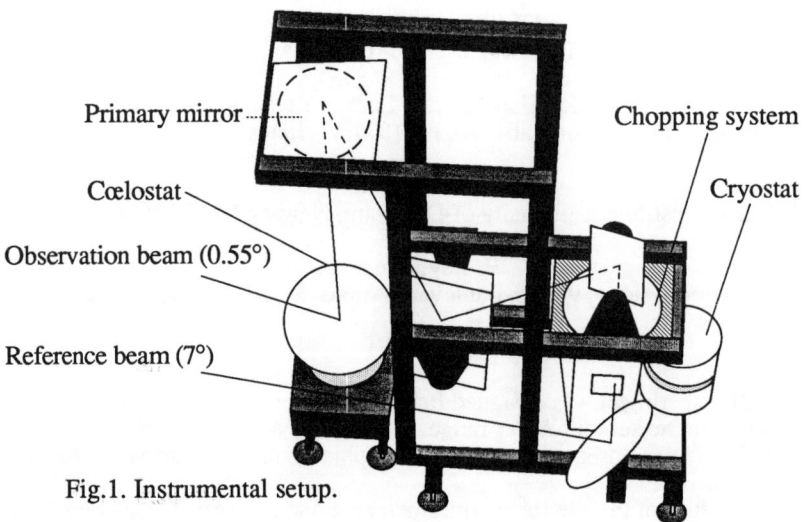

Fig.1. Instrumental setup.

can be sent on another blackbody at a different temperature thus providing an absolute calibration of the experiment. The off-axis configuration of the 45 cm diameter spherical mirror avoids the presence of any emissive parts within the beam. An oversized coelostat placed in front of the primary mirror carries out the tracking and the azimuthal scanning. This design confers to the optics a total emissivity of about 8 % due to the optical surfaces and the elements seen by diffraction. The experiment uses a photometer equipped with 3 composite bolometers cooled at 1 K in an ^{4}He cryostat. The photometric band is selected by combining a bolometer and its cold bandpass filter with an ambient temperature narrow bandpass filter mounted on a wheel. Photometric characteristics (effective wavelength, effective bandwith) of the channels for a 300K blackbody are : FA (990 µm, 410 µm), FB (1000 µm, 170 µm) and FD (870 µm, 40 µm).

The EMILIE experiment was set at Amundsen-Scott Station (United States) from mid-november to mid-december 1984. The experiment was located 8 km from the main station to avoid any pollution (humid and warm air from the generators and the buildings).

3. PHOTOMETRIC CHARACTERISTICS OF THE SOUTH POLE SITE

In order to calibrate our observations, we needed to evaluate the atmospheric transmission. The main component responsible for the atmospheric opacity at these wavelengths is the water vapor. The total precipitable water can be known by two methods. The easiest way is to use the meteorological balloon soundings made every 12 hours at the station which give the temperature, pressure and dew point altitude profiles. Another standard method is to use the absolute measurements of the sky emission together with an atmospheric numerical model. Then in both cases the model was used to derive the transmission from the total precipitable water. For all computations we used an average atmosphere at South Pole deduced from the meteorological soundings which is rather different from the usual U.S. standard atmosphere (Fig.2) but results show quite small differences using the two types of atmosphere. There is a very good agreement between the two determinations of the total precipitable water (Fig.3). During the observations, the precipitable water vapor content stayed around 0.5 mm at zenith and the temperature varied from -33 to -26 Celsius. This is to be compared with the best values at Mauna Kea Observatory, Hawaii, rarely below 1 mm at zenith.

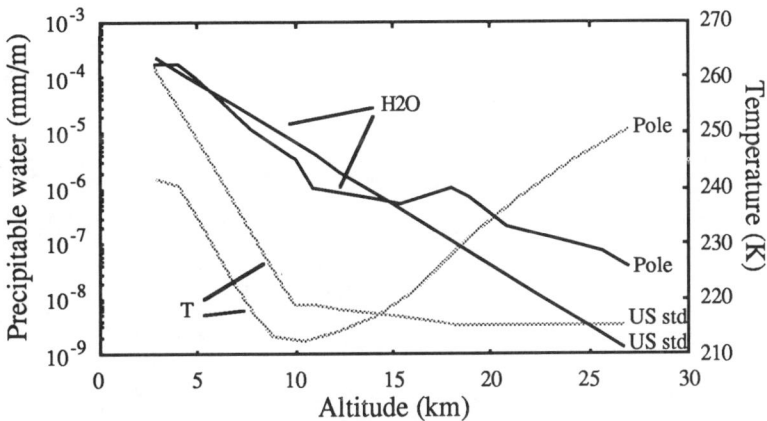

Fig.2. Average vertical profiles of the atmosphere at South Pole (water content and temperature) during the campaign compared to the standard model.

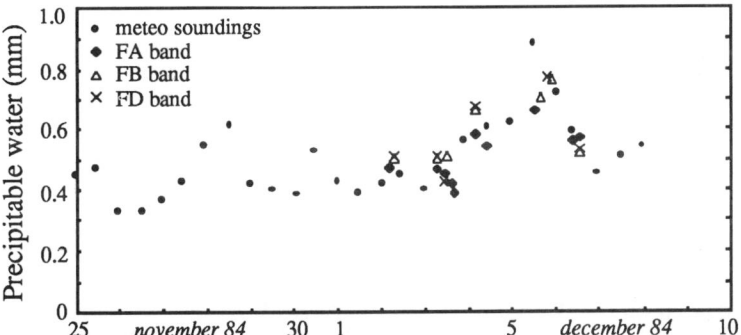

Fig.3. Precipitable water at South Pole deduced from meteorological balloon soundings and from the measurements of the EMILIE experiment.

An other remarkable feature obtained at the South Pole is the uniformity of the atmospheric emission at constant elevation and large angular scale (180°). In the FA band, this emission was found to be almost linear with a drift of 1 K when the total emission was around 80 K (Fig.4). A sample of the time variations of the atmospheric emission is given in Fig.5.

4. THE INTEGRATED GALACTIC EMISSION

Observations were carried out over the following galactic longitude ranges: $235° < l_{II} < 275°$, $330° < l_{II} < 360°$ and $0° < l_{II} < 6°$. Some long integration periods at high galactic latitude were spent to probe the ultimate sensitivity that could be achieved with this instrument. A signal was detected for the central regions of the galaxy and an upper limit (2.5 σ equal to $2.5 \cdot 10^{-8}$ Wm^{-2}sr^{-1} [λI_λ]) was set elsewhere. Data reduction and scientific interpretation of the results are presented extensively elsewhere[3]. The width at half maximum of the latitude profile averaged over the [-24°,+6°] longitude range is equal to 1.0° after deconvolution from the beam. We observed an a-priori empty region of the sky (high galactic latitude and no known submillimetre source in the vicinity) to determine the sensitivity of the experiment. A sum of 1000 identical 6° wide scans (a 1 hour integration) lead to a 1 σ dispersion over the 6° equal to 0.45 mK (equal to 0.37

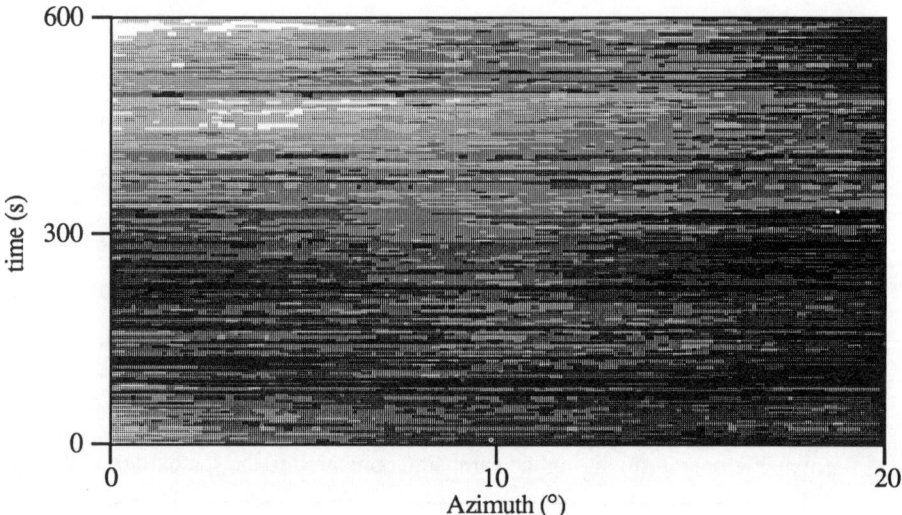

Fig.4. Relative sky brightness at 900 μm (FA band). Proportional grey scale from 0 mK (black) to 300 mK (white).

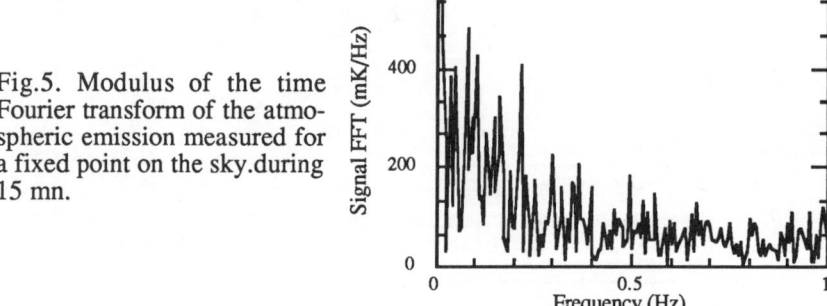

Fig.5. Modulus of the time Fourier transform of the atmospheric emission measured for a fixed point on the sky.during 15 mn.

10^{-8} Wm^{-2}sr^{-1} for a 300 K blackbody) after removal of a linear baseline. Pseudo periodicity in the remaining fluctuations indicates that the main limitation is not pure noise but systematic instrumental effects.

Acknowledgements. This experiment was supported by the Institut National des Sciences de l'Univers and by the National Science Foundation (Grant Number DPP-8120258).

REFERENCES

1. Pajot, F., Gispert, R., Lamarre, J.M., Peyturaux, R., Puget, J.L., Serra, G., Coron, N., Dambier, G., Leblanc, J., Moalic, J.P., Renault, J.C., Vitry, R., *Astron. Astrophys.* , **154**, 55 (1986)
2. Smythe, W.D., Jackson, B.V., *Appl.Opt.* **16**, 2041 (1977)
3. Pajot, F., Gispert, R., Lamarre, J.M., Peyturaux, R., Pomerantz, M.A., Puget, J.L., Serra, G., Maurel, C., Pfeiffer, R., Renault, J.C., *Astron. Astrophys.* in press.(1989)

SOUTH POLE SUBMILLIMETER ISOTROPY MEASUREMENTS OF THE COSMIC MICROWAVE BACKGROUND

Mark Dragovan
Joseph Henry Laboratory, Princeton University, Princeton, NJ 08544

Stephen R. Platt and Robert J. Pernic
The University of Chicago, Yerkes Observatory, Williams Bay, WI 53191

Antony A. Stark
AT&T Bell Laboratories, Holmdel, NJ 07733

ABSTRACT

Observations were made from the United States Amundsen-Scott South Pole Station during the austral summer of 1988-89 to search for spatial anisotropy in the submillimeter Cosmic Microwave Background. Three 30' x 30' regions of the sky were observed at 350 μm, 450 μm, and 600 μm with the University of Chicago 32-Channel Submillimeter Photometer and a 1.2-meter off-axis parabolic telescope, designed and constructed at AT&T Bell Laboratories. Reimaging optics gave each of the 32 bolometers in the array a 5-arc minute field of view. The search is sensitive to fluctuations on all angular scales between 5- and 30-arc minutes.

INTRODUCTION

Observations today reveal a structure-filled Universe. Galaxies, clusters of galaxies, even clusters of clusters of galaxies, are seen in every direction. This present-day clumpiness must have evolved from initial density enhancements in the early Universe. The photons that we can observe today from the Cosmic Microwave Background are the best probe of the physical conditions of the early Universe. The photons arrive undistorted, except for a red-shift due to the general expansion of the Universe, from the epoch of recombination. It was then that free electrons and protons recombined to form neutral hydrogen, leaving the Universe transparent to photons. Prior to this, the photons and baryonic matter were in thermal equilibrium. No direct observations can be made to probe the history of the Universe prior to recombination.

The microwave background radiation has a thermal spectrum with a temperature of 2.78 K, and its intensity peaks at a wavelength of 1.8 mm. Any structure in the Universe at recombination will cause the physical temperature of one region of the sky to be slightly different from a neighboring region. The expected contrast is low, however, with current limits approaching $\Delta T/T = 10^{-5}$. Isotropy measurements on both the Wein and Raleigh-Jeans portions of the spectrum are important. Long wavelength measurements are sensitive to primordial density fluctuations, while short wavelength measurements are sensitive to emission from protogalaxies.

We used a small (1.2 meter) submillimeter (SMM) telescope for isotropy observations of the Cosmic Microwave Background (CMB) from the United States Amundsen-Scott South Pole Station during the 1988-89 austral summer. Recent results by the Berkeley-Nagoya collaboration[1] indicate an excess flux in the submillimeter background spectrum. In order to constrain the origin of the excess, it is necessary to measure its isotropy. The South Pole site, combined with a sensitive array of liquid helium-cooled bolometers and a special purpose telescope, permitted observations to be made that would be impossible anywhere else in the world.

THE SOUTH POLE SITE

The South Pole provides a unique site for submillimeter observations. Such observations can be pursued only from extremely cold and dry sites, where the atmosphere is exceptionally stable and contains less than 1 mm of precipitable water vapor. The South Pole is at high altitude (roughly 10,800 feet pressure altitude, corresponding to 530 mm Hg pressure), and it is cold, with surface temperatures varying from -50 to -20 F during November and December. The high altitude reduces the pressure broadening of the H_2O and O_2 lines which are the primary sources of emission in the submillimeter atmospheric windows. The cold temperatures mean that most of the water vapor in clouds is frozen. Ice crystals do not have the rotational absorption bands characteristic of water vapor and are essentially transparent to SMM radiation. Measurements of the atmospheric emission at the South Pole indicate a precipitable water vapor content of 0.3 mm on a typical day, and ~1.0 mm on the worst. Variations in atmospheric emission, caused by water vapor clouds passing through the telescope beam, give rise to an additional source of noise ("sky noise"). The South Pole has roughly an order of magnitude less sky noise than Mauna Kea, Hawaii.

The South Pole also has a singular advantage over any other observatory location: a given patch of sky never rises or sets. Consequently, observations are limited only by the length of the stretches of clear, stable weather, rather than the length of time the source takes to transit. Since the elevation of a source is constant at the Pole, no corrections need be made for the change in atmospheric opacity with elevation. The only source of variable opacity is the change in weather conditions during the observations. In addition, since the zenith angle of a field is constant, variations of telescope side-lobe pickup from the warm Earth are minimized.

The combination of high atmospheric transmission, low sky noise, and long observing times on a given patch of sky make the South Pole the best available location for SMM observations. It is an especially good site from which to search for anisotropy in the CMB.

SUBMILLIMETER ISOTROPY MEASUREMENTS OF THE MICROWAVE BACKGROUND

Recent measurements of the SMM CMB spectrum indicate that there is an excess flux amounting to 10% of the total power in the background radiation. Measurements of its isotropy are necessary to place constraints on its origin.

Our observations were made during November and December 1988 from the US Amundsen-Scott South Pole Station using a specially designed 1.2-meter off-axis telescope, built at AT&T Bell Laboratories. The telescope was designed to have a low side-lobe response to minimize the effects of ground pickup. Additional shielding was used to reflect all scattered radiation back onto the sky.

The detector array was the University of Chicago Submillimeter Photometer, SCAMP I. Each of the 32 ^{3}He-cooled silicon bolometers is mounted within an integrating cavity and illuminated through a parabolic Winston concentrator. The square array of concentrators and detectors are in turn illuminated by a system of transfer optics which efficiently couple each detector to the telescope and reject unwanted thermal background radiation. Signals are amplified by LN_2-cooled JFET source-followers and room-temperature preamplifiers, then digitally sampled and phase-demodulated.

Three different patches of sky were observed, chosen from 100 µm IRAS maps for a minimum of thermal dust emission. Each field was observed at each of three wavelengths, 350 µm, 450 µm, and 600 µm. The combination of telescope and camera optics resulted in a 5-arc minute field of view for each of the 32 detectors, covering a patch of sky approximately the size of the full moon. The telescope was first pointed at the field of interest and then the sky was allowed to drift through the telescope beam. The use of drift scans eliminated the possibility of detecting any noise associated with the telescope drive electronics.

The entire data set consists of 40-150 10-15 minute scans at each wavelength for each of the three positions. With only a fraction of the data for one wavelength (600 µm) at one position reduced, we find the isotropy to be $\Delta I/I \approx 10^{-2}$, which corresponds to $\Delta T/T \approx 10^{-3}$. When all of the data is reduced, this limit should be approximately an order of magnitude lower, placing severe constraints on dust models for the origin of the excess.

ACKNOWLEDGEMENTS

We wish to thank M. Pomerantz of the Bartol Research Institute for his support and encouragement during the first incarnation of this experiment. Construction of SCAMP I was made possible by NSF grant AST 8513974. Travel and observing support is provided by NSF grant AST 8815628. SRP was supported by NASA Training Grant NGT-50073. This work was supported at the South Pole by NSF grant DPP87-17300.

REFERENCES

1. T. Matsumoto, S. Hayakawa, H. Matsuo, H. Murakami, S. Sata, A. E. Lange, and P. L. Richards, Ap. J. 329, 567 (1988).

ATMOSPHERIC TRANSPARENCY OVER ANTARCTICA FROM THE MID-INFRARED TO CENTIMETER WAVELENGTHS

John Bally
AT&T Bell Laboratories
HOH-L245, Holmdel, NJ 07733

ABSTRACT

The high altitude, extreme cold, and low water vapor column density make the Antarctic Plateau the best known site on the Earth for astronomical observations from 20 µm to 1 mm. I use estimates of total water vapor column density obtained from both in-situ atmospheric sampling and radiometric measurement of the sky brightness near 1 mm and a program written by Erich Grossman to estimate the range of useful wavelengths which can be observed from the Antarctic Plateau.

I. INTRODUCTION

There is increasing evidence that the atmospheric conditions on the Antarctic Plateau are the best in the world for astronomical observations in the mid-infrared, far-infrared, and millimeter wavelengths for which water vapor is the primary absorber. The Antarctic Plateau ranges in elevation from nearly 3,000 meters at the South Pole to over 4,000 meters near the geographic center of the continent. The center of the ice-shelf is located in a nearly permanent high pressure zone with air descending from high altitudes dominating the atmospheric circulation. These factors, combined with the extremely low surface temperatures (-80 to -20 C) result in very dry atmospheric conditions.

Two kinds of measurements provide evidence for a low total percipitable water vapor column over the Antarctic ice-shelf. Balloon-born in situ-measurements (radiosonde) have been conducted at both the South Pole (Smythe and Jackson 1977) and at the Vostok site operated by the U.S.S.R. at an elevation of 3488 meters near the geographic center of Antarctica (Burova et al. 1986). This data has been summerized by Townes and Melnick (1989) and demonstrate that over the South Pole, the water column ranges from a peak of 1.6 mm of H_2O to under 400 µm in the Austral summer and averages around 700 µm. In the Austral winter, the average amount of percipitable water vapor averages around 200 to 500 µm. At the somewhat higher Vostok site, Burova et al. (1986) report mean values around 400 μm in the Austral summer and around 200 μm in the Austral winter. However, radiosonde measurements may not be reliable in the low humidity, low temperature environment of the Antarctic atmosphere. Long detector equilibration times may result in an over-estimate of the H_2O content of the atmosphere at higher altitudes.

A second method is to make broad-band photometric or narrow-band spectroscopic measurements of the absorption of the atmosphere in a near infrared band containing H_2O lines or to make sky brightness measurements in the sub-mm or mm wavelength range. Optical measurements have been performed by Smythe and Jackson (1977) and Murcray et al. (1981). Dragovan et al. (1989) report sky brightness measurements at 700 μm in the sub-mm portion of the spectrum. The derivation of total H_2O column densities from this data is complicated since it requires a detailed understanding of the line formation process. The far line wings responsible for the broad band opacity depend on the details of the short-range interaction between H_2O and the other molecules with which it collides. The measurements indicate average H_2O content over the South Pole is sufficiently low to produce a zenith opacity of order 0.1 to 0.2 at 700 μm during the Austral summer. This is consistent with about 200 to 500 μm of percipitable H_2O. The available measurements suggest that sometimes during the Austral winter the percipitable H_2O column is lower than 100 μm and is rarely greater than 1 mm. It is possible that from a site at 4,000 meters elevation water column densities of order 50 μm may be realized.

II. MODELING THE TRANSMISSION OF THE ATMOSPHERE

I use the program written by Erich Grossmann (1989) to model the atmospheric transmission over the Antarctic Plateau. Atmospheric line parameters are obtained from the AFGL Atmospheric Absorption Line Parameters Compilation main data base which is described in Rothman et al. (1983) and includes the parameters for all lines produced by the 7 major constituents of the atmosphere which have strong lines (H_2O, CO_2, O_2, O_3 N2O, CO, and CH_4: N_2 does not have a dipole moment and does not contribute to the absorption). The program uses either a Lorentzian or the Zhevakin-Naumov (1963) line shape to compute the absorption in the line wings. Since at any particular frequency, the opacity is determined by the superposition of the absorption produced by the far line wings of many individual spectral lines, the computation of the spectrum requires that all lines within a pre-selected frequency range of each point on the spectrum be included. For any particular frequency, a range in excess of several hundred GHz must be searched for lines whose pressure broadened wings contribute to the opacity.

Figure 1 illustrates the transmission of the atmosphere computed with the Grossmann program over a frequency range of 0 to 1,800 GHz. I assume 100 µm of percipitable H_2O and a zenith distance of 0 degrees (looking straight up). A pressure altitude of 3800 m is assumed, which corresponds to a high site on the Antarctic Plateau. These conditions probably represent what can be obtained at the South Pole under the most favorable circumstances. Since there are uncertainties in the detailed line shapes, and the resolution of the calculated grid is finite, the spectrum is only a rough guide to what can be expected. At some frequencies, especially near narrow features, the calculated opacities may be in error by as much as 50%. The actual frequency of occurrence of conditions as good as these is hard to estimate and there is no substitute for actual on-site measurements.

Model computations indicate that about 85% of the spectrum at wavelengths longer than 300 µm can be observed most of the time. Observations at 350 µm may be as easy in Antarctica as observations at 3 mm at existing mm-wave observatory sites such as Kitt Peak. Several new windows open up at shorter wavelengths such as the region between 190 and 250 µm. Under the best conditions, it may be possible to observe the 158 µm, C^+ fine structure line, the most luminous line emitted by the interstellar medium.

Figure 2 illustrates the atmospheric absorption up to 20,000 GHz ($\lambda = 15$ µm) under the best Antarctic conditions (assuming 50 µm H_2O) which may be encountered a few percent of the time during Austral winter from the highest sites. Many new windows open at wavelengths below 100 µm and the atmosphere is relatively transparent at wavelengths between 17 and 60 µm, making mid-infrared observations possible. Between 18 and 60 µm, the atmosphere remains sufficiently transparent even under 300 µm of water vapor.

Existing data, combined with model calculations of atmospheric transmission suggests that Antarctica may have the worlds best conditions for observing at infrared to millimeter wavelengths. Further observations should be conducted to explore the statistics of water vapor column density over both the South Pole and other potential observing sites in Antarctica.

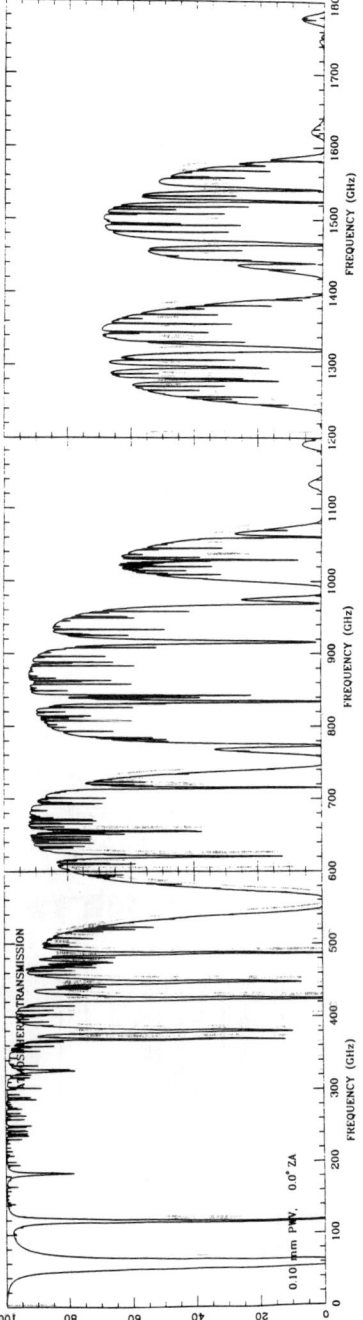

Figure 1. Computed transmission for 100 μm percipitable H$_2$O at an altitude of 3800 m covering the frequency range 0 to 1,800 GHz (λ = 167 μm).

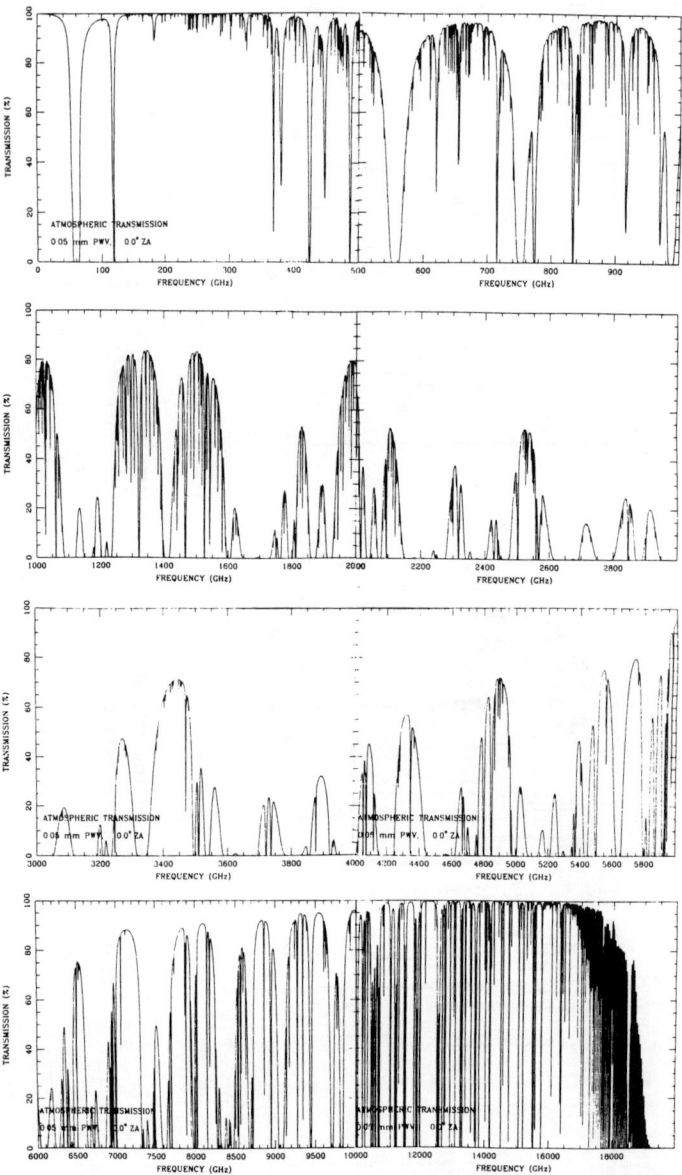

Figure 2. Computed transmission for 50 μm percipitable H_2O at an altitude of 3800 m, covering the frequency range 0 to 20,000 GHz (λ = 15 μm).

REFERENCES

Burova, L.P., Gromov, V.D., Luk'yanchikova, N.I., and Sholomitskii, G.B. 1986 *Soviet Astron. Lett.*,**12**,339.

Dragovan, M., Stark, A.A., Pernic, R., and Pomerantz, M 1989 *Applied Optics*, (in press).

Grossmann, E. 1989 (Dept. of Astronomy - Univ. of Texas, Austin: (private communication).

Murcay. D.G., Murcay, F.J., Murcay, F.H., and Barker, D.B. 1981 *Antarctic Journal of the U.S.*, **15**,199.

Rothman, L.S., Gamache, R.R., Barbe, A., Goldman, A., Gillis, J.R., Brown, L.R., Toth, R.A., Flaud, J.M., Camy-Peyret, C 1983 *Applied Optics*,**22**,2247.

Smythe, W.D. and Jackson, B.V. 1977 *Applied Optics*,**16**,2041.

Townes, C.H. and Melnick, G. 1989 *Atmospheric Transmission in the Far Infrared at the South Pole and Astronomical Applications*, (preprint).

Zhevakin, S.A. and Naumov, A.P. 1963 *Geomagnetism and Aeronomy*,**3**,537.

ON THE FUTURE OF SUBMILLIMETER ASTRONOMY AT THE SOUTH POLE

Antony A. Stark
AT&T Bell Laboratories; Holmdel, NJ 07733

ABSTRACT

The South Pole is the best accessible submillimeter-wave observatory site in the world. The Antarctic Submillimeter Telescope/Remote Observatory (AST/RO), a 1.5 m offset telescope for heterodyne spectroscopy, will be a step toward a permanent submillimeter observatory there. Realization of the site's potential will require the founding of a South Pole Observatory and construction of considerably larger instruments.

I. A UNIQUELY CLEAR SKY

The South Pole is unique among potential observatory sites for the consistent clarity of its submillimeter sky. Submillimeter-wavelength observations can be pursued only from extremely cold and dry sites, where the atmosphere contains less than 1 mm of precipitable water vapor. Most places on the earth's surface have a water vapor burden substantially higher than this, which renders the atmosphere opaque; roughly speaking, zenith optical depth in the submillimeter band is equal to the precipitable water vapor measured in millimeters (Zammit and Ade 1981). Measurements reported by Pajot (1989) and Dragovan et al. (1989) indicate that the broad-band submillimeter atmospheric opacity at the South Pole during the summer is typically $\tau_{zenith} \approx 0.2$. "Good weather" at Mauna Kea in Hawaii corresponds to $\tau_{zenith} \approx 1$. The South Pole in the Austral summer is consistently several times better than Mauna Kea. In the Antarctic winter, conditions are likely to be nearly as good at submillimeter wavelengths (but not the far-infrared) as the Kuiper Airborne Observatory, and some mid- and far- infrared bands may become sufficiently transparent for some types of observations (Bally 1989). Telescopes at the South Pole can make observations which are very difficult or impossible elsewhere, because the atmosphere above the South Pole is transparent to submillimeter wavelengths essentially all the time. This makes possible entirely new kinds of astrophysical observations.

II. THE SIGNIFICANCE OF SUBMILLIMETER OBSERVATIONS

Submillimeter observations are a means of approaching some of the important problems in astronomy, problems relating to the origin of the elements, the birth of stars, and the evolution of the Galaxy. All the atoms in the Earth were once part of the interstellar medium and participated in the dynamics of the Galaxy; the solar system formed in a dense molecular cloud, a result of complex chemical and physical processes. Submillimeter observations of the Galaxy probe the component of the interstellar medium which is an active participant in current star formation. There are hundreds of potentially observable spectral lines from dozens of different molecular and atomic species (Sutton et al. 1985), showing complex velocity and positional structure over many square degrees of sky. Of particular importance is the fine-structure transition of neutral carbon at 609µm, which traces the interface between atomic and molecular regions of interstellar clouds. This line may be pervasive at a low brightness level (Harwit, Houck, and Stacey 1986), covering much of the sky with weak spectral lines that are difficult to measure unless the transparency of the atmosphere is good. Large-scale surveys in the CI line may prove as important to understanding the interstellar medium as have large-scale surveys of HI and CO.

High-lying rotational lines of CO, such as the J=4→3 line at 650µm, are expected to be tracers of warm, moderately dense molecular gas associated with regions of active star formation, the dense portions of Giant Molecular Clouds, and interstellar shock waves propagating in dense molecular clouds. Submillimeter transitions of high dipole-moment molecules such as CS are the best available tracers of high density gas in protostellar condensations or gas in the immediate vicinity of recently formed stars. Dense star-forming regions are shrouded in cold dust which emits strongly at wavelengths longer than 100µm; submillimeter photometry can be used to measure the physical properties of protostellar clouds (e.g. Davidson 1987). For recent reviews of current work in submillimeter astronomy, see Phillips (1987), Genzel and Stacey (1985), and Melnick (1987).

The broad-band spectrum of most spiral and irregular galaxies peaks near 100μm wavelength, and falls off sharply through the submillimeter (e.g. Stark *et al.* 1989). At high redshifts the 100μm peak and the strong interstellar cooling lines of CII and OI will be shifted into a wavelength band at which the Milky Way as a whole and other foreground galaxies do not emit much continuum radiation. Sufficiently large and sensitive submillimeter telescopes may therefore be able to detect and study protogalaxies.

Submillimeter telescopes can also make a contribution to studies of molecules in the earth's atmosphere. High-quality spectra of stratospheric molecules can be made with about an hour of observing time, and inverted to obtain abundances of molecules like O_3 and CO as a function of altitude, with an altitude resolution of about 10 km and an accuracy of about 5% (Bevilacqua and Olivero 1989; Waters, Wilson, and Shimabokuro 1976). This is not nearly as accurate as direct, balloon-borne measurements, but remote measurements can be made at frequent intervals and in different directions on the sky. The molecules which are easiest to measure with submillimeter spectroscopy are those at very high altitudes where balloons cannot reach.

III. THE FEASIBILITY OF OPERATING TELESCOPES AT THE POLE

It has been possible to live and work at the South Pole since 1957; many of the facilities need for an observatory are already there. The enormous logistical difficulties of working on the Antarctic plateau have largely been obviated by the National Science Foundation's support of a permanently staffed Pole station. A number of individual Principal Investigator experiments (EMILIE, Dragovan's 1986-87 experiment, and the 1988-89 Cosmic Microwave Background Radiation [CMBR] experiments) have demonstrated the feasibility of temporarily installing and operating 1 meter class telescopes at the Amundsen-Scott South Pole Station during the Austral summer.

Instruments for the South Pole do require special care and preparation. They must, of course, be capable of operating at low temperature and air pressure: these requirements are less stringent, however, than those on a balloon gondola. Care must be taken to seal against blowing ice crystals ("diamond dust"). Equipment must be reliable because replacement parts are weeks or months away. Design of stable foundations may present problems (Stark and Gress 1989).

In some ways telescope operation at the South Pole is easy. For the CMBR campaign, a C-130 cargo airplane was backed up to the site, and equipment unloaded with forklifts on the spot. Weather on the Antarctic plateau is usually benign, aside from the extreme cold. It never rains, nor are there any insects. The South Pole Station staff are friendly and able people who are willing to help with problems.

IV. THE AST/RO TELESCOPE

AST/RO is a 1.5-m class telescope scheduled for permanent installation at the South Pole in late 1992, the result of a successful proposal to the NSF by A. A. Stark, J. Bally and R. W. Wilson of AT&T Bell Laboratories, T. M. Bania and A. P. Lane of Boston University, and K.-Y. Lo of the University of Illinois. The immediate scientific goals of this instrument are heterodyne spectroscopy of galactic molecular clouds and molecular lines in the earth's stratosphere at wavelengths near 600µm, but with some modification it is potentially a general-purpose telescope for the millimeter, submillimeter and far-infrared. Plans for the telescope will be described here in some detail because of this potential for its use by the reader. It should be stressed that some parts of the system — the beam chopper, Nasmyth focus and upgraded carbon-fiber primary — are not part of the original project but are enhancements currently being considered.

A telescope aperture of 1.5 meters yields a beamsize of 80" (λ/600µm) This beamsize is large enough to allow large-scale mapping programs, yet is small enough to map distant clouds in the Galaxy and to just resolve hundreds of external galaxies; it is not small enough, however, to study distant galaxies or to study protostellar regions or the center of our Galaxy in any detail. All of the optics in AST/RO are offset for high beam efficiency and to avoid the creation of inadvertent reflections and resonances. The receiver is in a coudé room under the mount, in a spacious, shirt-sleeve environment. To improve the reliability of the system, there will be a two-fold redundancy of the receiver components, cryogenic systems, and spectrometer. The data acquisition system will also be able to reduce the data and make maps. The entire system will be highly automated to reduce to a minimum the need for operator intervention.

Construction of a 1.5 m class submillimeter telescope is much simpler than construction of a larger instrument, because the primary reflector is a single piece, rather than a collection of panels. The primary reflector will be small enough and light enough to be lifted and put into place by a few people. The mount can be carried to the South Pole in a C130 aircraft without disassembly and then unloaded and moved into place using a forklift. This is considerably simpler than the heavy construction work, assembly and alignment of multiple panels that will be required for a larger telescope.

Because the telescope will be smaller than most submillimeter telescopes, diffraction effects at the edges of the optical surfaces are relatively more important. The optical elements are therefore completely off-axis: there are no aperture blockages anywhere in the system. This allows all the optical surfaces to be "large", in the sense that the beam illumination at the mirror edge is small. In particular, the secondary mirror is oversized without blocking the primary. Also, there are no inadvertent resonant cavities, where the beam reflects back and interferes with itself, greatly reducing the problems of "baseline ripple" found in on-axis systems, particularly on-axis heterodyne systems where leaking local-oscillator power and receiver noise can be reflected back into the receiver.

Since the secondary mirror is off-axis, there is considerable freedom in its size, orientation and focal length, which can be put to good use. AST/RO is a modified Gregorian, an optical scheme developed for the AT&T Bell Laboratories Cosmic Microwave Background Experiment by Dragovan (1988): the secondary is a concave ellipsoid, which is followed by a flat tertiary. The advantage of a Gregorian is that the concave secondary images the surface of the primary onto a plane at the surface of the flat tertiary, a "real exit pupil". Suppose the tertiary mirror is tilted by a mirror mover. This will tilt the wave front at the primary, steering the beam, without changing the illumination or spillover at the primary since the secondary images the field distribution at the tertiary onto the surface of the primary. The secondary must be oversized, because its illumination changes. Beam chopping between source and reference positions on the sky can therefore be accomplished by relatively small movements of the flat tertiary, and the sidelobe and backlobe response will be very nearly the same at both source and reference.

Present plans call for a primary reflector made of carbon fiber and epoxy with a vacuum-deposited aluminum surface, having a surface roughness of 6 μm and an rms figure of about 10 μm. The diameter of the primary may be as large as 1.7 m, with a focal length of 1.2 m. The Gregorian secondary will be a prolate spheroid, with one focus 0.1 m from the vertex and another focus about 1.5 m from the vertex. The angle between the line of foci of this spheroid and the axis of rotation of the (off-axis) paraboloidal primary can be set so as to cancel lowest order coma (Dragone 1982). This provides a diffraction-limited field-of-view which is about 3° in diameter at 3 mm wavelength and about 30' in diameter at 200μm. The chopper can make use of this field-of-view, because it does not change the illumination of the primary. The telescope will have a Nasmyth focus, the beam passing through an elevation bearing having a 0.3 m diameter hole. This focus will be almost identical in its optical properties to the bent Cassegrain focus on the Kuiper Airborne Observatory, except for a larger field-of-view. Array detectors of various types could be used at this focus.

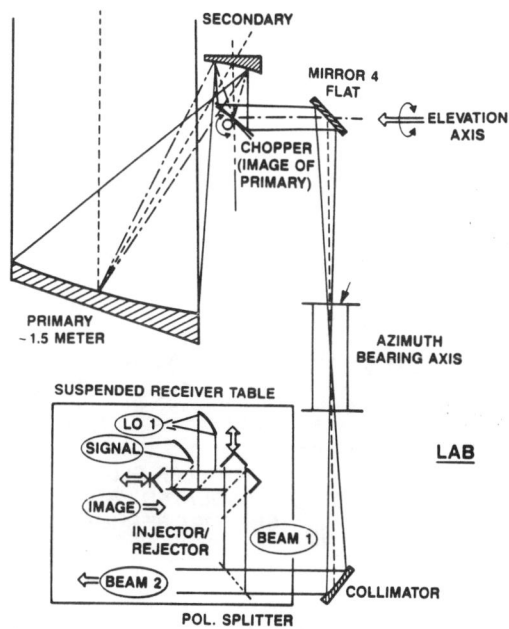

Fig. 1. Optics scheme for the AST/RO telescope. The telescope is offset along both the elevation and azimuth axes, so that a coudé focus can be achieved with only four mirrors. A Nasmyth focus results from removal of the fourth mirror. The (flat) chopper rotates around a line which lies in the plane of the exit pupil. The cable wrap and bearing configurations are schematic.

The heterodyne detector and calibration systems will be nearly as large as the telescope itself. A coudé-type arrangement is used, where the detector and calibration system are stationary with respect to the Earth. Figure 1 shows how the addition of the fourth and fifth mirrors sends the beam through the elevation and azimuth axes and into the receiver in the "coudé-room" beneath the mount. Because the azimuth shaft must be hollow to permit passage of the beam, a hollow-shaft encoder is used. The axes will each have two DC torque motors with about 150 foot-lbs of torque, acting in opposition on external bull gears. This instrument is designed to be mounted on the roof of a laboratory. The roof must be capable of supporting the approximately 2500 lb weight of the telescope and variable 100 lb wind loads with less than 10" flexure. This means, essentially, that the support structure must have a steel frame anchored in the ice.

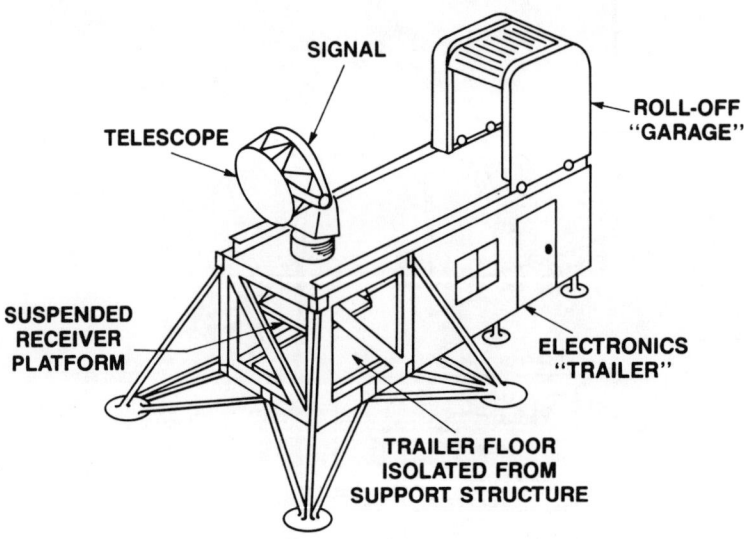

Fig. 2. Support structure (walls removed) and lab for the AST/RO telescope.

It will be necessary to cover the antenna during snowstorms, to prevent snow accumulation. This will be done with a roll-away shed, approximately 8'×8'×8'. Figure 2 shows the telescope with its support structure and cover. The laboratory space underneath the antenna holds a rack of electronics, including the AOS spectrometer, space for the computer terminal and printer, laboratory bench space, storage for tools, and room for the receiver (about 5'×5' of floor space); all this will fit in a modest-sized room. The compressors for the refrigerators are about 100 cubic feet, but are noisy and should be in a separate soundproofed closet that has ducts and a blower to carry the heat where it will be useful.

Potentially the operation of this instrument is completely automatic. The 7-m antenna at Bell Labs is largely automated now; as long as it is not necessary to retune the receiver, it will run unattended for long periods. The AST/RO telescope will never be completely unattended, since the South Pole base is permanently staffed.

V. THE BENEFITS OF A SOUTH POLE OBSERVATORY

An observatory is needed to coordinate efforts and to provide facilities which are beyond the resources of a single group. In the present state of affairs, each experimental group performing astronomical observations from the South Pole must provide its own facilities, shipping entire laboratories to and from the Pole using air transport resources which are expensive and limited. Common facilities should include:

1. electronics shop with standard test instruments: oscilliscope, voltmeter, spectrum analyzer, etc.
2. accurate time and frequency standards
3. means of data communication
4. cryogenics shop with liquid helium and nitrogen supply including liquefier, liquid helium transfer equipment and storage dewar
5. vacuum equipment: leak tester, high capacity roughing pumps and diffusion pumps
6. mechanical shop with precision lathe, milling machine, drill press and an assortment of tools
7. general electronic and mechanical supplies such as screws and resistors.

During the 1988-89 CMBR campaign, each of the three groups shipped most of these items in both directions in order to be sure they would be available if needed.

The observatory could also organize the recruitment and training of technical personnel. Increased numbers of technically proficient winter-over personnel will be needed. With coordination, technical workers can be cross-trained to share jobs, providing greater flexibility and back-ups in work assignments. It seems likely that situations will arise where, say, two different experiments each have programs that require the attention of two people, but only at long intervals or in special situations. Instead of supporting four winter-over people, effective organization could result in two or three people being assigned to both experiments.

Also needed are mechanisms for common use of telescopes. The AST/RO telescope will potentially be useful for a variety of possible observing projects, but no mechanism exists for proposing or carrying out relatively short observing runs, or indeed for using it in any capacity but the 600μm heterodyne spectroscopy for which it was originally conceived. If an observatory existed, it would be the natural entity to solicit proposals and have them refereed, and to organize efforts so that major campaigns are not needed for each experiment.

Much of the science at submillimeter wavelengths requires a large telescope with a small beam. Many interesting objects including protogalaxies and protostars are smaller than 10 seconds of arc, and are best observed with a 10 meter class single-dish instrument or an interferometer. It is hoped that in future, submillimeter astronomy at the South Pole will include such capabilities.

VI. CONCLUSION

The South Pole is an exceptionally good site for submillimeter astronomy. Most of the facilites needed for an observatory are already there. In the next few years, the AST/RO telescope will be installed, and will give some permanent capability for submillimeter observations. Founding of a South Pole Observatory could enhance the capability of AST/RO and other experiments, by organizing services to avoid unnecessary duplication of effort. A South Pole Observatory would also pave the way for major instruments to come.

BIBLIOGRAPHY

Bally, J. 1989 this conference.

Bevilacqua, R. M., and Olivero, J. J. 1989 *J. Geophys. Res.*, in press.

Davidson, J. A. 1987 *Ap. J.*, **315**, 602.

Dragone, C. 1982 *IEEE Trans. Ant. Prop.* **AP-30**, 331.

Dragovan, M. 1989 *Appl. Optics*, in press.

Dragovan, M., Stark, A. A., Pernic, R. and Pomerantz, M. 1989 *Appl. Optics*, in press.

Genzel, R., and Stacey, G. J. 1985 *Mittg. Der. Astr. Gesellschaft,* **63**, 215.

Harwit, M., Houck, J. R., Stacey, G. J. 1986 *Nature,* **319**, 646.

Melnick, G. 1987 in Proc. IAU Symposium No. 120, *Astrochemistry*, ed. M. S. Vardya and S. P. Tarafdar, (Dordrecht: Reidel), pp. 137-152.

Pajot, F. 1989 this conference.

Phillips, T. G. 1987 in *Interstellar Processes*, ed. D. J. Hollenbach and H. A. Thronson, Jr., (Dordrecht: Reidel), p. 707.

Stark, A. A., Davidson, J. A., Harper, D. A., Pernic, R., Loewenstein, R., Platt, S., Engargiola, G., and Casey, S. 1989 *Ap. J.*, **337**, 650.

Stark, A. A. and Gress, J. 1989 this conference.

Sutton, E. C., Blake, G. A., Masson, C. R., and Phillips, T. G. 1985 *Ap. J. Suppl.*, **58**, 341.

Waters, J. W., Wilson, W. J., and Shimabokuro, F. I. 1976 *Science,* **191**, 1174.

Zammit, C. C., and Ade, P. A. R. 1981 *Nature,* **293**, 550.

Millimeter and Sub-millimeter Photometry from Antarctica

J. B. Peterson

Joseph Henry Laboratory, Princeton University,
Princeton, N. J. 08544

Introduction

Over the past five years five groups, from Italy, France and the United States, have attempted photometric observations from Antarctica at wavelengths near one millimeter. These observations have all been made with small, portable telescopes, so the sciences goals of the projects have been limited to large beam width observations. Observations of emission by cold dust in the Galactic plane, in nearby dust clouds, and in local group galaxies have been attempted. In addition several instruments used in Antarctica have been designed to search for anisotropy in the 2.7K cosmic background radiation (CBR). With the recent report (Matsumoto et. al. 1988) of isotropic sky brightness in excess of the 2.7 K planck spectrum, at wavelengths near 500 μm and 700 μm, there has been increased interest in the use of Antarctica as a sub-millimeter observing site. The isotropy of the sky brightness at these wavelengths provides one of the best tests of models of this sub-millimeter excess, but these observations are very difficult from the standard mid-latitude sites.

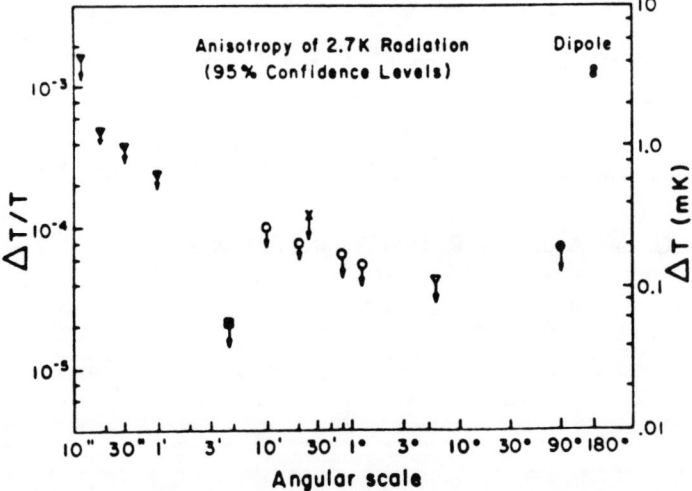

Figure 1. Limits to the anisotropy of the CBR. All measurements shown have been made at frequencies below the peak of the CBR spectrum.

© 1989 American Institute of Physics

Figure one is a plot of the upper limits to the anisotropy of the cosmic background radiation (Wilkinson 1987). No experiment has yet reached the $\Delta T/T \sim 10^{-5}$ level. A diverse set of instruments was used to make the measurements including the VLA, radio telescopes, and balloon-borne microwave radiometers. Progress has been slow in the last few years, in this field The VLA measurements, at lower frequency than the other measurements, are limited by radio point source confusion. Radio telescope observations are limited by atmospheric noise, by pickup of Earth emission, and some are now limited by radio point source confusion. Balloon-borne instruments are constrained to short integration times and, since balloon-borne telescopes used to date scan the sky, they spread that integration time over many patches. In addition, all the measurements shown in figure one have been made below the peak of the cosmic background spectrum, in part because, above the peak, sky noise would severely limit ground based measurements made at mid-latitude sites.

Atmospheric Emission Windows

Figure two is a plot of the surface brightness of astrophysical sources along with an estimate of atmospheric emission from the south pole. The plot shows that as the water vapor content of the atmosphere decreases, atmospheric emission decreases in the windows near one millimeter. In addition transparence improves as the water column decreases. Measurements in the 700 μm band (Dragovan, et. al., 1989) indicate that a total precipitable water content $\sim 300\mu$m is typical at the south pole in summer. Balloon soundings from the pole and from Vostok (Burova et. al. 1986) indicate that days with water content below 100μm are common in winter. At mid-latitude sites, the shortest wavelength photometry window commonly used is at 350 μm. Note that on the polar plateau, several usable windows, with $\sim$50% transparence, appear at wavelengths below 350 μm, when the water content falls below 100 μm.

A number projects in submillimeter and millimeter wavelength photometry can reach better sensitivity levels at dry Antarctic sites than at mid-latitude sites.

Cosmic background anisotropy tests have been done in the past at frequencies below the peak, where low atmospheric emission has made this work possible from mid-latitude sites. As photometer sensitivity improves these observations have moved up in frequency to avoid interference from radio point sources. In addition the report of excess emission at frequencies above the peak has rekindled interest in observations on the Wein side of the CBR spectrum. Observations at these frequencies require very dry sites.

Figure two includes an estimate of the spectrum of interstellar Galactic dust emission, in a slab model, calculated for Galactic latitude 15°. The estimate is made using data from the IRAS 100μm band, assuming a slab model dust

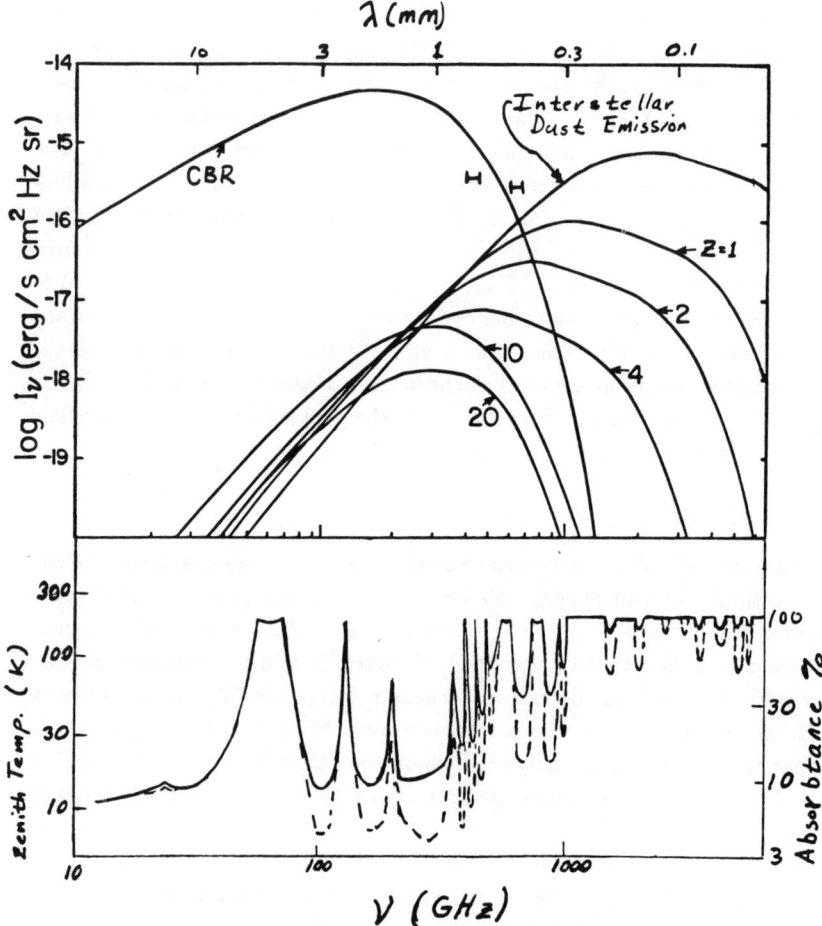

Figure 2. Brightness of astrophysical sources shown in the top panel include: the CBR, emission by 20 K dust in the Galaxy at 15° Galactic Latitude, and dust emission by a galaxy like our own, seen at high redshift (curves labled by redshift). The error bars show the sub-millimeter excess reported by Matsumoto et. al. The lower panel shows a calculated estimate of atmospheric emission and absorbtance for 300 μm H_2O column, and 100 μm H_2O column (dashed curve). The windows available in the Antarctic winter are well placed to study the sub-millimeter excess, and dust emission by high redshift galaxies.

distribution, with dust emissivity $\epsilon \sim \nu^2$ (Draine and Lee, 1984). Atmospheric emission is so strong near the peak of the dust emission spectrum that most measurements of Galactic dust emission have been done from aircraft, balloons or spacecraft. The reduced atmospheric emission available in Antarctica has made observations of Galactic dust emission possible from the ground.

Also plotted in figure two is an estimate of the surface brightness of a galaxy like our own, seen at high redshift (Wright and Peterson 1989). Note that, above redshift 5 this dust emission brightness is centered near wavelength 1 mm. In fact for $\lambda > 1$mm high redshift galaxies are brighter than low redshift galaxies. This wavelength range is unique; at these wavelengths, if galaxies can be observed through the atmosphere at all, they can be seen all the way back to the redshift at which they formed. To date the sensitivity of millimeter and submillimeter observations has not been adequate to study high redshift dust emission. New bolometer designs, optimized for these atmospheric windows are under development, and if these bolometers can be used with ~ 10 m aperture telescopes on the polar plateau, the new field of dust cosmology may begin the provide important constraints to models of galaxy evolution.

Sky Noise

Because, in the near-millimeter windows, the atmospheric emission zenith temperature is larger than the antenna temperature due to astrophysical sources, photometry is accomplished using switched beam observations. That is, the telescope beam is rapidly switched between nearby patches of sky, and a difference is taken. As water vapor clouds pass through the beams, the fluctuating difference of atmospheric emission along the two lines of sight limits the sensitivity of these observation. Sky noise should decrease as the water content of the atmosphere decreases but, sky noise is also dependent on the invisible spatial structure of the water vapor clouds, and on the degree of atmospheric turbulence above the telescope.

In the near-millimeter windows, atmospheric emission is mostly due to the far wings of the strong, pressure broadened, water line at 560 μm and the wings of water lines at shorter wavelengths. So, most sky noise is contributed by the first pressure scale height above the telescope. At a stable, low wind speed site, like the south pole, we can expect less turbulence in this layer than at a wind swept mountain top site.

During November 1988 through January 1989, John Ruhl of the Princeton group, operated a sky noise monitor (Ruhl and Wilkinson 1989) for the 3 mm window, at the south pole. The monitor used a room temperature receiver, which was much less sensitive than the cryogenic detector used, also at 3 mm, for a CBR isotropy test. Because the monitor used a larger beam throw, 7.5°, and used a gradient sensitive beam switching pattern, it was possible to measure sky noise, above receiver noise, about half the time. The same monitor was used at another flat CBR site, Saskatoon, Saskatchewan, by Dave Wilkinson, in March 1989.

Data from the monitor are shown in figure 3. The sky temperature difference, between beam positions, was measured by switching at 26 Hz and synchronously detecting using a lockin. Lockin integration time constant 10 s was used. The results presented are peak to peak fluctuations in the lockin output, measured over one hour intervals. The data from Saskatoon have median sky noise 80 mK, with most one hour periods well above receiver noise. At the pole the median measured noise was 50 mK, only slightly above receiver noise alone. The data are sufficient to conclude that the south pole, in summer, is a better site for photometry at 3 mm, than Saskatoon in late winter. In addition, we can attempt to extrapolate data from the monitor, using a Kolmogorov $1/f^2$ power spectrum (Levin, et. al. 1987) for water vapor spatial structure, to the smaller throw angles used for CBR isotropy tests. We conclude from this extrapolation, that using a gradient insensitive sensitivity pattern, in the 3 mm window, isotropy observations with beam throw less than 1°, at the pole, in summer, can reach isotropy limits well below 10^{-5}.

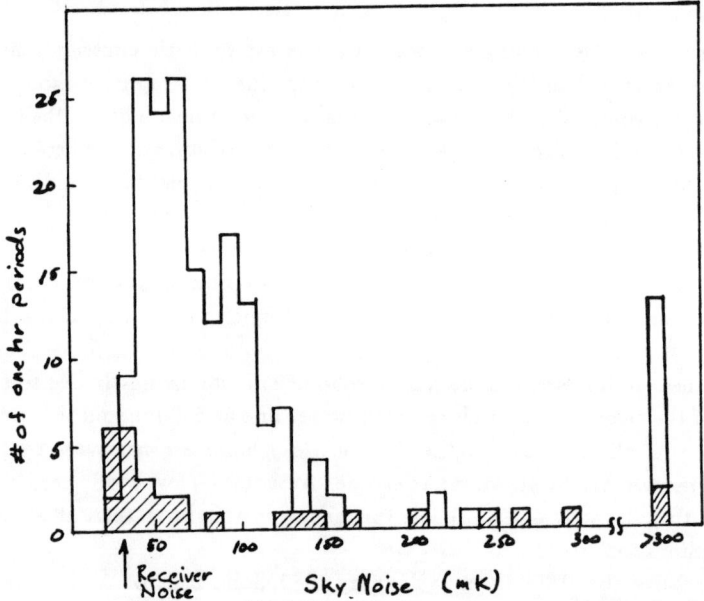

Figure 3. Sky noise in the 3 mm window at the south pole, in summer (cross hatched) and at Saskatoon, Saskatchewan, in late winter. The data are discussed in the text.

So far, we have not used the monitor at a mountain top site. However, the University of Chicago 32 pixel array of bolometers for the sub-millimeter windows has been used at the south pole and at Mauna Kea. Although no controlled comparison was made between the two sites, at the pole, the Moon was used for calibration at 10° (seven air masses), in the 350 μm window. Sky noise at Mauna Kea would have made this calibration impossible.

Published Results

There have been two published photometry results from antarctic observations. Pajot et. al. (1986) measured dust emission in the Galactic plane, in a broad spectral band from 800 μm to 4 mm, from the south pole. The instrument made a comparison between a narrow beam, which was scanned through the plane and a broader reference beam.

Andreani et. al. (1989) used a 1 m telescope with a hydraulically driven chopping primary at the Italian base at Terra Nova Bay. They measured emission at 1 mm and 2 mm, from dust at low galactic latitudes, from the Chamaeleon cloud, from an IRAS cirrus cloud and from the Large Magellanic Cloud . They find a larger flux from these objects than expected from the IRAS 100 μm images, indicating a larger mass of cold 10K dust than models predict.

Current Projects

In the 1988-1989 summer season there were three groups studying CBR isotropy at the south pole.

A group from University of California, Santa Barbara, led by Lubin, brought two instruments, a switched horn radiometer for 13 and 20 mm wavelength, with a 7° beam width, and a 1 m telescope for 3 mm wavelength with a 30 arc-minute beam width. Both instruments used receivers, the 13 and 20 mm receivers had HEMT first rf amplifiers, and the 3 mm receiver used an SIS mixer.

A group from AT&T Bell Labs, led by Dragovan, brought a 1.2 m off-axis telescope and a 32 pixel, 300 mK bolometer array. Using filters for the 350 μm, 450 μm, and 650 μm wavelength windows, they searched for anisotropy of the sub-millimeter excess using 5 arc-minute beam widths.

A group from Princeton, led by Peterson, used 50 mK bolometers in the 3 mm wavelength window, with a 1.2 m cassegrain antenna, with 9 arc-minute beam width, and a 20 cm refracting telescope, with 1.8° beam width.

No group has published results from this season to date, but the list of instruments includes some of the most sensitive photometers available today, and covers a wide range of wavelengths and angular scales. If these instruments reached the sensitivity possible with these state-of-the-art detector systems, new limits to CBR isotropy will be set, using observations from Antarctica.

Future Programs

In the 1989-90 season a group led by Smoot from Berkeley plans to attempt a measurement of the spectrum of the CBR from the pole. This project is a continuation of their White Mountain measurements (Smoot, et. al. 1985) from 3 mm to 100 cm wavelength.

The Italian group plans to return to the Antarctic coast with a 2m telescope for millimeter wavelengths in the 1989-90 season.

In ~1992, the Astro project, a 1.8 m sub-millimeter telescope, will be installed at the pole. This Bell Labs-University of Illinois-Boston University project is designed primarily to observe line emission, but may also be useful for broad band photometry. Astro is designed to operate in the Austral winter.

If, in the future, large aperture (~10m) telescopes become available on the Antarctic plateau, observations from Antarctica can include studies of continuum emission by QSOs, studies of dust emission by high redshift galaxies, and studies of star forming regions.

Since the 15 m James Clark Maxwell telescope on Mauna Kea opened, it's time allocation committee has been force to turn away 3/4 of requests for telescope time. The polar plateau seems to be ~4 times drier in summer, ~ 10 times drier in winter than Mauna Kea. A large aperture telescope on the plateau would allow even more sensitive measurements in the windows used at mid-latitudes, and make possible, from the ground, observations in shorter wavelength windows that must now be done from aircraft, balloons, or spacecraft.

References

Andreani, P., Dall'Oglio, G., Martinis, L., Piccarillo, L. Pizzo, L., Rossi, L., Venturino, C., 1989, submitted to Ap. J.
Burova, L. P.,et. al., 1986, Sov. Astron. Lett., **12**,339.
Dragovan, M., Stark, A. A., Pernic, R. and Pomerantz, M.,1983, Applied Optics, in press.
Draine, B. T. and Lee, H. M., 1984, Ap. J. **285**,89.
Levin, S. M., et. al., 1987, Radio Sci, **22**,521
Matsumoto, et. al., 1988, Ap. J. **320**,567.
Pajot, F., et. al., 1896, in Space-Borne Sub-Millimetre Astronomy Mission, 189
Ruhl, J. and Wilkinson, D. T., 1989, in preparation
Smoot, G. F., et. al., 1985, Ap. J. Lett. **291**,L23.
Wilkinson, D. T., 1987, Proceedings of the 13th Texas Symposium on Relativistic Astrophysics, Ulmer, P. M.(ed), 209.
Wright, G. A. and Peterson, J. B., 1989, in preparation

This work was supported by a grant from the N. S. F.

INFRARED ASTRONOMY IN ANTARCTICA

D. A. Harper
University of Chicago
Yerkes Observatory, Williams Bay, WI 53191

Infrared observations are crucial to resolving many of the most important problems in modern astrophysics. Infrared light can penetrate dense clouds of dust to reveal the birth of stars and planetary systems and the structure of galactic nuclei. The infrared spectrum also contains the fossil record of the visible universe as it appeared when the first galaxies were born. A number of direct observational tests of theories of galaxy, star, and planet formation are possible if sufficient sensitivity can be achieved.

Astronomers have appreciated for some time that Antarctica should be a good infrared site. However, the difficulty of traveling to and working in the Antarctic (particularly during the long polar night), the existence of relatively easy access to good infrared sites at mid-latitudes (e.g., Mauna Kea in Hawaii), and limitations in the performance of infrared detectors have discouraged the development of astronomical facilities in Antarctica. Recent technological advances, however, promise to radically alter this situation. Future advances in the sensitivities of infrared measurements will depend much more strongly on improving site quality rather than detector performance.

Among Earth-based sites, Antarctica offers unique opportunities to circumvent environmental limitations. There is increasing evidence that atmospheric conditions on the Antarctic Plateau are the best in the world for astronomical observations at infrared, submillimeter, and millimeter wavelengths. The elevation of the polar ice cap ranges from nearly 3,000 meters above sea level at the South Pole to over 4,000 meters near the geographic center of the continent. The center of the Plateau is located in a nearly permanent high-pressure zone in which air descending from high altitudes dominates the atmospheric circulation. These factors, combined with the low surface temperatures (-80 to -30 C), result in extremely dry conditions and open up new atmospheric "windows" for studying the Universe. At near-infrared wavelengths, which lie on the exponential side of the ambient-temperature Planck radiation function, the low temperatures also result in a dramatic reduction in thermal background radiation.

During the past several years, a number of submillimeter and millimeter-wavelength experiments conducted during the Austral summer have confirmed many of the expected benefits of Antarctic atmospheric conditions and sparked a resurgence of interest in both summer and year-round observing. Coupled with the recent revolution in the sensitivities and sizes of infrared detector arrays and advances in instrument automation, these efforts have provided fresh incentive for the development of new astronomical initiatives at the South Pole and other high-altitude sites in the continental interior. In the following, I will review some of the potential advantages of Antarctica as a site for infrared astronomy and touch briefly on a few of the scientific investigations which could be pursued there.

High Atmospheric Transmission

Bally[1] and Townes and Melnick[2] have recently summarized current knowledge of transmission and water-vapor content of the Antarctic atmosphere. It is clear from submillimeter and millimeter measurements made during the last few Austral summers that the atmospheric opacities are consistently much lower than observed at good mid-latitude sites such as Mauna Kea, Hawaii. The measurements indicate that the average water-vapor content over the South Pole is sufficiently low to produce a zenith opacity of order 0.1 to 0.2 at a wavelength of 700 μm during the Austral summer[3]. According to the model calibration adopted by Bally[1], this corresponds to about 200 to 500 μm of precipitable water vapor. Available measurements suggest that at times during the Austral winter the precipitable water vapor is lower than 100 μm and is rarely greater than 1 mm.

The model-atmosphere computations, when calibrated by the submillimeter sky-dip observations, indicate that about 85% of the spectrum at wavelengths longer than 200 μm can be observed most of the time. Observing at 350 μm (or through the even poorer atmospheric window at 35 μm) from Mauna Kea can be frustrating, since the sky usually transmits only 5 to 20% of the radiation and is highly variable, making calibration extremely difficult and creating "sky noise" which limits the sensitivity of measurements to many times less than the value set by the basic thermal "photon noise." From Antarctica, observations at 35 μm or 350 μm may be as easy as observations at a wavelength of 3 mm at existing millimeter-wave observatory sites such as Kitt Peak, where measurements can be performed 80% of the time through a steady atmosphere with less than 30% attenuation. At a given atmospheric transmission, the South Pole has the additional advantage of the greater stability which comes from the absence of the strong diurnal effects which plague observations at temperate latitudes.

Observations between 40 and 320 μm are impossible from any ground-based site at temperate latitudes, where the water vapor burden is almost always greater than 500 to 1000 μm. When the amount of precipitable water vapor drops below 200 to 500 μm, this spectral range develops many new "holes" which can be exploited in much the same way that astronomers now use the 350-μm and 650-μm submillimeter windows from Mauna Kea. The new windows between 35 and 100 μm are especially significant, since at the South Pole it is feasible to build a much larger telescope (with greater angular resolution) than possible in the near future for any airborne or orbital observatory.

Low Background Emission

Perhaps the greatest potential for gains in sensitivity over other ground-based sites exists in the near-infrared. The sensitivity of imaging at wavelengths of 1 to 4 μm is limited by the high background light from airglow emission from OH radicals at altitudes of 80-90 km, from atmospheric thermal emission (at the longer wavelengths), and, ultimately, by diffuse celestial light such as the zodiacal light (scattered and thermally re-emitted sunlight from dust within the Solar System). Since the near-infrared spectrum lies on the Wien side of the 300 K blackbody spectrum radiated by room-temperature objects, a small decrease in ambient temperature lowers the thermal background light by a large factor. In a spectral band between 2.27 and 2.45 μm, fortuitously devoid of OH airglow emission, the net background flux during the Austral winter may approach the limit set by the zodiacal light. Since this is also the minimum in the brightness of the zodiacal light, the background light in this band would be fainter than at any other wavelength, including visible wavelengths. Because of the lower telescope temperature, a South Pole telescope would even have a lower background than the Hubble Space Telescope. Cooled space telescopes like SIRTF

(NASA's planned Space Infrared Telescope Facility, a 1-m class instrument) will be able to reach the zodiacal-light limit, but for 2.4-µm observations, future large Antarctic telescopes will enjoy a significant advantage in aperture and cost.

Computations of thermal emission from the atmosphere in the 2.4-µm window made by Lubin[4] show quantitatively the gains which may accrue from working in the cold conditions at the South Pole. Figure 1 compares the expected thermal emission from the polar atmosphere and from the sky above Mauna Kea, an excellent mid-latitude site. The Mauna Kea calculations were based on a surface temperature of 2.6 C and a total amount of precipitable water of 1200 µm. For the South Pole data, the corresponding values were -59 C and 260 µm. The parameters chosen for the computations were intended to simulate good but not exceptional conditions at the two sites.

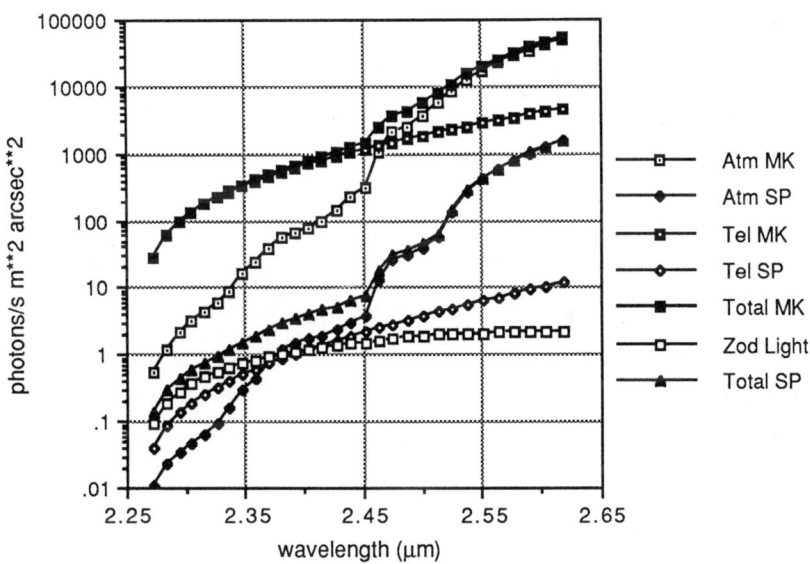

Figure 1 A comparison of integrated background emission for telescopes at the South Pole and Mauna Kea, Hawaii.[4] The y-coordinate is the value of the definite integral of the photon flux density from 2.27 µm to the upper limit given by the x-coordinate. The important contributers to the integrated background are atmospheric thermal emission (Atm), the instrumental thermal emission from the telescope and camera window (Tel), and the extraterrestrial background, which is dominated by zodiacal light. The atmospheric and zodiacal light values represent fluxes incident on the entrance aperture of the telescope.

In addition to the atmospheric thermal emission, Figure 1 shows the expected instrumental contribution to the thermal background (assuming a telescope emissivity of

5%), the estimated zodiacal light background for a position near the ecliptic pole, and the sums of the three components. The zodiacal light contribution was estimated from models presented by Murdock and Price[5]. These models suggest that the zodiacal light will be a factor of two to four times brighter than at the ecliptic pole over much of the anti-solar hemisphere. Note that the data presented are integral background fluxes, in photons/s m^2 arcsec^2, from a lower wavelength limit of 2.27 µm to a progressively larger long-wavelength limit given by the x-coordinate.

Several conclusions can be drawn from Figure 1. (1) Given detectors with sufficiently low dark current, the potential advantage of the South Pole site is *extremely* large. The average ratio of integral fluxes at Mauna Kea to those at the South Pole is ~220 over a range of passbands from 2.27–2.33 µm to 2.27–2.45 µm. (2) The ratio of the telescope's thermal emission to atmospheric emission is larger at Mauna Kea than at the Pole. This results from the fact that the polar atmosphere has a pronounced temperature inversion, so that the polar sky flux is dominated by emission from altitudes at which the temperature is somewhat higher than the surface temperature. Thus, the comparison between the two sites would favor the South Pole even more strongly if the assumed instrumental emissivities were ~10-15% (values more characteristic of those actually realized on existing telescopes than the value of 5% used in the computation). In fact, it should be easier to preserve the quality of mirror coatings in the cold, dry, polar conditions, so the *practical* advantage of the South Pole telescope will be larger still. (3) For the South Pole telescope, the computed contributions of zodiacal, instrumental, and atmospheric emission are comparable. In addition to the basic uncertainties in the atmospheric model or the zodiacal light, differences will arise because of variations in atmospheric conditions, zenith angle, ecliptic latitude, solar elongation, and mirror quality. Consequently, optimal combinations of telescope and instrument design, filter passband, pixel size, and observing strategy can be determined only through actual experience at the Pole. Early on-site measurements will be essential to the efficient planning of future instrumentation and observing programs.

Better on-site measurements of atmospheric emission are particularly important. Hoffmann, Frey, and Lemke[6] made photometric measurements of near-infrared airglow emission in several spectral bands from balloon altitudes. In a 2.35–2.45 µm band, they found that the zenith surface brightness was $<6 \times 10^{-11}$ W/cm^2 µm sr, corresponding to an integrated 2.27–2.45 µm flux of <33 photons/s m^2 arcsec^2. Ito et al.[7] reported seeing an excess airglow emission during a balloon flight in a passband with bandwidth of $\Delta\lambda/\lambda = 0.1$ centered at 2.4 µm, but they also stated that the excess may have been caused by a shift in the passband of their filter due to cooling to nitrogen temperature and to the very fast f/0.8 focal ratio of their telescope, resulting in the inclusion of known OH lines from the spectral region shortward of 2.27 µm. The upper limit from the Hoffmann, Frey, and Lemke results is approximately 4 times the estimate presented in Figure 1 for the total background flux for a South Pole telescope. Taking this as a worst case, the South Pole telescope would still enjoy an advantage in background compared to a mid-latitude telescope of more than a factor of 50.

Measurements by Baker et al.[8] using a rocket-borne spectrometer flown through an extremely bright aurora did not reveal any strong emission lines in the 2.27-2.45 µm band. As in the case of the airglow, however, more sensitive measurements are needed to determine whether this result holds at levels comparable to the brightness of the zodiacal light.

In summary, working in the cold conditions on the Antarctic Plateau will enable an enormous gain in sensitivity over what can be achieved at warmer sites. The difference in the thermal background at a wavelength of 2.4 μm may be a factor of 220 or more (based on the comparison above between a night at the South Pole with 0.26 mm of precipitable water vapor and a temperature of -59 C, and a night at Mauna Kea with 1.2 mm of water and a temperature of 2.6 C). In this case, the warmer telescope suffers a penalty of a factor of 15 in signal-to-noise ratio. If all else is held equal, the difference in the noise translates to the following gains for background-limited observations, in the sense of what a warmer telescope would require compared to the colder telescope:

(1) Source flux brighter by 2.9 magnitudes

(2) Telescope diameter larger by a factor of 15

(3) Integration time longer by a factor of 220.

For very deep images, there is an additional and perhaps even more important benefit of lower background. Experience with optical CCD cameras has shown that it is relatively easy to calibrate an image well enough to detect detail at an intensity level which is about 1% of the background. However, even with techniques such as drift scanning and the use of "median sky flats," it has proven very difficult to go below ~0.1% of the background. In this case, the South Pole telescope might enjoy an advantage in signal-to-noise ratio proportional to the decrease in background, an improvement of a factor of 220 in limiting flux density (5.9 magnitudes).

From the background fluxes in Figure 1, one can calculate limiting fluxes (per pixel) for various combinations of telescope aperture, pixel size, integration time, and detector performance. The results are presented in Figure 2, for three different pixel sizes and for telescope diameters ranging from 0.6 to 8 meters. Parameters used in the calculations included a signal-to-noise ratio of 5, a total system efficiency of 0.5, a bandwidth of $\Delta\lambda/\lambda = 0.08$, a detector dark current of 2 electrons/s, a readout noise of 50 electrons, a full-well capacity of 100,000 electrons, an integration time of 5000 seconds, and zodiacal light and atmospheric emission intensities a factor of two larger than the levels shown in Figure 1 (which were intended to represent the minimum values found at the ecliptic pole and zenith, respectively).

Examples of Astronomical Gains

The study of galaxies at optical wavelengths has in recent years been advanced significantly by the advent of sensitive CCD detector arrays and the development of techniques to measure signals as small as one part in 10^4 of the sky background. It is now evident that information available at near-infrared wavelengths may be crucial to the study of galaxy evolution in general, and to the detection and identification of primeval galaxies in particular (see, e.g., Lilly and Cowie[9]).

The high background at these wavelengths, however, has limited the utility of these measurements. For example, the typical background photon flux in the K band at a mid-latitude site is over 200 times that in the V band. Since many galaxy spectra are approximately "flat" in photon flux for a constant fractional bandwidth, this means that it is difficult or impossible to see the same objects in the deepest current infrared pictures.

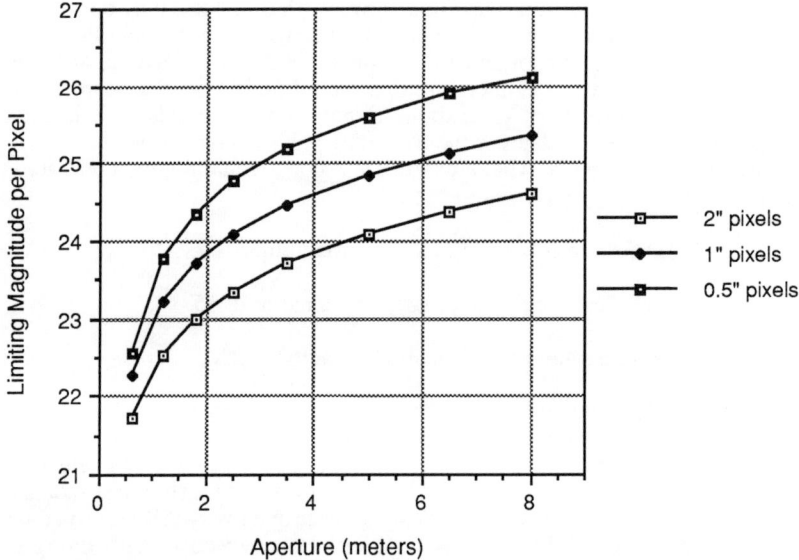

Figure 2 Limiting K magnitudes (per pixel) for South Pole telescopes for an integration time of 5000 s and a statistical significance of 5σ. See the text for a detailed description of the assumptions upon which the calculations were based.

Since the near-infrared background may be over a factor of 200 lower during the Antarctic winter than at temperate latitudes, going to the South Pole can put infrared and optical measurements on an equal footing. Indeed, for objects at high redshifts, it is the infrared light which corresponds to the spectral region we know best, while the visual light corresponds to the rest-frame ultraviolet, which can vary widely from galaxy to galaxy. Thus, counts of galaxies in infrared images should be much less sensitive to the presence of hot young stars than counts in optical images. *Comparisons* of deep infrared images with deep optical images will be a powerful tool for studying these evolutionary effects, including identifying the youngest galaxies. The color distribution provides a constraint on the redshift distribution, and the infrared counts and the optical-infrared colors together provide a better constraint on cosmological models than classical tests.

Detectors good enough to take advantage of the dark Antarctic sky are now becoming available. As readout noises drop to fewer than 50 electrons and the sizes of focal-plane arrays increase to levels comparable to optical CCD cameras, their application to the study of faint galaxies is an increasingly compelling prospect.

Near-infrared observations from Antarctica can, of course, serve a much broader scientific program than faint galaxy studies. For example, near-infrared surveys can be an extremely valuable source of statistically significant numbers of rare objects. To cite just one example, brown dwarfs may be a thousand times rarer than primeval galaxies at a given magnitude (see, e.g., Lilly and Cowie[9]), yet a 2.5-m telescope equipped with a wide-field infrared camera would be able detect them in the general field by the thousands or tens of thousands. Such instrumentation would also be invaluable for studying heavily obscured star formation regions in our own and other galaxies.

Exciting new observations will also be possible at far-infrared and submillimeter wavelengths. Of particular interest are the high-angular resolution observations which would be possible with large (10-m class) telescopes or interferometric arrays. Such observations are essential for understanding the structure and energetics of protostars, star-forming regions, and active galactic nuclei. Until very large telescopes are available in orbit or on the Moon, Antarctica offers the only prospect for achieving arcsecond or better resolution at wavelengths between 35 and 350 µm.

Acknowledgements

I would like to acknowledge John Bally, Patrick Espy, David Hall, Mark Hereld, Richard Kron, and Daniel Lubin for their help in gathering and preparing the material presented above.

References

1. Bally, J., in these proceedings
2. Townes, C.H. and Melnick, G., "Atmospheric Transmission in the Far Infrared at the South Pole and Astronomical Applications," preprint (1989).
3. Dragovan, M., Stark, A.A., Pernic, R., and Pomerantz, M., *Applied Optics*, in press (1989).
4. Lubin, D., Masters thesis, University of Chicago (1988).
5. Murdock, T. L., and Price, S. D., *Astron. J.*, **90**, 375 (1985).
6. Hoffmann, W., Frey, A., and Lemke, D., *Nature*, **250**, 636 (1974).
7. Ito, K., Matsumoto, T., and Uyama, K., *Publ. Astron. Soc. Japan*, **28**, 427 (1976).
8. Baker, K. D., Baker, D. J., Ulwick, J. C., and Stair, A. T., Jr., *J. Geophys. Res.*, **82**, 3518 (1977).
9. Lilly, S.J. and Cowie, L.L., in *Infrared Astronomy and Arrays*, eds. C.G. Wynn-Williams and E.E. Becklin (Honolulu:University of Hawaii, 1987), p. 473.

Geospace Plasmas

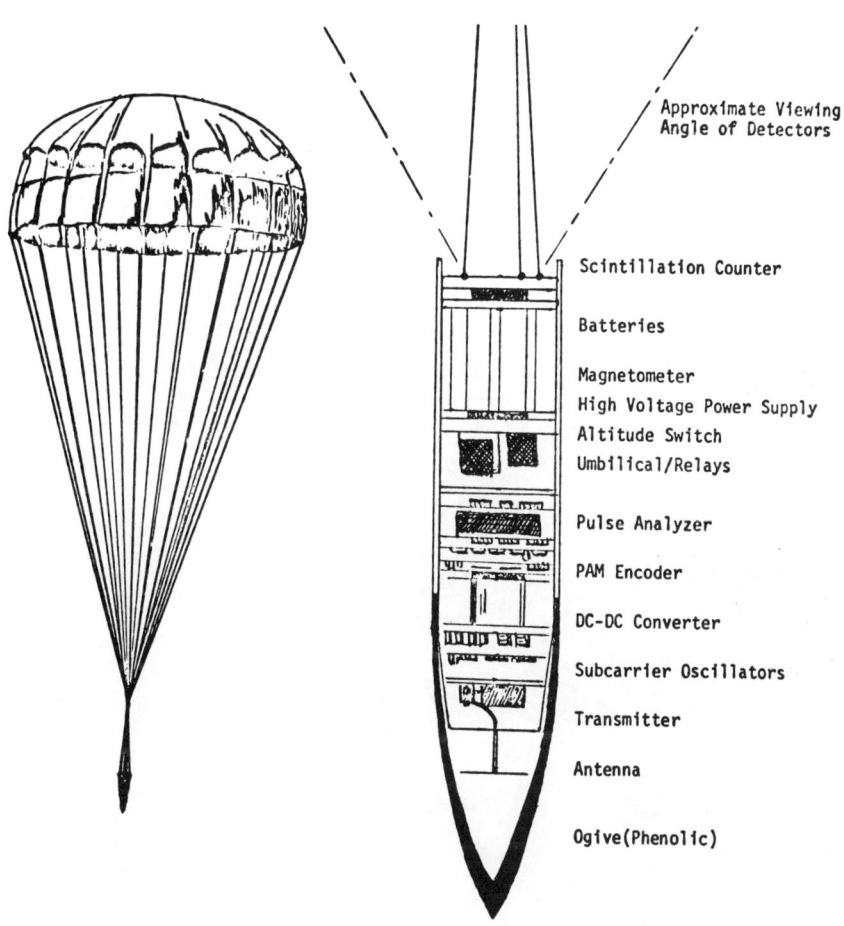

GEOSPACE PLASMAS

R.L. Arnoldy

Institute for the Study of Earth, Oceans and Space, University of New Hampshire

INTRODUCTION

By Upper Atmospheric Research I am referring to what might better be called Solar-Terrestrial Physics or Space Plasma Physics. the major objective of this research is how the energy carried from the sun by the solar wind is coupled to the Earth System and is manifest as such geophysical phenomena as magnetic storms, ionospheric disturbances and auroral displays. The ultimate objective over the next decade or two will be to seek an understanding how this energy is deposited in the various components of the Earth System such as the atmosphere, hydrosphere, etc., and how it affects the interactions between these components and the underlying processes that structure life on Earth.

The solar wind plasma is an electrically charged gas in which the atoms have been stripped of one or more of their electrons. Because of its high electrical conductivity it carries some of the solar magnetic field with it as it thermally expands into space from the sun. The interaction between the Earth System and the solar wind is principally via the magnetic field of the Earth and the plasma and trapped radiation it contains. A cartoon of this interaction is given in Figure 1 showing how the solar wind is deflected by the Earth and how it in turn distorts the magnetic field of the Earth creating a cavity which contains it, called the magnetosphere. The solar wind energy absorbed by the Earth System, 10^{11}-10^{12} watts, which is about 1% of the energy incident on the magnetosphere, is stored in the magnetosphere in the form of trapped radiation and current systems that distort the Earth's magnetic field. In addition significant energy, on the order of a few times 10^{11} watts, is transported through the magnetosphere and deposited in the neutral atmosphere during each magnetic substorm. The magnetic substorm results in acceleration of particles in the magnetosphere, the dumping of some of these particles into the atmosphere producing aurora, and the intensification of ionospheric current systems. The apparent explosive release of energy by the magnetosphere during a substorm, a 'magnetospheric hiccup', is the primary means of transporting the solar wind energy to other components of the Earth System.

THE IMPORTANCE OF THE POLAR REGIONS

The polar regions are important in the study of terrestrial Space Plasma Physics (Magnetospheric Physics) for several reasons, First, referring back to Figure 1, the polar regions directly sample the solar wind at the polar cusps. The polar cusps separate the Earth's magnetic field lines which close within the dayside magnetosphere from the open tail field lines which connect to the solar wind. At the polar cusps, the shielding effect of the Earth's magnetic field may virtually vanish. Here the solar wind plasma can penetrate directly into the magnetosphere and release charged particles into the ionosphere along the dayside auroral oval. Satellites have detected strong and complex plasma flows in the cusps. Intense plasma wave activity is found in the cusps where currents link the electrodynamics of the ionosphere to those currents that must exist on the sub-solar surface of the magnetosphere creating the boundary between the solar wind and the Earth's magnetic field. Although various spacecraft have probed the polar cusp regions

THE EARTH'S MAGNETOSPHERE
(Meridian Sectional View)

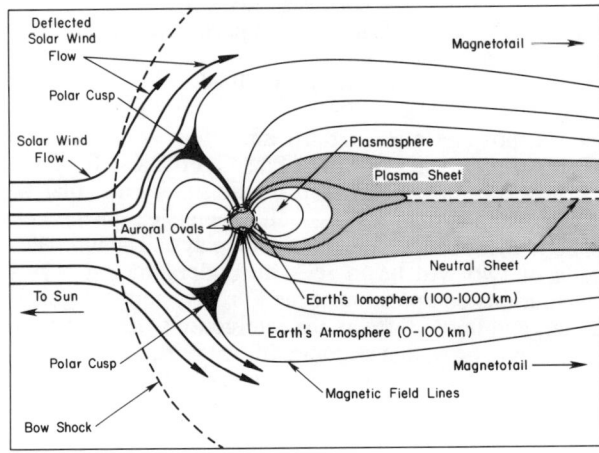

Fig 1. Schematic of the matgnetosphere depicting its interaction with the solar wind.

very little is known about them. Previous measurements have been inadequate due to limited instrumentation, telemetry capability, and spatial coverage. The structure of the cusp entry region is highly turbulent, and the entry process is likely to be very complex.

A second reason for the importance of the polar regions is that they are connected via magnetic field lines to the entire boundary between the magnetosphere and the solar wind, the magnetopause. After the solar wind is decelerated and heated in passing thorough the bow-shock, it expands and flows around the dayside magnetopause where momentum and energy coupling between it and the magnetosphere begins. The strength of these coupling is dependent on the direction of the solar wind magnetic field. A southward directed solar wind field is a favorable configuration for energy transfer with the Earth's field by a process known as magnetic reconnection. By this process the solar wind magnetic field becomes topologically connected to the Earth's magnetic field, the magnetosphere is opened and the solar wind plasma and the Earth's plasma intermix. However, since the solar wind plasma is moving rapidly along the magnetopause, plasma entry to the magnetosphere is spread out over great distances and in fact most of it occurs on the night side of the Earth. The magnetic lines of force of the solar wind magnetic field that have merged with the Earth's field are distended on the night side of the Earth and form the geomagnetic tail. The solar wind plasma on these field lines is subject to a variety of plasma instabilities and forms the plasma mantel in the high-latitude tail. Since the polar regions are geomagnetically connected to this process of opening and transporting the Earth's field to the tail, information about the process is propagated to the polar regions by magnetohydrodynamic (MHD) waves. Although spacecrafts have perhaps identified the fundamentals of the coupling of solar wind energy to the

Earth, a more thorough monitoring of the process to better understand it by means of polar region MHD studies is important. I will elaborate on this more later.

A third reason for the importance of the polar regions in magnetospheric studies is that they are the recipients of the solar wind energy that has been stored in, or transported through, the geomagnetic tail and funnelled to the auroral ovals during magnetic storms or substorms. The details of how this transport and/or release of energy occurs is a major topic of research in Magnetospheric Physics. It is generally believed that as the Earth's magnetic field lines (identified only by the plasma gyrating about them) are opened on the dayside and folded back into the tail they release their stress by reconnecting near the midplane of the magnetic tail (see Figure 2). This reconnection process and the mixing of flowing plasma from both hemispheres will generate large-scale plasma turbulence and produce a rapid heating of the plasma. Solar flares on the sun may be a similar release of energy stored in solar magnetic fields to energetic particles and electromagnetic radiation. Observationally, very hot plasma and energetic particle beams are measured in the Earth's tail when the magnetic field in the tail suddenly becomes less stressed, known as a substorm. Substorms represent a major transitory loss of stored tail energy as well as a major source for the injection of energy into the near-Earth magnetosphere and ionosphere.

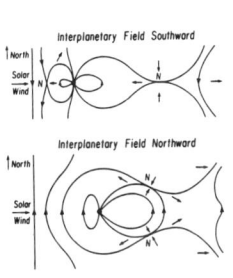

Fig. 2. Magnetic reconnection according to Dungey, J.W., Phys. Rev. Lett., 6, 47, 1961.

SPECIFIC POLAR SCIENCE PROGRAMS

There are three platforms from which scientific observations have and will continue to be made in the polar regions. These are the ground, balloons, and sounding rockets. In my discussion of these platforms I will direct my attention to the Antarctic since it is the preferred pole for several reasons. First, because the Earth's magnetic dipole is off-centered, the auroral oval experiences twenty-four hours of darkness during the Austral winter while this is not the case during the Arctic winter. As a result, dayside aurora can be best studied by Antarctica. Second, Antarctica has no territorial boundaries (or population) which put restrictions on scientific endeavors. Finally, during the Austral summer it is theoretically possible to deploy field sites giving a good local-time coverage of the auroral oval and the polar cap, this cannot be done in the Arctic because of limited land mass coverage.

I. GROUND EXPERIMENTS

The underlying scientific justification for much of the ground work that I will discuss for the Antarctic is that with multiple sites one can make simultaneous measurements related to a large portion of the magnetosphere to an extent that cannot now, nor in the foreseeable future, be accomplished by satellites. Coupled with northern high latitude measurements, multiple site Antarctic data can provide valuable insight into the topology of the magnetosphere and the plasma physics processes that occur at the important boundaries of the magnetosphere. I will now discuss briefly several types of ground observations that could accomplish this 'magnetospheric mapping'.

A. OPTICAL MEASUREMENTS

Figure 3 gives a view of the aurora obtained by the University of Iowa imager on the DE-1 spacecraft. Occasionally these global views of the aurora will have a sun-aligned arc crossing over the pole forming what looks like the Greek letter theta, hence the name Theta Aurora. There is no doubt that such data is invaluable in formulating a global view of the magnetosphere. Just how the Theta Aurora maps to the geomagnetic tail remains a mystery; understanding its topology will for sure be a major milestone in Magnetospheric Physics. Another view that we have of the auroral oval is from DMSP spacecraft at altitudes of 1000 km. An example of these data is given in Figure 4 where the spatial resolution is considerably improved over the DE-1 images. From such data one now starts to get a better view on how transpolar or sun-aligned arcs are connected to the nightside aurora. However, the DMSP photos are generated using the orbital motion of the spacecraft, hence the photo in Figure 4 was compiled over a time period of a few tens of minutes. Anyone who has witnessed an aurora is impressed with the dynamics of the aurora with time variability up to several Hertz, the resolution of the camera systems, and probably well beyond this. Figure 5 presents data from an intensified TV system (courtesy of University of Alaska) having a 30 km field of view. The pencil rays evident in Figure 5 have scale sizes on the order of hundreds of meters to a fewkilometers and it is the motion of these rays that provides much of the dynamics of the aurora. The data in Figure 5 for 'quiet arc'

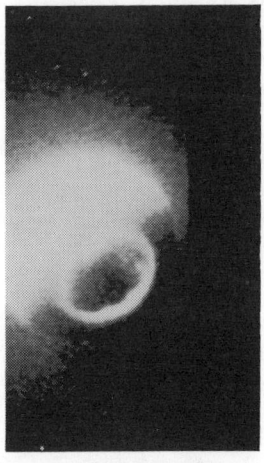

Fig. 3. Auroral image from Dynamic Explorer spacecraft, courtesy L. Frank, U.

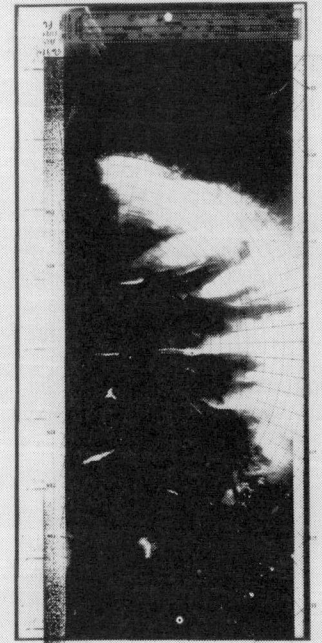

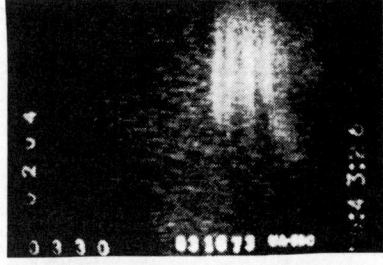

Fig. 5 Image-intensified, ground TV image of the aurora, courtesy, T. Hallinan, Geophysical Institute of Alaska.

Fig. 4. Auroral image from a DSMP spacecraft.

which is very inappropriately named. Clearly the Physics of the aurora lies in understanding this fundamental structure of the aurora which cannot easily be done from orbiting spacecraft. Ground and rocket experiments give us our best shot. What I view as an ultimate data set is to image a large portion of the auroral oval from the ground with a spatial and temporal resolution that can uncover the Physics involved. Answers to such questions as: What is the structure of the transpolar arc in Theta Aurora? Is there any coherency in the dynamics of the aurora at different local times? will provide information about the boundaries and the interactions between different boundaries from which the auroral energy originates.

While the polar cap is usually a region dominated by soft electron precipitation, satellite observations have shown that discrete auroral forms may have characteristic energies and fluxes comparable to those seen simultaneously in the oval. A commonly used diagnostic technique uses the ratio of optical emissions at two wave lengths to determine the rate of auroral energy deposition and the mean electron energy. Using a network of ground-based observing sites, it is possible to apply the method to obtain a map of the energy deposition rates over much of the polar cap.

B. MAGNETIC FIELD MEASUREMENTS

As discussed earlier, the fundamental question of Magnetospheric Physics is how the energy released during substorms flows through the magnetosphere from the solar wind. Is it stored in the magnetotail and suddenly released by a trigger? or, Is it continuously transported through the magnetosphere once some input threshold is exceeded The answer to this controversy lies in having a global index giving the occurrence of substorms. The AE index, which is the envelope of all the horizontal components from a network of ground magnetic observatories, is presently obtained from northern hemisphere sites which do not provide adequate coverage at the high latitudes because of the lack of suitably situated land masses. Proponents of the direct coupling or 'driven' model argue that as the consequence of this limited coverage, high latitude substorms are not included in the present AE index, hence the occurrence rate of substorms is underestimated. Antarctica can provide the high latitude magnetic field data required to perhaps resolve this issue.

When the interplanetary magnetic field is directed northward, the geomagnetic activity in the auroral zones is generally low and the auroral oval and polar cap shrink in size. However, during these periods, the polar CAMP magnetic field as measured by satellites is found to be quite disturbed which cannot be very well studied from the ground because of our limited observatory coverage at high latitudes. With magnetometers suitably located around an invariant latitude circle near 78° to 80° we will be able to study the interaction of the solar wind and the Earth's magnetosphere during period of northward interplanetary magnetic field as reflected in polar cap magnetic fields and obtain a reliable measure of 'quiet-time' global activity.

In addition to the polar cap network of observing sites, a meridional array extending down the Antarctic peninsula will tie the polar cap measurements to auroral oval and subauroral latitudes. The simultaneous observation of aurora and magnetic disturbances produced by auroras from multiple stations on the same meridian would permit the monitoring of the expansion of the aurora in response to the onset of magnetic substorms. The initial poleward expansion of the aurora has been well correlated with the thinning of the plasma sheet in the Earth's distant tail prior to reconnection of magnetic field lines in the tail. Following reconnection the plasma sheet again refills and there is believed to be a sudden poleward 'leap' of the aurora at this time. Refilling or thickening of the plasma sheet would theoretically correspond to the plasma reaching progressively higher latitude field line which would map on the ground to a poleward movement. The timing of the poleward motions from multiple stations could test the validity of such a model.

In addition to having the Antarctic array, it is essential that conjugate measurements be attempted in the Arctic. Such experiments are being performed between Sondre Stromjford, Greenland and Frobisher Bay, Canada in the Arctic and South Pole, Antarctica. Somewhat to our surprise we are finding that there is a high degree of simultaneity in magnetic pulsations between Sondre Stromfjord and South Pole which are located at about the same magnetic latitude but are separated in longitude by over 30°. Figure 6 gives an example where bursts of magnetic pulsation caused by auroral particle entry into the atmosphere are very similar in the two hemispheres. The fact that this type of simultaneity occurs about 85% of the time at latitudes where the magnetic field is quite distorted by the solar wind is unexpected. A clue to the

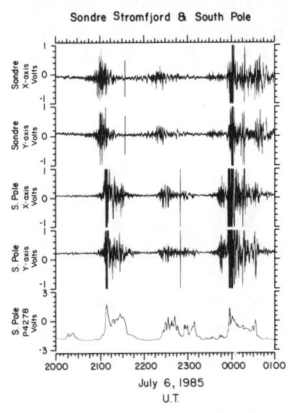

Fig. 6. Opposite hemisphere, high latitude, irregular magnetic pulsations.

simultaneity of the pulsations seen in Figure 6 is provided by DMSP auroral image given in Figure 7 South Pole is located under a three-arc system and presumably Sondre Stromfjord in the northern hemisphere is beneath a similar arc system. A time shift in Figure 6 between the wave events at the two opposite hemishere sites suggests a westward propagation of the auroral arcs. Either the Earth's magnetic field is sufficiently distorted that South Pole and Sondre Stromfjord are magnetically conjuage or the auroral arcs are very wide in longitudinal extent. The latter seems more plausible. In addition, the data of Figure 6 also requires that the source of electrons bombarding the atmosphere to produce the aurora and the magnetic pulsations illuminates both hemispheres with the same variability. This suggests

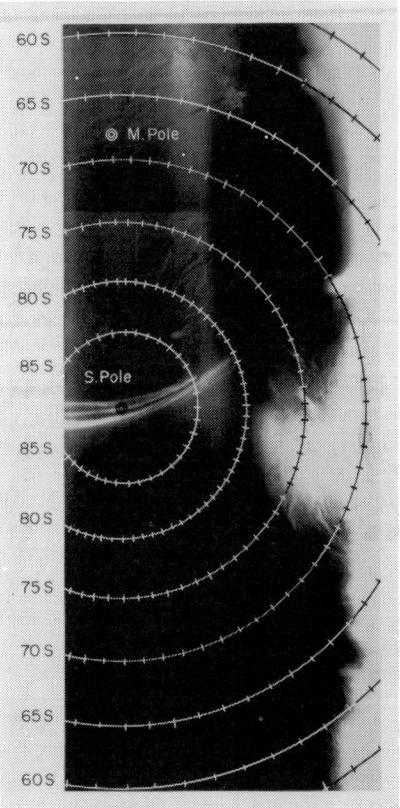

Fig. 7. DMSP satellite image of the aurora conjugate to the magnetic pulsations given in Figure 6.

that the source mechanism for these high atitude arcs is in the geomagnetic tail and is little influenced the local ionospheric conditions since there is no strong seasonal dependence on the simultaneity observed.

Besides using ULF waves to probe the large-scale auroral morphology, ground magnetic pulsations might also provide a means to measure the inherent time variability of the auroral beam itself not possible from moving observatories such as satellites or rockets. Clearly the human eye and image-intensified TV show that the aurora is very dynamic (turbulent) and not at all a DC phenomenon. Recent coordinated ground measurements made in the Antarctic at subauroral latitudes (Siple Station, L = 4) have shown that these ground pulsations (Pi 1) are very closely correlated with short bursts of auroral light above a very low level of auroral background as seen in the sample of data presented in Figure 8a. The magnetic pulsations measured by an induction antenna have a distinctive asymmetry which suggests that the signal is produced by the magnetic field of a sudden turn-on or enhancement of an ionospheric current followed by a much slower decay in this current system. These pulsations are therefore the result of ionospheric conductivity enhancements due to the correlated auroral particle precipitation measured by a riometer (not shown in Figure 8a) and the auroral photometer. This type of particle precipitation is generally termed "microburst", it is quite energetic since it is often recorded by ground riometers and at balloon altitudes as x-rays, and finally, is typical of early morning or dayside subauroral zone precipitation. At auroral latitudes and higher, similar ground ULF Pi pulsations are correlated with photometer fluctuations as seen in Figure 8b. However, these induction antenna signals are not asymmetric and often are correlated only with auroral light bursts and not with riometer absorption. The magnetic and auroral light fluctuations given in Figure 8 show an incredible detailed phase correlation. However, the auroral light fluctuation amplitudes for this event are only a few kilo Rayleighs superimposed on a background auroral signal of several tens of kilo Rayleighs. The suggestion here is that these high latitude, ground Pi signals may indeed be a measure of magnetospheric Alfven waves that are also very clearly associated with particle precipitation measured optically on the ground. The The role of Alfven waves in auroral acceleration mechanism is one of current great interest.

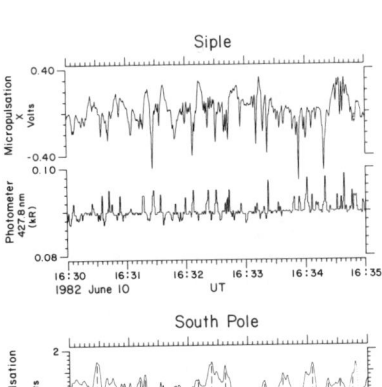

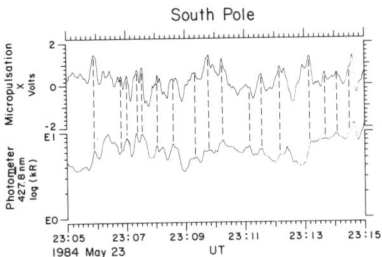

Fig. 8. Magnetic pulsations and pulsating aurora at the equatorward and poleward edge of the auroral oval.

C. ENERGETIC PARTICLES

Charged particles entering the Earth's atmosphere can be sensed from the ground by the ionization they create in the process of being absorbed by the atmosphere. Radio waves from galactic sources between 30- and 50 MHz are absorbed in passing through the ionosphere to ground receivers by this enhanced ionization. Riometers (an acronym for relative ionospheric opacity meters) indirectly measure particle precipitation to the atmosphere by this technique and by comparing absorption at different frequencies can get a crude spectrum of the radiation. All the advantages of making multiple site measurements in the Antarctic that have been discussed above apply equally well to the riometer technique. However, the Antarctic has an additional uniqueness for particle measurements. Because the Earth's magnetic field is anomalously weak in the South Atlantic, charged particles trapped in the Earth's magnetic field mirror closer to the atmosphere here than anywhere else in the world. Consequently any interaction the particles may have that directs their motion more along the magnetic field may result in their loss to the atmosphere in this region of the world. The South Atlantic and the Antarctica peninsula is therefore a sink for trapped particles, hence arrayed measurements of particle precipitation would radially map regions of particle interactions (mostly with electromagnetic or hydromagnetic waves) in the magnetosphere.

A significant improvement in the riometer technique for examining cosmic noise absorption was made with the installation of a 64-element phased array riometer system at South Pole, Antarctica in January, 1988. The use of antenna arrays producing multiple narrow beams is necessary in order to examine the spatial scale of regions of energetic (> few keV) electron precipitation, which are coincident with cosmic noise absorption activity, and to resolve time fluctuations from spatial variations. The use of such antenna arrays allows an image to be made of the overhead precipitation pattern not unlike that obtained by all-sky cameras or TV. This technique makes possible precipitation imagines during daylight and overcast conditions when photometry is not possible.

D. PLASMA TRANSPORT OVER THE POLAR CAPS - CONVECTION

As dayside magnetic field lines presumably connect to the interplanetary magnetic field lines they are swept over the polar caps along with the plasma attached to them. This plasma transport, due to an electric field drift, can be measured by several techniques. At ionospheric F-region altitudes the electron drift can be measured by the Doppler shift of UHF radar scattered by the electrons. At lower altitudes, the electric field transporting plasma over the polar caps can be sensed by the Doppler shift of HF radar scattered off plasma irregularities. Since the drifting plasma will impart a drift velocity on neutral atoms by collisions, polar cap transport in the form of thermospheric neutral winds can be detected optically by measuring the Doppler shift of radiation emitted by the excited neutral atoms. Finally, 6300Å emission by drifting plasma patches can be measured from the ground by image-intensified TV systems. Polar cap transport is an important indicator of the interaction between the magnetosphere and the solar wind as seen in Figure 2. If the interplanetary field is directed southward than reconnection occurs on the dayside and magnetic field energy and plasma are transported to and deposited in the geomagnetic tail. In contrast to this, for northward interplanetary field directions the interplanetary field is draped over the poles and little transport occurs. The details of these interactions, however, are not fully understood even though we have statistical images of the transport process from satellite data and some ground sites. It would be desirable to obtain a body of simultaneous measurements of neutral wind and ion drift along the auroral oval and in the polar cap. Simultaneous data are required from all locations in order to separate local time and universal time effects, and the frequency of the measurements should commensurate with changes in substorm phenomena and response times of ion-neutral interactions. This ideal is beginning to be realized with the recent installation of an HF radar at Halley Bay, Antarctica. The radar is a joint Anglo-

American project. It is magnetically conjugate to an identical radar operated by the Johns Hopkins Applied Physics Laboratory at Goose Bay, Labrador. The two radars have large conjugate fields-of-view with an overlapping area in excess of 3 million km.^{2} The two radars provide a unique tool for studying conjugate phenomena, including the conjugate formation of ionospheric irregularities, pulsations, and convection. Preliminary analysis of the data set has already dramatically demonstrated the effect of the interplanetary magnetic field direction on the plasma transport over the polar caps.

E. MAGNETIC RECONNECTION - FLUX TRANSFER EVENTS

The location of South Pole Station near the magnetospheric cusp allows the possibility of searching for specific ionospheric signatures of processes that may be occurring at the Earth's magnetopause. In particular, a topic of high current interest involves the signatures that may exist in the ionosphere of magnetic field reconnection events (flux transfer events) on the dayside magnetosphere boundary. These instances of sporadic reconnection of geomagnetic and interplanetary magnetic field lines should be evident in currents induced in the ionosphere under the magnetospheric cusp.

Data from the South Pole and its conjugate region in the northern hemisphere, near Iqaluit on Baffin Island, have been used by Lanzerotti et al., 1977 to search for evidences of convected field-aligned currents in the ionosphere that may be produced by flux transfer events. Shown in Figure 9 is a typical example of a conjugate signature that has been suggested to be an example of such a current system produced by reconnection on the dayside magnetosphere. Also shown are magnetic field data acquired at Sondre Stromjford, Greenland, approximately 2 hours earlier in local magnetic time than Iqaluit. Of particular relevance for the convected field-aligned current is the signature in the vertical magnetic component, which at each station shows a sharp upward increase in the magnetic field intensity with a time width at half-amplitude of ~ 5 min. The dotted lines on each of the magnetic traces represent a model ionospheric Hall current loop convected over the stations, with the parameters of the Hall loop and its location with respect to a given station determined by the vertical magnetic field signature. For a loop of ~ 200 km diameter in the ionosphere and at an altitude of about 100 km, the vertical magnetic signatures suggest a field-aligned current intensity of the order of 2×10^5 amperes. This event occurred when the interplanetary field was directed southward. The increase in the vertical magnetic field at each of the stations means that the field-aligned current is incident into the ionosphere in both the northern and the southern hemispheres.

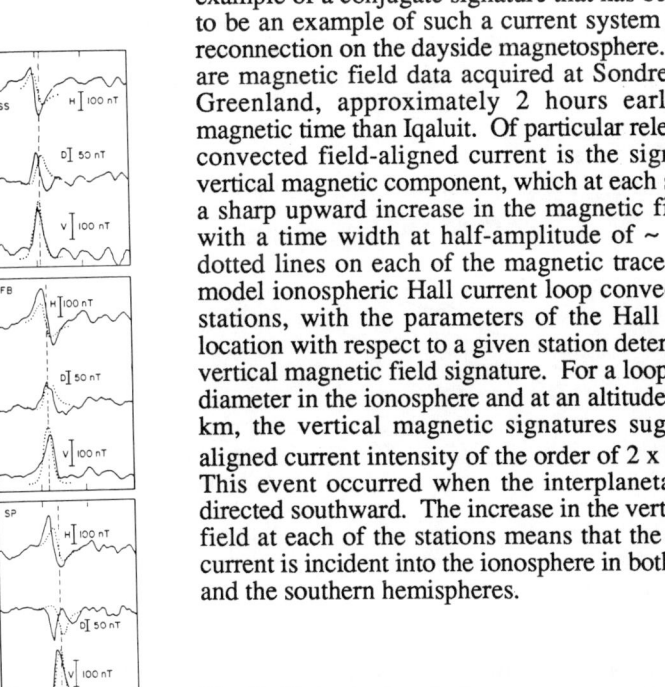

Fig. 9. Opposite hemisphere measurements of a ground magnetic signature of a flux transfer event after Lanzerotti et al., Geophys. Rev. Lett., 13, 1089, 1986.

The distinctive magnetic signature of an FTE at a polar cap magnetic station, essentially a one-cycle Pc 5 wave, can be used to infer the signs of the B_y and B_z components of the interplanetary magnetic field. In addition, ground measurements at or near the cusp can monitor the rate of reconnection at the dayside magnetopause in a manner not possible with rapidly moving spacecraft, if indeed the Pc 5 "FTE events" are ground signatures of this reconnection. A large quantity of data currently exists from both flux gate and induction magnetic antennas located in opposite hemispheres at conjugate and near-conjugate sites. This suite of data will greatly expand our knowledge of dayside reconnection and conditions under which it occurs.

Recent studies have indicated, however, that the ground Pc 5 "FTE" events do not occur on field lines that map to the boundary of the magnetosphere. A very fortunate conjugate VIKING satellite measurement at 13,000 km altitude and Sondre Stromfjord, Greenland, ground measurement of such an FTE is given in Figure 10. The satellite (moving poleward) is magnetically conjugate to Sondre Stromfjord at 1331 U. T. when the FTE is evident on the ground. The satellite magnetometer and electric field sensors (not shown in Figure 10) measure an Alfven wave at this time. As the satellite moves poleward, further magnetic fluctuations occur some of which can be identified as signatures of field-aligned currents. When almost 6 degrees of magnetic latitude poleward of the ground FTE, charged particle detectors aboard the satellite identify the cusp by soft electron precipitation as indicated by the dashed lines in Figure 10. Clearly, the Sondre Stromfjord FTE occurred equatorward of the Cusp. Before entering the Cusp and about three minutes after the Sondre Stromfjord FTE, the satellite ion detectors measured energy dispersed, presumably, magnetosheath ions. This time delay is consistent with injection on the Sondre Stromfjord field line at a distance of 10 Re near geomagnetic equator. The high frequency Pc 1 magnetic pulsa-

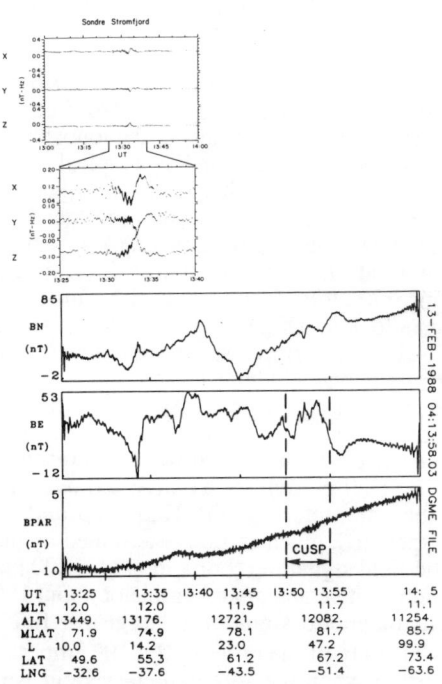

Fig. 10 Conjugate ground magnetic the signature of a flux transfer event and VIKING satellite magnetic signature.

tions riding on the ground FTE was observed in space as well and might be related to the ion injection mechanism. It is not clear at this time if such events indeed do represent recently reconnected geomagnetic field lines to the interplanetary field. The ground data is important in helping to understand the spacecraft data in that it helps order space and time. Clearly the ground FTE is a temporal event. Likewise the satellite conjugate event is probably also temporal.

I would like to end this section of ground Antarctic experiments by showing recent ground auroral photometry data that can be very important in helping us understand the transport of solar wind energy into the magnetosphere. Figure 11

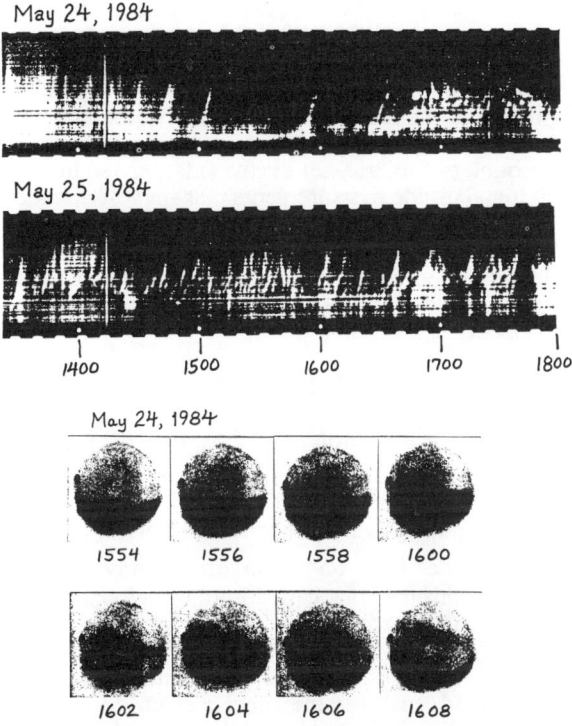

Fig. 11. Ground Keogram and all sky TV images of the aurora at South Pole, Antarctica from Rairden and Mende, J. Geophys. Res., 94, 1402, 1989.

gives data from two different optical instruments that are sensing the aurora at South Pole. In the U. T. time interval from 1400 to 1800, South Pole is at dusk local time and generally passes under the auroral oval from being equatorward of it at local noon to moving poleward of it at local midnight. The top two traces are data from a meridional slit (keogram) camera. The keogram meridian is aligned such that the top edge of the film is toward the magnetic pole and the bottom edge toward the equator, while a motor drives the film across the camera slit at a rate of one inch per hour. The keogram is a record of the motion of the intersection of auroral features with this slit and gives a good time-resolution of the poleward and equatorward movement of arcs within the auroral oval. Most of the auroral features in Figure 11 are vertical streaks slanted to the right which means auroral motion from equatorward to the pole. The bottom two traces are intensified all-sky images of the aurora. This camera photographs the sky through a 30 Å wide filter at 6300 Å. This filter eliminates much of the unwanted sky backgrounds and emphasizes the soft electron precipitation and high ion density regions. The camera field of view is 160 degrees full angle centered at zenith, which is a circle of diameter 1600 km at the assumed emission height of 200 km altitude. The magnetic pole is at the top of each image and magnetic east to the right. The keogram field of view is essentially a vertical line up the center of the all-sky camera field of view. The correlation of the all-sky camera images with the keogram data is necessary to determine the longitudinal extent of the keogram features and the component of their motions perpendicular to the magnetic meridian. For example, the all-sky images in Figure 11 pertain to the isolated vertical keogram streak on May 24, 1984, showing it is indeed an auroral arc of great longitudinal extent moving rapidly poleward.

The amount of north-south motion of the aurora at high latitudes as given in the data of Figure 11 is indeed astonishing. For some hours as many as two dozen auroral forms have moved predominantly poleward. The relationship of this dusk sector motion to flux transfer events is largely an unanswered question. There are several high latitude measurements that suggest that the boundary layer in the 1400-1600 local time sector is an important source for high latitude, early afternoon auroral arcs. These are field-aligned currents, enhanced F-region electron densities obtained from incoherent scatter radar measurements, and ground-based observations of enhanced 4.5 kHz hiss presumably associated with precipitating soft electrons.

BALLOONS

Because of the unique meteorology of the Antarctic, it is possible during the Austral summer to have high altitude balloons stay aloft for several days to weeks trapped in a high altitude wind vortex about the geographic pole. A balloon launched from McMurdo, at an invariant latitude of 80°, will move through a range of invariant latitudes from 65° to 85° in one traversal about the pole, or sample the auroral oval plus the polar cap. Multiple balloons aloft at the same time at different longitudes will therefore provide the same type of geophysical imaging of these important regions of the magnetosphere that we have discussed above using an array of ground sites. Balloons offer some unique measurements not possible from the ground, such as atmospheric electric fields which at 100,000 feet might be representative of ionospheric/ magnetospheric fields and the x-ray spectrum of energetic particle precipitation which is easier to unfold to give the primary particle spectrum than the ground measurements using riometers. Much of our present knowledge of the precipitation of particles from the trapped radiation has come from balloon x-ray measurements.

Several balloon campaigns have been conducted in the Antarctic. The most recent in December, 1985-January, 1986 resulted in over 460 hours of observation at float altitude. The balloons were instrumented by the U. of Houston and the U. of Maryland three axis double probe electric field detectors and x-ray scintillation counters. Data analysis is still in progress but phenomena observed include diurnal modulation of the convective electric field conjugate to northern hemisphere radar electric field measurements, very large vertical electric fields, the electric field associated with possible flux transfer events, intense x-ray precipitation in the cusp with fluxes up to 125 keV, and strong ULF fluctuations in the electric field.

Ground-Level Detection of LF and MF
High Latitude Terrestrial Emissions

Robert F. Benson

Laboratory for Extraterrestrial Physics
NASA - Goddard Space Flight Center
Greenbelt, MD 20771

ABSTRACT

Reports of ground-level detection of radio noise associated with auroral phenomena in the frequency range f ~ 100 kHz date back to the time period of the International Geophysical Year (1957-58). The main observational challenge is to confirm that such emissions are indeed related to auroral phenomena rather than to man-made interference, solar radio bursts or other signals detected as a result of unusual propagation conditions. Once confirmed, there are two theoretical problems: namely, to determine the generation mechanism and the propagation conditions that enable ground-level detection. Two brief (two to three weeks each) experimental campaigns were recently conducted (one in Alaska and one in Norway) using 4 commmercial communication receivers in an attempt to confirm the existence of such waves. The Alaskan experiment yielded signals attributed to auroral processes on nearly 1/2 of the nights monitored. A preliminary investigation of the data from Norway, where local interference was greater, has identified only one such event. The results emphasize the need for continuous monitoring over a wide frequency range in a high-latitude low-noise environment. The proposed automated geophysical observation in Antarctica would provide ideal platforms to conduct such an experiment.

INTRODUCTION

Reber and Ellis[1] apparently were the first to report the ground-level detection of high latitde terrestrial signals with f ~ 100 kHz. They observed signals down to 520 kHz in their low-noise cosmic noise experiment. Dowden[2] observed signals at 180 kHz which he attributed to the aurora. Jorgenson[3], Morgan[4], and Sato and Hayashi[5] also reported the ground-level detection of high-latitude terrestrial signals with $f \lesssim 100$ kHz. The reviews of Ellyett[6] and LaBelle[7] discuss these observations and many others (including those at much higher frequencies).

Most scientific ground-level routine monitoring of the high-latitude radio environment has either been in the frequency range $f \lesssim 10$ kHz (in search of ducted whistler (W) mode waves) or $f \gtrsim 10$ MHz (as a by-product of riometer cosmic radio wave absorption experiments). Interest in the frequency range between these extremes was stimulated by the theoretical work of Wu et al.[8] indicating that W mode waves propagating downward along the

direction of the earth's magnetic field B can be generated in the
low-altitude auroral region by the cyclotron maser instability.
(By contrast, the Cerenkov process, considered to be the
generating mechanism for W mode auroral hiss observed by
spacecraft in the topside ionosphere[9,10], generate W mode waves
propagating oblique to B which have difficulty reaching the
ground[9].) Waves propagating downward, along B, are only
restricted by the constrain $f < f_H$, where f_H is the electron
gyrofrequency[11], and the parallel propagation restriction can be
relaxed when ionospheric irregularities and electron collisions
are considered[12]). In an attempt to detect such waves and to
determine the maximum emission frequency, experiments were
performed in Alaska and Norway. This frequency is an important
parameter in the Wu et al.[8] mechanism yielding information on the
minimum altitude of wave generation because it is related to f_H.
The Alaskan results[13] will be briefly reviewed in the next section
before presenting some of the preliminary findings[14] from the
Norway experiment.

OBSERVATIONS

Benson et al.[13] used four Realistic (Radio Shack) phase lock
loop synthesized communication receivers with 10 m horizontal
ground-level monopoles near Fairbanks, Alaska to search for
aurorally generated emissions with $f \gtrsim 150$ kHz (the low frequency
limit of the receivers) during the spring of 1986. On 7 out of 15
nights of operation, signals attributed to ionospheric origin were
observed on one or more of the following frequencies; 150, 291,
500 or 700 kHz. A one-to-one correlation with geophysical
phenomena was not always evident but the strongest emissions
ocurred near midnight magnetic local time (MLT).

This experiment was repeated in January, 1989 in Andenes,
Norway in conjunction with the flight of the Oedipus A rocket from
the Andøya Rocket Range. This rocket consisted of a dual payload
connected by a conducting tether. One of the payloads included a
radio receiver covering the frequency range from 25 kHz to 5 MHz.
This receiver was derived from the WISP (Waves in Space Plasmas)
Space Shuttle program[15]. The goal of the ground-based radio
observations was to relate topside and bottomside observations of
auroral radio emissions with $f > 100$ kHz. The equipment used was
the same as in the Alaskan experiment except that a 16 bit analog
to digital converter and a personal computer replaced the
multichannel chart recorder.

A preliminary investigation of ambient noise levels was
conducted on the island of Andoya in September, 1988. Three sites
operated by the Andøya Rocket Range were tested and the science
buidling at the rocket range was found to be the guietest in the
evening hours due to the shielding of signals from European radio
stations provided by the adjacent mountains. During the January,
1989 experimental campaign, however, interference from the

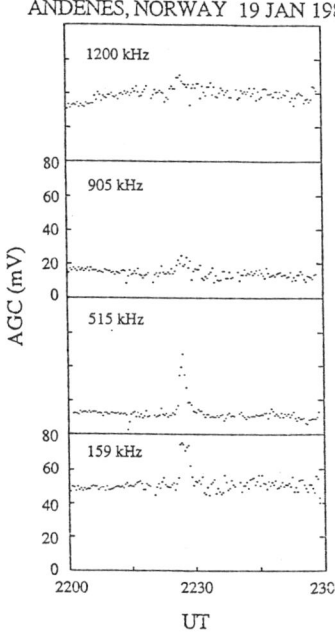

Fig. 1. Signal strength levels from ground-based receivers.

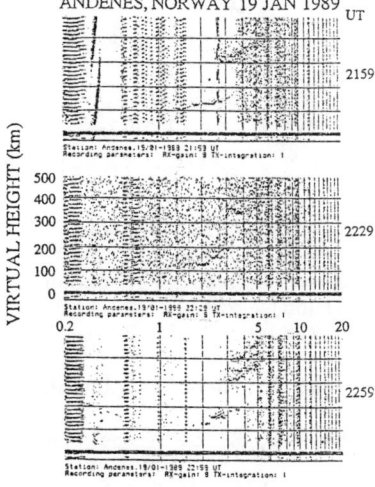

Fig. 2. Consecutive ionograms.

Norwegian telemetry transmmitter caused a problem. The ground-based receivers were operated during this period for approximately 93 hours when this transmitter was off the air. A preliminary scan of these data indicate that auroral radio noise of the type observed near Fairbanks during the spring of 1986 was less common at Andenes, Norway in January, 1989.

To date, one broad-band noise event has been identified in the January, 1989 date. Signals were detected from 159 to 1200 kHz using the 4 ground-based receivers (Figure 1), from 200 kHz to at least 3.5 MHz on the Andenes vertical ionosonde (figure 2) and even on the Andenes 32.5 MHz riometer. The individual points in Figure 1 correspond to averages of 24 s of raw data. The raw data (0.2 s resolution) indicate that the noise event persisted for about 3 min centered near 2227:30 UT. The ionogram recorded at 2229 UT in Figure 2 reveals a broad-band speckled noise pattern covering the entire virtual height scale over the frequency range from 200 kHz to about 3.5 MHz, whereas the adjacent ionograms (which are tyical of those preceding and following 2229 UT) shown this type of noise band only in the frequency range beyond the point of strong E and/or F layer echoes and with the greatest intensity up to about 10 or 12 MHz. The 32.5 MHz riometer record (not shown) recorded a large signal enhancement that started just before 2227 UT and persisted for about 1 min.

The above event occurred during a time period of magnetic disturbance near midnight MLT. The most intense events observed during the Alaskan experiment were also observed near midnight MLT. These results are consistent with satellite observations of the diurnal variation of the invariant latitude of the maximum intensity of 200 kHz auroral hiss[16] (Λ = 65° for Fairbanks and 66.3° for Andenes).

DISCUSSION

As pointed out earlier[13], the lowest W mode emission frequency from the cyclotron maser instability using the parameters of Wu et al.[8] would be about 350 kHz. Thus this model will not explain the 159 kHz emissions observed in Norway. In addition, the model as presented will not explain the emissions observed above the maximum ionospheric value for f_H, i.e., it cannot explain the ionosonde observations that extend at least to 3.5 MHz or the riometer observations at 32.5 MHz. Additional theoretical work is called for to explain emissions over such a broad frequency range. Additional observations are also required on a long term nearly continuous basis in order to better determine the geophysical conditions required for the emissions to be detected on the ground and to determine the exact frequency spectrum. Measurements of this type are planned for Alaska and have been proposed for inclusion into the Antarctican automated geophysical obseratories[17].

REFERENCES

1. G. Reber and G. R. Ellis, J. Geophys. Res., 61, 1 (1956).
2. R. L. Dowden, Nature, Suppl. 184, 803 (1959).
3. T. S. Jorgensen, J. Geophys. Res., 73, 1055 (1968).
4. M. G. Morgan, J. Geophys. Res., 82, 2377, (1977).
5. N. Sato and K. Hayashi, J. Geophys. Res., 90, 3531 (1985).
6. C. D. Ellyett, J. Atmos. Terr. Phys., 31, 671 (1969).
7. J. LaBelle, J. Atmos. Terr. Phys., in press (1989).
8. C. S. Wu, D. Dillenburg, L. F. Ziebell and H. P. Freund, Planet Space Sci., 3, 499 (1983).
9. D. W. Swift and J. R. Kan, J. Geophys. Res., 80, 985 (1975).
10. J. E. Maggs, J. Geophys. Res., 81, 1707 (1976).
11. H. G. Booker, Cold Plasma Waves (Martinus Nijhoff, Boston, 1984), Fig. 13.1.
12. R. A. Helliwell, Whistlers and Related Ionospheric Phenomena (Stanford University Press, Stanford, 1965), p.27.
13. R. F. Benson, M. D. Desch, R. D. Hunsucker and G. L. Romick, J. Geophys. Res., 93, 277 (1988).
14. R. F. Benson and M. D. Desch, EOS (Trans. Am. Geophys. Union), 70, 441 (1989).
15. H. G. James, T. R. Darlington, C. H. Hersom, R. S. Gruno and J. V. Gore, Proc. IEEE, 75, 218 (1987).
16. T. Laaspere, W. C. Johnson and L. C. Semprebon, J. Geophys. Res., 76, 4477, (1971).
17. J. LaBelle, personal communication, 1989.

GEOSPACE PLASMAS: REPORT OF THE WORKING GROUP

T. J. Rosenberg
Institute for Physical Science and Technology
University of Maryland, College Park, Md 20742

INTRODUCTION

The Geospace Plasmas Working Group (GPWG) had a short formal program of four speakers. Discussion on a variety of topics relating to the conference theme followed the formal presentations. This led to several recommendations which were presented on the last day of the conference along with a brief description of the Antarctic geospace science community for the benefit of those not familiar with the activities and interests of this group.

FORMAL PROGRAM

The speakers and their topics were:

Dr. J. T. Lynch: *A Proposal for an International Station in Antarctica*

Mr. J. P. Katsufrakis: *The Use of Blue-Ice Airfields in the Support of a New Wave Injection Facility on the Hollick-Kenyon Plateau*

Dr. W. B. Gail: *The Role of the Winter-Over Scientist in Antarctic Astrophysics*

Dr. R. F. Benson: *Ground-Level Detection of LF and MF High Latitude Terrestrial Emissions*

Dr. Lynch, the Program Manager for Polar Aeronomy and Astrophysics at NSF presented several reasons why the time is appropriate now to seriously consider building a truly international station in Antarctica. Such a station would demonstrate to the world that international scientific cooperation is alive, and has been growing and prospering, after nearly 30 years under the Antarctic Treaty. Furthermore, Antarctica is a logical stepping stone to manned exploration of the Solar System. In this context, it should be noted that the Outer Space and Moon Treaties are based on the Antarctic Treaty and share many common features; long winter-over isolation of small science and support crews is quite similar to what is expected in space; and the physical environment and scientific objectives can closely emulate a station on the Moon or Mars. Lastly, 1991 is the 30th anniversary of the Antarctic Treaty and 1992 has been declared International Space Year. Nothing, in Dr. Lynch's estimation, could be more appropriate for these occasions than the banding together of nations to study space from the last unoccupied land on Earth.

Dr. Lynch went on to make two specific suggestions for an international Antarctic station.

A. An Extremely Low Frequency/Very Low Frequency (ELF/VLF) Wave Injection Facility (WIF). This station, a follow-on to Siple station, could provide a unique opportunity to study the interaction of artificially generated electromagnetic waves with particles trapped in the inner magnetosphere. The thick ice makes it possible to construct reasonably efficient broadband antennas with which to launch the waves. The station would be located near the plasmapause (L = 4.0 to 5.5) near the base of the Antarctic peninsula (see Fig. 1). The magnetically conjugate area is accessible in Canada, which is important for studying the effects of the waves.

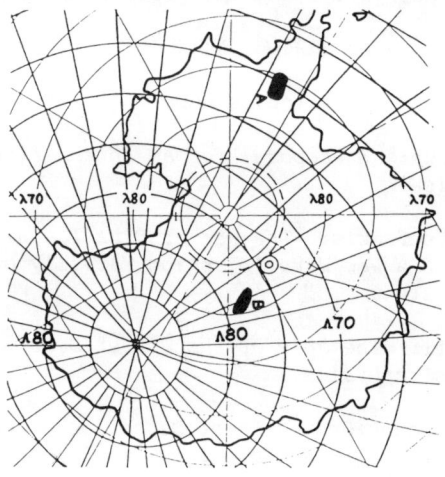

Fig. 1. The approximate locations of suggested new Antarctic international stations are shown by the solid areas denoted A and B. A—ELF/VLF wave injection facility (WIF); B—High altitude, high geomagnetic latitude observatory.

A workshop held at Stanford University in April 1989 formulated the science and engineering requirements and laid the foundation for the international resource sharing that would be essential in establishing such a facility. A report is in preparation to be submitted to NSF (U. S. Inan, personal communication).

B. A high altitude, high geomagnetic latitude observatory primarily for astronomy and aeronomy, to be located above the 4,000 meter altitude contour on the East Antarctic ice sheet (see Fig. 1). This station could closely approximate a lunar base (the pressure altitude of 16,000 to 17,500 ft might require a pressurized station). It would certainly be the lowest water vapor (and coldest) observatory on Earth, which would make it highly desirable for infrared, submillimeter and millimeter astronomical measurements. Its geomagnetic location would be in an area of the magnetosphere that will be the focus of attention of the International Solar Terrestrial Physics Program (ISTP) during the 1990s.

Mr. Katsufrakis, formerly of Stanford University and a key person in the design and operation of Siple station, discussed an alternative method of logistical support of a new WIF (see Lynch plan A above).

Construction and logistical support of Siple station were carried out by LC-130 ski-equipped Hercules aircraft operating out of McMurdo, with the attendant burdens on Squadron VXE-6 and the logistics budget. An alternative method, which shows great promise, is the one being considered for the construction of a new complex at South Pole Station. This method employs the use of a satellite blue-ice airfield located in the area between Mount Howe ($87°22'S$, $149°30'W$) and d'Angelo Bluff ($87°18'S$, $154°00'W$) within 250 nautical miles of South Pole Station. Standard heavy wheeled transport aircraft would be used to fly construction materials, supplies, fuel, and passengers from such locations as New Zealand, South America, and McMurdo Station. Passengers and priority cargo would be flown on to the Pole by ski-wheel shuttle, with the fuel and bulk cargo being transported by fast sled train.

The experience gained in this evolution can be directly applied to the construction and support of a new WIF located on the Hollick-Kenyon Plateau at $77°36'$ South Latitude and $93°44'$ West Longitude. The commercial blue-ice airfield at Patriot Hills ($80°20'S$, $81°25'W$) in the Ellsworth Mountains is recommended to be used in conjunction with the Mount Howe-d'Angelo Bluff blue-ice airfield at the United Kingdom Base at Rothera on Adelaide Island. Construction materials, supplies, and fuel flown into the Patriot Hills airfield by heavy lift wheeled transport aircraft would then be delivered to the new WIF site by fast sled train over a distance of 250 nautical miles (Fig. 2).

Wheeled aircraft that might be used include the C-130 Hercules, maximum payload approximately 22 tons; the C-141 Starlifter, maximum payload approximately 35 tons; and the KC-10A, maximum payload of approximately 85 tons. As a comparison, the LC-130 ski-wheel Hercules supporting Siple Station carried approximately 10.5 tons using wheel takeoff from the ice runway at Williams field and approximately 7 tons using ski takeoff. Carrying this payload required a refueling stop at Byrd Surface Camp on the return leg of the flight.

Sled trains operating in the Antarctic after 1992 will be able to use the Global Positioning System (GPS) satellite constellation (21 satellites) for navigating during the traverse. Locating the new WIF west of the Ellsworth Mountains should provide better aircraft operations weather in a region of less snow precipitation than experienced at Siple Station.

Fig. 2. Patriot Hills (A) and the sled train traverse of a possible WIF site (+) at 77°36' South, 93°44' West.

The new WIF design should incorporate wonder-arch covered tunnels for food and fuel storage. The life support spaces, the power house, the laboratories, the dispensary, etc. should be constructed of light weight materials and should be located on top of elevated platforms. Each platform should be a square which greatly reduces drifting. The platform should also have the capability of being raised periodically.

Dr. Gail, now at The Aerospace Corporation and a former winter-over at South Pole station, presented his views on the role of the winter-over scientist. Since his contribution is included elsewhere in these Proceedings, there is no need to elaborate further on his presentation here. Perspectives and suggestions along similar lines to those of Dr. Gail were also received, following the conference, from Mr. E. J. Wollack, a more recent (1988) winter-over at South Pole. Both Dr. Gail and Mr. Wollack felt that the creation of a Science and Technology Center could lead to improvements in all aspects of the winter-over program.

It will suffice to emphasize that the success of scientific activities in Antarctica relies heavily on the abilities and dedication of the individuals entrusted to manage the projects during the long period of isolation. Care in the selection of personnel, adequate training by experienced staff, clear lines of responsibility, meaningful interaction with principal investigators and continuity in the winter-over program are required to ensure such success.

Dr. Benson of the Goddard Space Flight Center described radiowave measurements made in Alaska and Norway that suggest the possibility of observing impulsive auroral noise emissions of a few hundred kHz from the ground. Such emissions are thought to be indicative of plasma processes occurring above the ionosphere. He suggested that new measurements made in the Antarctic polar cap, possibly from automatic geophysical observatories, may help to identify and elucidate the phenomena, leading to improved theoretical understanding.

Dr. Benson's contribution is also included elsewhere in these Proceedings and thus does not require further elaboration here. It is of interest to note, however, that Dr. Benson was also an Antarctic winter-over. His stint occurred in 1957 at South Pole at the dawn of the modern era of polar upper atmosphere research, the International Geophysical Year. In extemporaneous remarks following the conference banquet, he offered some historical perspectives on the people and working environment of those days which provided an interesting contrast to the much more recent experiences of the other former winter-overs in attendance.

DISCUSSION TOPICS

1. Long range plans for geospace plasma research in Antarctica (e.g., use of automatic geophysical observatories, long-duration ballooning, new international manned stations);
2. The Bartol S & T Center concept;
3. What the GPWG thought the functions of a Center should be; and
4. How the Geospace Plasmas community could both contribute to and benefit from the realization of such a Center.

GEOSPACE SCIENCE IN ANTARCTICA

Research on geospace plasmas deals with the interaction of the solar wind with the geomagnetic field and the energization and transport of plasmas within the terrestrial magnetosphere. Viewed in a broad context, knowledge gained from the study of plasma processes in the Earth-Sun system will contribute to understanding the characteristics of other astrophysical systems.

As discussed by R. L. Arnoldy in his overview lecture (included elsewhere in these Proceedings) the polar regions, and Antarctica in particular, play an important role in the study of geospace plasma phenomena. Although many aspects of space plasma processes can be studied in situ by rocket-and satellite-borne instruments, these methods generally are unable to obtain a clear separation between spatial and temporal effects in the phenomena observed. The polar regions present a window through which groundbased instruments can observe the geospace environment and perform this separation. Antarctica is uniquely situated for this purpose because it provides a land base without territorial boundaries, lengthy periods of darkness (for studying the aurora), and an extremely low background of manmade noise in all frequency ranges of interest.

Antarctic geospace science is very diverse. Many techniques

are employed, mostly in the form of principal investigator-class instruments. However, some facilities-class instruments, e.g. coherent scatter radars, wave injection transmitters, are also used. Geospace science involves "all" of Antarctica as well as sites at geomagnetically conjugate locations in the northern hemisphere.

The Antarctic geospace community has functioned in many respects like an unstructured S & T center. Meetings are held annually (2-3 day workshops) where data are exchanged, collaborations are initiated on analysis projects, science topics are discussed and new programs are planned.

The needs of the geospace community as they relate to what an S & T Center might be expected to accomplish include

1. the hiring and training of winter-over scientists;
2. improvement in satellite communications and data transfer;
3. centralization and coordination in the use of acquired data (e.g., archiving, on-line access); and
4. general advocacy of long-range plans.

The geospace community could be expected to contribute to a Center

1. experience in operations and planning;
2. results of geospace plasma studies pertinent to other astrophysical objects; and
3. background information of concern to astrophysicists (e.g., auroral luminosity, charged particle/x-ray environment).

RECOMMENDATIONS

The GPWG recognizes that there are compelling scientific arguments for the creation of a Center for Astrophysics Research in Antarctica, and enthusiastically supports its formation. We believe that such a Center would enhance the opportunities and support for astrophysical research in all areas that can benefit from the Antarctic location.

The GPWG recommends that the proposal to establish a Center clearly define the scope of the Center's activities and the responsibilities it intends to undertake on behalf of its constituent subgroups (e.g., geospace plasmas, neutrino astronomy, etc.). Consistent with Bartol's conceptual plans, we believe that such a Center should provide an intellectual focus for research as well as become a source of technical and logistical expertise.

The GPWG recommends that the Center have an Advisory Committee with scientific representation from each of the subgroups.

Antarctic Ballooning

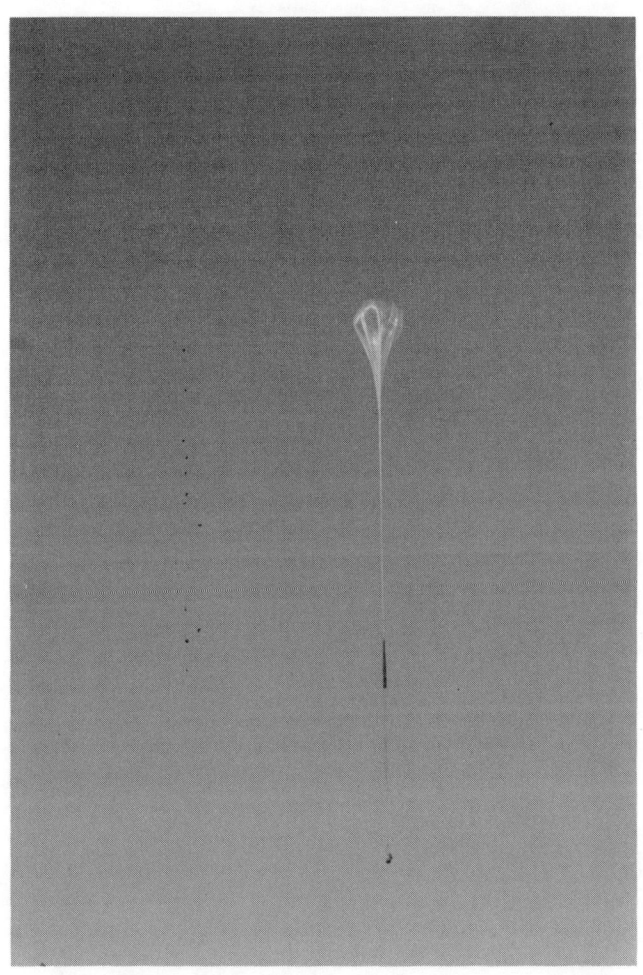

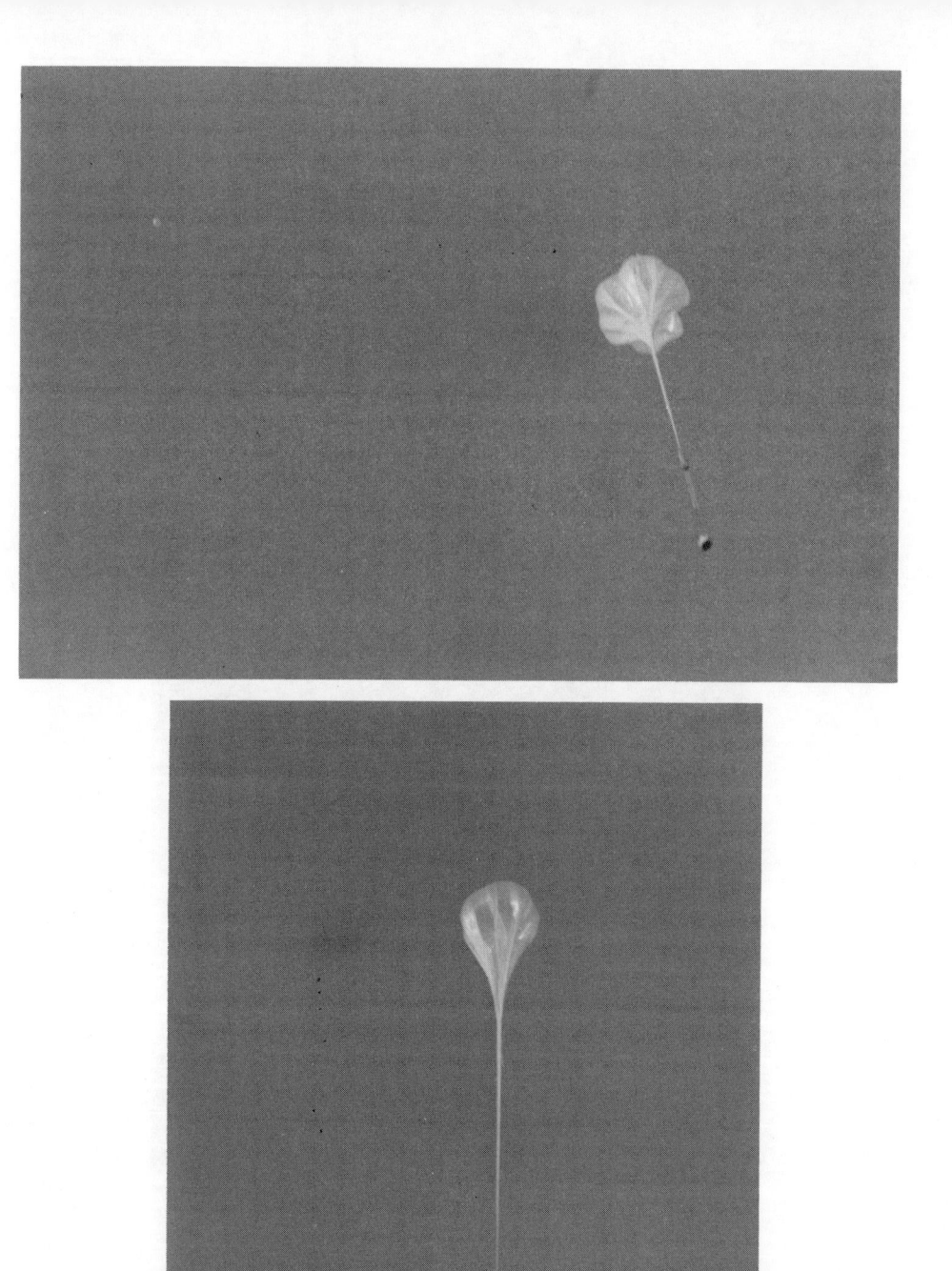

COSMIC AND SOLAR GAMMA-RAY, X-RAY AND PARTICLE MEASUREMENTS FROM HIGH ALTITUDE BALLOONS IN ANTARCTICA

R. P. Lin
Space Sciences Laboratory, University of California, Berkeley, CA 94720

ABSTRACT

For measurements of cosmic and solar gamma-rays, hard X-rays, and particles, Antarctica offers the potential for very long, 10–20 day, continuous, twenty-four-hour-a-day observations, with balloon flights circling the South Pole during austral summer. For X-ray/gamma-ray sources at high south latitude the overlying atmosphere is minimized, and for cosmic ray measurements the low geomagnetic cutoff permits entry of low rigidity particles. The first Antarctic flight of a heavy (~2400 lb.) payload on a large (11.6×10^6 cu. ft.) balloon took place in January, 1988, to search for the gamma-ray lines of ^{56}Co produced in the new supernova SN 1987A in the Large Magellanic Cloud. The long duration balloon flights presently planned from Antarctica include those for further gamma-ray/hard X-ray studies of SN 1987A and for the NASA Max '91 program for solar flare studies.

I. INTRODUCTION

Antarctica potentially offers a number of significant advantages for balloon-borne measurements of cosmic and solar gamma-rays, hard X-rays and particles. The most important advantage is that very long observing times, perhaps even tens of days, may be obtained with long duration balloon flights circling the South Pole during austral summer. For X-ray and gamma-ray sources in the southern sky, continuous twenty-four-hour-a-day observations are possible. For sources near the South Pole the hard X-ray/gamma-ray observations would have the further advantage of minimal overlying atmosphere so photon transmission through the atmosphere is maximized. For cosmic ray measurements Antarctica is advantageous since the low geomagnetic cutoff permits entry of low rigidity particles.

Until recently, however, because of the perceived severe environmental and logistical problems, no serious effort has been made to fly state-of-the-art balloon-borne instruments in Antarctica. Such instruments typically weigh up to one to two tons and are carried by $\sim 11-28 \times 10^6$ cu. ft. balloons to altitudes of ~35–40 km.

On February 23, 1987, the supernova 1987A was first detected in the Large Magellanic Cloud (LMC), a companion galaxy to the Milky Way located ~52,000 parsecs away. This is the first nearby supernova detected in over 380 years, and thus it offers a truly unique opportunity to study in detail the supernova explosion, the process of explosive nucleosynthesis, and possibly the formation of a neutron star. Eventually, SN 1987A may well have even greater impact on astrophysics than the Crab Nebula, the best studied previous supernova remnant.

Supernova explosions are believed to produce most of the heavy elements in our universe, via the process of explosive nucleosynthesis. Gamma-ray line spectroscopy provides one of the best ways to study this process, since many of the heavy elements are first produced as an unstable nucleus which then decays, with the emission of gamma-ray lines of characteristic energies, until a stable nucleus is reached.

SN 1987A is located at 69° south declination, only 21° from the pole. In early 1988 an ~2400 lb. instrument designed to make high spectral resolution measurements of the gamma-ray lines of SN 1987A was successfully launched on an 11.6 million cu. ft. balloon from McMurdo, Antarctica.[1,2] After ~3 days' flight the payload was cut down because of electronic problems, but not before collecting significant data on the supernova. This flight opened the door to serious consideration of Antarctica as a scientific ballooning site for X-ray and gamma-ray astronomy. Exploratory flights are planned for this coming austral summer (1989-90), and very likely for the next as well, with a possible full-scale balloon campaign for the next solar maximum (1991-92 and beyond).

In section II, I review our current understanding of the advantages and problems of research on balloons in Antarctica. Section III discusses the scientific objectives for high energy gamma-ray and hard X-ray astrophysics. The advantages of Antarctica for balloon-borne cosmic ray research are summarized in other papers in these proceedings.[3]

II. BALLOONING IN ANTARCTICA

For high quality gamma-ray and hard X-ray measurements, heavy ($\gtrsim 10^3$ lb.) instruments must be carried to altitudes of ~35–40 km (overlying atmosphere ~2.5–6 g/cm^2). Most scientific balloon flights last from ~10 hours to ~two days, limited either by the balloon drifting out of range or by the loss of altitude over time. In the past several years there has been significant interest and experimentation with long duration balloon flights (LDBF). These LDBF's take advantage of the strong, stable, zonal winds which flow approximately along latitude lines in the summer time at mid-latitudes; the balloon is launched from one continent and allowed to drift to another for cutdown (see Figure 1).

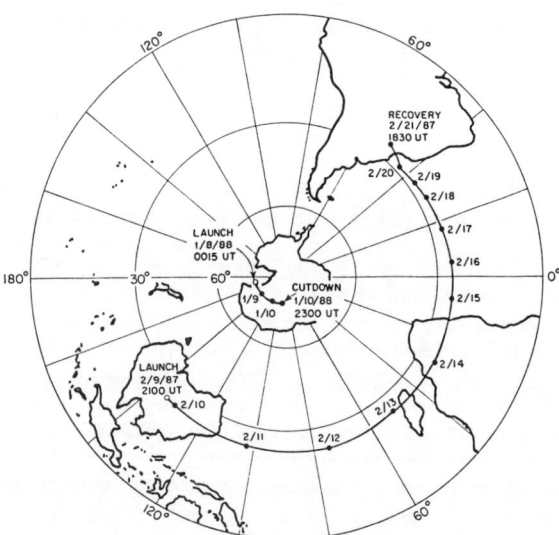

Figure 1. The trajectories for two long duration balloon flights, at mid-latitudes[4] (~23°S, Alice Springs, Australia to Brazil), and in Antarctica[2] (~78°S, McMurdo). In the mid-latitude flight the balloon slowed down as the float altitude decreased.

For standard zero-pressure balloons the temperature of the gas, and therefore the balloon altitude, are controlled by the radiation received from the Sun and the

Earth. Thus the balloon is at high altitude during sunlight hours but drops during nighttime (Figure 2). If the balloon initially reaches a high daytime float altitude it will remain above the tropopause in its day-night excursions under normal conditions without ballast drops. Then, in this simple RAdiation COntrolled balloON (RACOON) mode,[5] flight durations are limited only by balloon lifetime and gas losses. The daytime float altitudes monotonically decrease, due to loss of lift with time. Calculations based on standard atmospheric models indicate that ~2.5–3% lift is lost in each day-night cycle. For a seven-day flight (six nights) this implies that 15–18% of total (payload plus balloon) weight is required in ballast (1050–1300 lbs. out of an ~3×10^3 lb. payload on a 28 million cu. ft. balloon) to maintain the initial daytime float altitude of ~126–130K ft. through the flight. At present LDBF's have been generally limited to the southern hemisphere for political reasons.

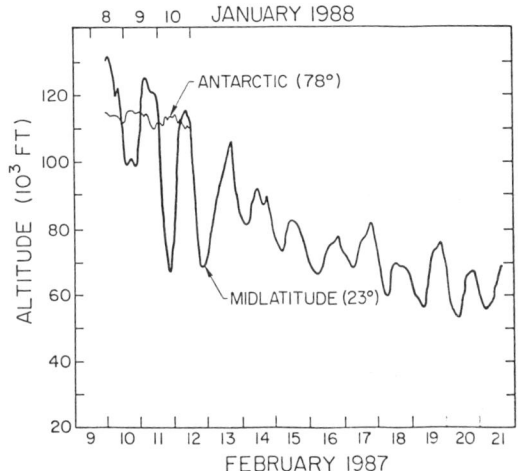

Figure 2. The altitude profile for the two balloon flights.[2,4] The large diurnal variations for the mid-latitude flight are due to the day-night cycling. The Antarctic flight was in sunlight at all times but the Sun angle varied through the day.

An attractive alternative is provided by flights in Antarctica during its summer. An 11.6 million cu. ft. balloon with an ~2390 lb. payload was successfully launched in January 1988 from near McMurdo (78° latitude).[1,2] After 65 hours the payload was cut down because of a power supply failure. The balloon drifted westward approximately along the 78° latitude line and came down near Vostok, about 80° west of McMurdo.

Figures 1 and 2 also show the trajectory and altitude of this flight. Because of the 24-hr./day sunlight the diurnal altitude excursions are much more limited. The lift loss of less than 0.5%/day implies that 15% of total payload weight in ballast would last 30 days, compared to seven days at mid-latitudes. Thus, much longer flights are possible. From the 1988 flight and sparse wind data, it is expected that one circumnavigation of the South Pole will take ~10–20 days' flight time. There are no problems with crossing international boundaries and essentially no risk of accidental landing in a populated area.

For sources in the southern sky an important advantage of Antarctic balloon flights is the possibility of continuous 24-hr./day observations. The left panel of Figure 3 shows the angle from the zenith for a variety of X-ray/gamma-ray sources as a function of local time for McMurdo (~78° latitude). The right panel plots the transmission of photons of various energies through the atmosphere as a function of zenith angle for a balloon at ~40 km altitude. For near-polar sources such as SN 1987A, LMC, or GX301-2, Antarctic balloon observations are clearly superior to mid-latitude observations, particularly at low energies. For sources at low declinations, however, the photon transmission through the atmosphere would be lower than at mid-latitudes, but by only ~10–25% (averaged over 12 hours for mid-latitude, and 24 hours for Antarctica) at gamma-ray line energies (see Figure 4).

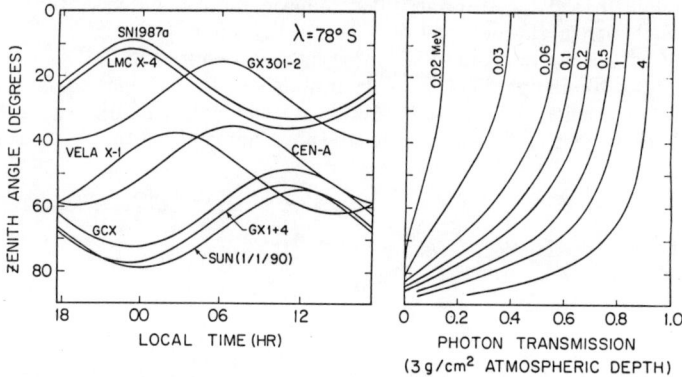

Figure 3. The angle from the zenith versus local time for various southern hemisphere X-ray/gamma-ray sources is plotted in the left panel for observations from McMurdo. The transmission for photons of various energies is plotted on the right versus zenith angle for observations from ~40 km altitude.

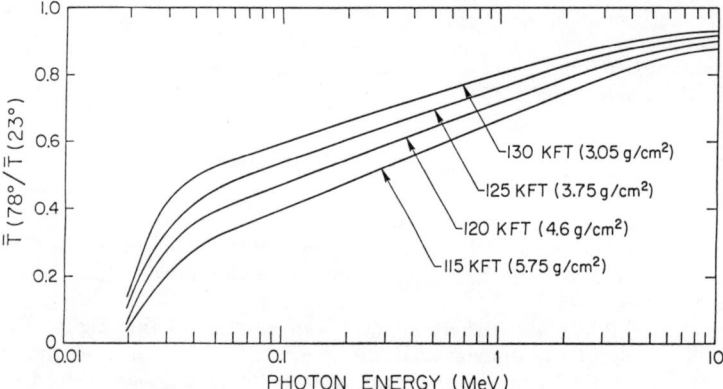

Figure 4. The ratio of photon transmission, averaged over the 12 hr/day observing period for 23° latitude and 24 hr/day observing period for 78° latitude, for a source at 23° declination and balloon float altitudes ranging from 115 to 130 kft.

Given the 24-hr./day observing time at Antarctica compared to ~12-hr./day at mid-latitudes, however, more photons are collected per flight day in Antarctica than at mid-latitudes, at energies above ~50 keV. Taken with the much longer flight durations which are potentially available, Antarctica is clearly superior both for long continuous observations for high sensitivity and for catching rare transient events such as cosmic gamma-ray bursts or large solar flares.

The plans for hard X-ray/gamma-ray astronomy in Antarctica have concentrated on balloon launches from McMurdo where extensive support facilities are available. With the wind at float altitudes flowing typically along latitude lines, however, an intriguing possibility is to launch balloons from the South Pole. The balloons should then stay near the South Pole, and a single ground station could provide continuous line-of-sight telemetry. At present, however, only small $1-2 \times 10^5$ cu. ft. balloons have been hand-launched from the South Pole. Equipment and techniques would need to be developed for launch of large balloons and heavy payloads.

It should be noted that the hard X-ray/gamma-ray background at balloon float altitudes, which is produced predominantly by cosmic ray interactions with the Earth's atmosphere, will be substantially (several times) higher in the Antarctic since the geomagnetic cutoff is very low. In addition, terrestrial energetic electron precipitation, from auroral and polar cap sources, will also increase the hard X-ray background occasionally. On the other hand, the low geomagnetic cutoff will allow low rigidity cosmic rays to enter, and studies of such particles, as well as of terrestrial radiation, are important scientific objectives in their own right.[3,6]

There are a number of practical aspects to scientific ballooning in Antarctica which have yet to be explored: How stable are the wind patterns at 35–40 km altitude? What are the chances for recovery? What communication channels are available for commands and data telemetry? What is required for balloon launches and recovery on the snow and ice in Antarctica? Exploratory flights of large ~3000 lb. payloads on 28 million cu. ft. balloons are presently planned for 1989-90 and 90-91 austral summers to answer some of these questions.

III. SCIENTIFIC OBJECTIVES

From the previous discussion it is clear that Antarctic LDBF observations could be advantageous for the study of essentially all hard X-ray/gamma-ray sources in the southern sky. Here we describe some of the gamma-ray/hard X-ray spectroscopic studies of two objects, SN 1987A and the Sun, as examples of the types of astrophysics that can be done.

Supernova 1987A

The January 1988 balloon flight from McMurdo was aimed at the detection and measurement of the gamma-ray lines from ^{56}Co produced by nucleosynthesis in the new supernova 1987A. Supernova 1987A appears to be a type II supernova, which is thought to occur in a massive (>8 $M_\odot$) giant stars. Theoretical models, such as 10 HMM of *Pinto and Woosley*,[7] for SN 1987A have provided generally good agreement with observations. Model 10 HMM starts with an ~20 $M_\odot$ star on the main sequence that has lost 4 $M_\odot$ in its evolution. It has a hydrogen shell of 10 $M_\odot$ and it produced 0.075 $M_\odot$ of ^{56}Ni (the progenitor of ^{56}Fe) in the explosion. ^{56}Ni decays with a half-life of 6.1 days to ^{56}Co with the production of positrons and gamma-rays, but these most likely cannot escape, since the Ni is created in the innermost part of the supernova shell and the shell is still very dense at this early stage. ^{56}Co decays to ^{56}Fe with a half-life of 77 days, by electron capture or positron emission, producing gamma-rays at 847 and 1238 keV (as well as an assortment of other energy photons).

Since July 1987 the supernova appears to be powered predominantly by the radioactive decay of ^{56}Co. The bolometric light curve shows an exponential decay with time constant characteristic of ^{56}Co and the intensity expected for 0.075 M$_\odot$ of ^{56}Co produced in the initial explosion, provided most of the decay gamma-rays and positrons are absorbed and thermalized in the supernova shell. As the ^{56}Co decays away and the supernova shell becomes transparent to ^{56}Co gamma-rays, other longer-lived radioactive isotopes with lower gamma-ray energies, such as ^{57}Co, may take over.[8]

^{57}Ni decays with a half-life of 1.5 days to ^{57}Co. ^{57}Co decays to ^{57}Fe with a 270-day half-life by electron capture, and emits either two photons of 122 keV and 14 keV energy (89% of the time), or a single 136 keV photon.[9] If ^{57}Co were produced in the explosion with ^{57}Co/^{56}Co ratio equal to the solar ^{57}Fe/^{56}Fe abundance ratio of 0.0243 by mass, then heating by ^{57}Co should begin to dominate over ^{56}Co after ~1200 days,[8] leading to an exponential decay of the supernova bolometric luminosity with the ~270-day half-life of ^{57}Co.

Figure 5 shows the calculated flux for the ^{56}Co 847 keV line and the ^{57}Co 122 keV lines as a function of time[10] for the 10 HMM model. For the solar ratio the 122 keV line should peak at $\sim 5 \times 10^{-5}$ (cm^2 s)$^{-1}$ at ~1000 days. *Woosley and Pinto*[10] indicate that the ^{57}Co/^{56}Co ratio could be a factor of 2 higher, but probably not lower, than the solar value.

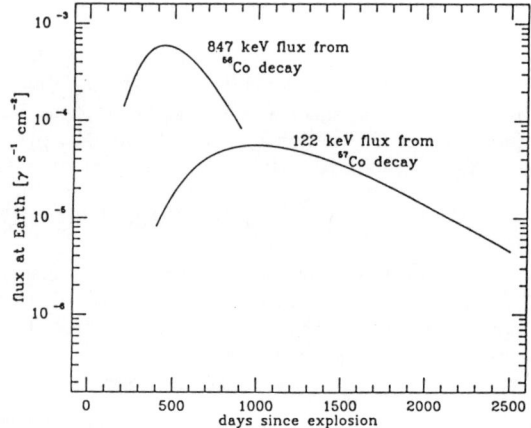

Figure 5. The predicted light curve from model 10 HM for the 847 keV ^{56}Co and 122 keV ^{57}Co gamma-ray lines for SN 1987A.[10]

The ratio of ^{57}Co to ^{56}Co provides the ratio of ^{57}Ni to ^{56}Ni during explosive silicon burning, and therefore gives important information on the neutron excess, defined as $N = \Sigma (N_i - Z_i) \dfrac{X_i}{A_i}$ where N and Z are the neutron and proton numbers of the nucleus with mass fraction X and nucleon number A.

Accurate determination of the line center and line profile (which depend on Doppler shift and broadening) will give information on the expansion velocity and geometry of the Fe shell. Since the ^{57}Co gamma-ray line emission is expected to take place on the innermost surface of the debris and therefore passes through all of the expanding shell, characteristics of the shell can be deduced by measurement of the same gamma-ray line at different times, and of lines of different energy (and different absorption coefficients) at the same time, and of the ratio of the line to the continuum at energies just below the line.

At even later times heating by ^{44}Ti gamma-rays (68, 78 and 1157 keV energy, and decay half-life 78 years) could take over from ^{57}Co if other energy sources are absent. However, a collapsed object is expected to be formed by the type II supernova explosion, most likely a neutron star, although a black hole is also a possibility. Depending on its rotation rate, magnetic field, and the accretion rate of infalling material, the neutron star could be a bright hard X-ray emitter similar to X-ray pulsators such as Her X-1. If the magnetic field is several times 10^{12} Gauss, cyclotron line features similar to those reported for Her X-1 may be present in the hard X-ray energy range. Recently, a half ms period optical pulsation has been reported for SN 1987A (C. Pennypacker, private communication, 1989). This pulsed emission was apparently detected only on one observing run out of several, suggesting possible transient obscuration of the optical emissions. The hard X-ray emission should be much more penetrating.

The spectral region ~30–150 keV should be particularly interesting for studying SN 1987A. Because the photoelectric opacity is high below 20 keV the continuum spectrum of an accreting neutron star would probably be peaked near 30 keV. Cyclotron features, if present, should be at energies of a few tens of keV. Clearly, high resolution hard X-ray/soft gamma-ray line and continuum measurements of SN 1987A over a several-year period beginning in 1989-90 are crucial. These measurements are best done from Antarctica because SN 1987A is at an angle of only 21° from the South Pole. Thus atmospheric attenuation will be minimal and SN 1987A can be seen 24 hours/day for 10–20 days on a long duration balloon flight.

Solar Flares

The ability to release energy impulsively and accelerate particles to high energies is a common characteristic of cosmic plasmas at many sites throughout the universe, ranging from magnetospheres to active galaxies. These high-energy processes play a central role in the overall physics of the system at each site where they are observed. Nowhere can one pursue the study of the basic physics of these processes better than in large solar flares, which are the most powerful natural particle accelerators in our solar system. Here, the acceleration of electrons is revealed by hard X-ray and gamma-ray bremsstrahlung; the acceleration of protons and nuclei is revealed by nuclear gamma-ray lines and neutrons. The accelerated particles, notably the electrons with energies of tens of keV, probably contain a major fraction of all the released flare energy, thus indicating the fundamental role of the high-energy plasma processes.

Gamma-ray line emission from flares results from interactions of accelerated ions with the ambient solar atmosphere (Figure 6). The strongest lines are at 2.223 MeV, from neutron capture on ^{1}H; at 0.511 MeV, from positron annihilation; and at 6.129, 4.438, 1.634, 1.369, 1.779 and 0.847 MeV, from ^{16}O, ^{12}C, ^{20}Ne, ^{24}Mg, ^{28}Si and ^{56}Fe deexcitations, respectively.[11,12] In addition, reactions of alpha particles with He nuclei produce lines at 0.429 and 0.478 MeV from ^{7}Li and ^{7}Be.[13,14]

Gamma-ray line measurements to date have been obtained almost exclusively by NaI scintillation detectors with energy resolution insufficient to resolve any of the lines (Figure 7). Liquid nitrogen-cooled germanium detectors can resolve all the lines (except the 2.223 MeV) and thus provide detailed information on directivity of the accelerated particles (through Doppler shifts); elemental abundances of the solar atmosphere; and the temperature, density and ionization state of the positron annihilation region, etc. In addition, Ge detectors will provide accurate measurements of the bremsstrahlung continuum, which can be unfolded to obtain the accelerated electron spectrum. High resolution is required to resolve the superhot component[16] — probably the hottest true thermal plasma produced in the flare —

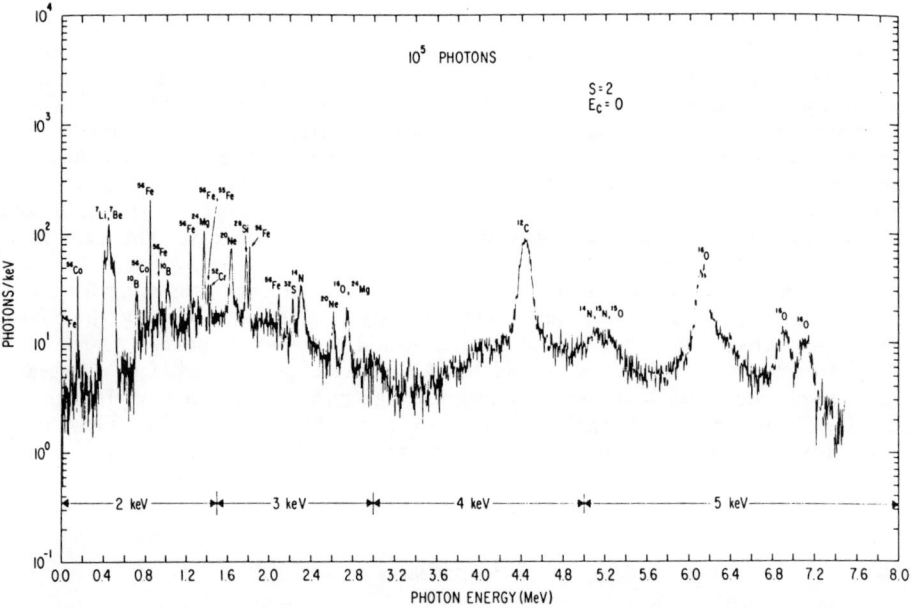

Figure 6. Calculated nuclear de-excitation gamma-ray spectrum from energetic particle reactions.[15]

Figure 7. The spectral resolution as a function of photon energy for a two-segment germanium (Ge) detector is compared to the resolutions of the hard X-ray and gamma-ray spectrometers on the Solar Maximum Mission (SMM). The typical widths expected for gamma-ray lines in solar flares are also shown. Note that none of these lines were resolved with the Gamma-Ray Spectrometer (GRS) on SMM, but all except the neutron-capture deuterium line at 2.233 MeV, with a predicted width of <0.1 keV, will be resolved with cooled Ge detectors. Similarly, the broken line, indicating the energy resolution to resolve the spectrum from the superhot plasma at a temperature in excess of 30×10^6 K, shows that this component was not resolved with the SMM instruments but can be clearly resolved with cooled Ge detectors.

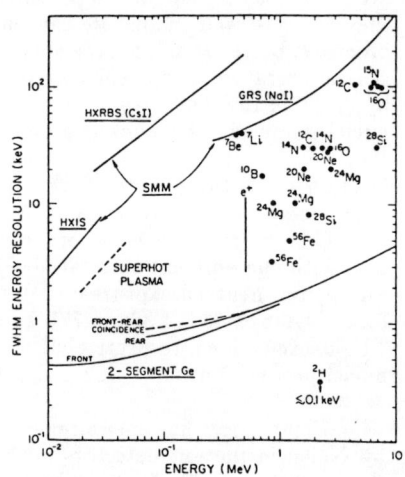

and to resolve the sharp breaks in the spectrum[17] which appear to be intimately related to the acceleration mechanism.

Large solar flares, which produce copious fluxes of gamma-rays and hard X-rays and thus are amenable to detailed study, occur infrequently and unpredictably. Thus long observing times are required to catch such flares. Recently, NASA started the Max '91 LDBF program for detailed study of solar flares in the upcoming solar activity maximum. Powerful new instruments were selected for hard X-ray and gamma-ray imaging and spectroscopy of solar flares. These are to be carried on LDBF's to maximize the likelihood of observing a large solar flare. The present plan is for these payloads to be flown in Antarctica, beginning in 1991-92, if the practical problems with ballooning in Antarctica can be resolved.

ACKNOWLEDGMENTS

This research has been supported in part by NSF grants ATM-8402231 and DPP-8717481 and NASA grant NAGW-516 and contract NAS8-37687.

REFERENCES

1. A. C. Rester, R. L. Coldwell, F. E. Dunnam, G. Eichorn, J. I. Trombka, R. Starr, and G. P. Lasche, accepted for publication, Astrophys. J. (1989).
2. J. Ground, K. Dallas, R. Cowie, and W. R. Thorn, preprint AFGL-TR-0265 (1988).
3. A. Westphal et al., these proceedings (1989).
4. R. P. Lin, D. W. Curtis, J. H. Primbsch, P. R. Harvey, W. K. Levedahl, D. M. Smith, R. M. Pelling, F. Duttweiler, and K. Hurley, Solar Phys. *133*, 333 (1987).
5. V. Lally, Proc. XXIV COSPAR Conf., Ottowa, *1*, p.1.4 (1982).
6. J. Benbrook, these proceedings (1989).
7. P. A. Pinto, and S. E. Woosley, Nature *333*, 534 (1988).
8. S. E. Woosley, P. A. Pinto, and D. Hartmann, submitted to Astrophys. J. (1989).
9. D. D. Clayton, Astrophys. J. *188*, 155 (1974).
10. S. E. Woosley, and P. A. Pinto, in Nuclear Spectroscopy of Astrophysical Sources, ed. N. Gehrels and G. Share (Amer. Inst. Physics, New York, 1988), p.98.
11. R. Ramaty, B. Kozlovsky, and R. E. Lingenfelter, Space Sci. Rev. *18*, 341 (1975).
12. R. Ramaty, and R. J. Murphy, Space Sci. Rev. *45*, 213 (1987).
13. R. J. Murphy, R. Ramaty, D. J. Forrest, and B. Kozlovsky, in 19th Internat. Cosmic Ray Conference Papers, *4*, 249 (1985a).
14. R. J. Murphy, D. J. Forrest, R. Ramaty, and B. Kozlovsky, 19th Internat. Cosmic Ray Conference Papers, *4*, 253 (1985b).
15. R. Ramaty, in The Physics of the Sun, ed. P. A. Sturrock, T. E. Holzer, D. Mihalas, and R. K. Ulrich, Vol. II, (D. Reidel Publ. Co., Dordrecht, Holland, 1986), p.291.
16. R. P. Lin, R. A. Schwartz, R. M. Pelling, and K. C. Hurley, Astrophys. J. Lett. *251*, L109 (1981).
17. R. P. Lin, and R. A. Schwartz, Astrophys. J. *312*, 462 (1987).

LONG-DURATION BALLOONING AT MID-LATITUDES AND IN ANTARCTICA

W. Vernon Jones
Space Physics Division, Code ES, NASA Headquarters
Washington, DC 20546

ABSTRACT

After a series of catastrophic balloon failures in the early-mid 1980's the high altitude research balloon program now enjoys unprecedented success with only two balloon failures in more than 100 flights, and increasing demands are being placed on the program. Four test flights from Australia to South America have led to the design of a new long-duration flight support system with global tracking, command, and telemetry for both mid- and high-latitudes. The successful flight of the Gamma Ray Advanced Detector by the Air Force Geophysics Laboratory has resulted in numerous requests for flight opportunities in Antarctica. The first test flight for the joint NASA-NSF program to support balloon flights in Antarctica is scheduled from McMurdo in December 1989. After summarizing the current status of scientific ballooning, this paper describes the recent long-duration test flights, the development efforts underway for future long-duration operations, and plans for near-term flights in Antarctica..

CURRENT STATUS OF THE NASA BALLOON PROGRAM

The National Aeronautics and Space Admimistrastion (NASA) Balloon Program conducts approximately 50 high altitude balloon flights per year in support of the astrophysics, space physics and upper atmosphere research activities. The program is based on five standard balloons having the characteristics and capabilities given in Table I. Of the three basic balloon sizes, two have heavy-lift (H) versions. i.e., greater effective wall-thickness through the use of more caps. The maximum nominal float altitude of 130 kft is available with only one balloon, but a range of altitudes can be accommodated with varying weights within the balloon design limits. The weight allocation for science instruments also depends on the flight duration and the time of day the payload is launched.

The program was plagued with numerous catastrophic balloon failures between 1983 and 1985.[1] The flight program was completely shut down in November 1985, extensive tests of the film quality and the film sealing methods were undertaken, and a flight-test program was initiated. As of June 1989, there have been only two balloon failures out of about 100 flights since the recovery plan was put into effect in early 1986.[2]

The unprecedented flight success led to re-evaluation in 1988 of restrictions due to concerns about ballooning safety. It was hoped that the new success rate might lead to some reduction in the restrictions on launches from the National Scientific Balloon Facility (NSBF) in Palestine, Texas. However, the new safety considerations led to even greater restrictions on Palestine launches. In particular, the risk for turnaround flights from Palestine was determined to be in excess of the accepted

© 1989 American Institute of Physics

Table I. Characteristics and nominal weights for the standard balloon sizes.

BALLOON VOLUME (MCF)	BALLOON WEIGHT (LBS)	SUSPENDED WEIGHT (LBS)	NSBF SYSTEMS (LBS)	NOMINAL ALTITUDE (K FT)	10-12HR FLIGHT[a,c] BALLAST (LBS)	10-12HR FLIGHT[a,c] SCIENCE (LBS)	20-36HR FLIGHT[b,c] BALLAST (LBS)	20-36HR FLIGHT[b,c] SCIENCE (LBS)
12	1,710	2,100	405	120	220	1,475	695	1,000
12H	3,220	5,500	650	100	525	4,325	1,575	3,275
23	3,750	4,300	600	120	475	3,225	1,450	2,250
28	3,355	3,000	500	130	375	2,125	1,150	1,350
28H	4,555	5,500	650	120	600	4,250	1,825	3,025

NOTES:

a. The 10-12 hour flight assumes a morning launch.
b. The 20-36 hour flight assumes either a morning or evening launch.
c. The weights listed are nominal; the maximum and minimum payload weights would vary with the ballast allocation.

casualty expectation, 1×10^{-6}. Due to increased population density in Texas, only flights with trajectories to the west (i.e., after the spring turnaround in May and before the fall turnaround in September) will be permitted from Palestine in the future.

The new restrictions actually result from a better understanding of the risks associated with the flights. Although the current balloon reliability has greatly reduced the risk involved during the launch phase by eliminating the catastrophic failures during ascent, it has essentially no effect on the risk during the cut-down/impact phase. During turnaround the wind directions are *a priori* unknown and variable, and payload impacts, which can occur anywhere within a 300 mile radius of Palestine, must be assumed to occur in the least desirable location. Planned flight terminations require that the predicted impacts be in sparsely populated regions, and cutdowns are not permitted if towns or cities are within the impact dispersion areas for either the balloon or the payload/parachute.

Fort Sumner, New Mexico has been designated a semi-permanent NSBF site for conducting turnaround flights in the western U.S., in order to mitigate the loss of Palestine for that purpose. Flights during the period after spring turnaround and before fall turnaround can, of course, still be carried out from Palestine. However, due to the high velocity winds in mid-summer only the first few weeks after spring turnaround and the last few weeks before fall turnaround offer viable opportunities for most investigations. The relative locations of Palestine and Fort Sumner is such that they can be used as convenient down-range telemetry stations for flights launched at the other site.

In order to estimate the requirements to be placed on the balloon program during the next several years, a limited survey of flight requirements for the major disciplines was carried out in September 1988.[2] The general trends gleaned from the survey can be summarized as follows:

(1) There will be an increase of approximately 50% in the total number of flights requested over the next 3 - 5 years.
(2) There will be an increase in the average, requested flight duration (about 90% of the investigators would like > 40 hr).
(3) There will be an increase in the number of investigations requesting long-duration (> 1 week) exposures, including solar

physics (Max '91 Program), cosmic physics, and high energy astrophysics payloads.
(4) There will be an increase in the number of investigations requesting exposures from the southern hemisphere, including upper atmosphere and high energy astrophysics payloads.
(5) There will be an increase in the number of investigations requesting exposures at high latitudes, including ionospheric payloads at the north and south poles, flights from McMurdo, and flights from Canada.

If these more demanding scientific requirements and increasing numbers of flights are to be met, near-simultaneous launches and launches from remote sites will likely become the operational standard. This new mode of operations will, of course, require additional resources and place heavy demands on the NSBF launch personnel.

RECENT LONG-DURATION FLIGHTS AT MID-LATITUDES

Two long-duration campaigns, one in January-February 1987 and the other in January-February, 1988, have tested the feasibility of long-duration balloon flights between Australia and South America. Each campaign consisted of two flights, for a total of four missions, which had their payloads recovered in South America, the first in Paraguay and the other three in Brazil. Brazil is the planned termination area for future semi-global flights launched from Australia.

One flight in each long-duration campaign was a passive, nuclear emulsion chamber for studying ultra-high energy cosmic rays, including a search for the postulated, but as yet unconfirmed, quark-gluon plasma state of matter.[3] Each flight carried enough ballast for six day-night transitions, and, with the exception of one night during the second mission, each remained above 110 kft for the flight duration. Figure 1 shows the altitude profile for the 1987 flight; the profile for the 1988 flight was similar. Figure 2 shows the balloon trajectory, which deviated little from the launch latitude.

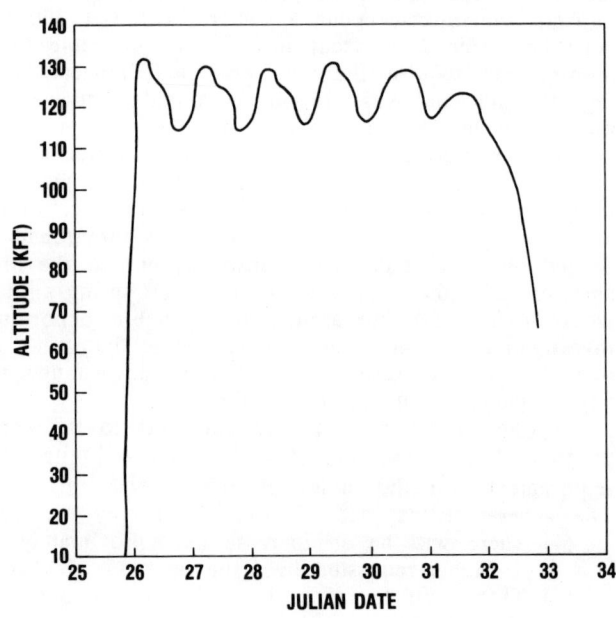

Fig. 1. Altitude profile of 1987 Australia-Paraguay flight (courtesy NSBF).

The second flight in the 1987 and 1988 long-duration campaigns was a considerably more sophisticated hard X-ray/low energy gamma ray experiment.[4] Both flights of this instrument experienced difficulties, but they provided valuable data on long-duration ballooning operations. The 1987 flight had all the ballast dropped during ascent. As shown in Fig. 3 its maximum daytime and nighttime altitudes decayed essentially throughout the 12-day flight. The 10-day flight in 1988 experienced difficulties with the initial extension of the solar panels, which led to power supply problems that impacted the ability to track the balloon. Consequently, the altitude profile of the flight is not well known. This flight also experienced ballast problems; the ballast-drop command apparently did not function properly during the early days of the mission, and the entire ballast was expended early in the flight.

In spite of the ballast problems on two flights, all four

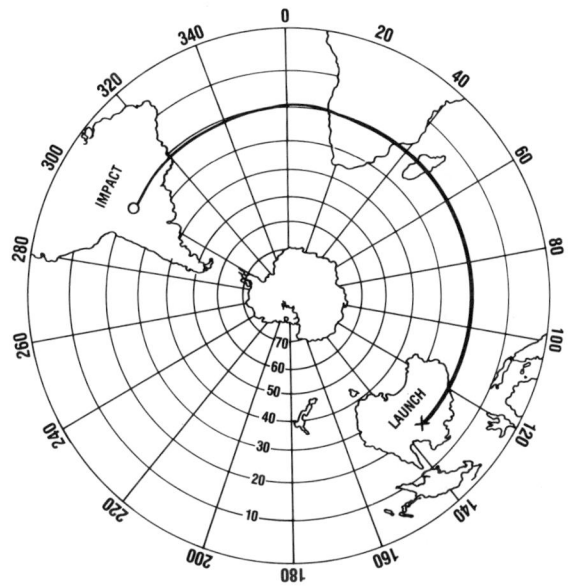

Fig. 2. Approximate trajectory of 1987 Australia-Paraguay flight (courtesy NSBF).

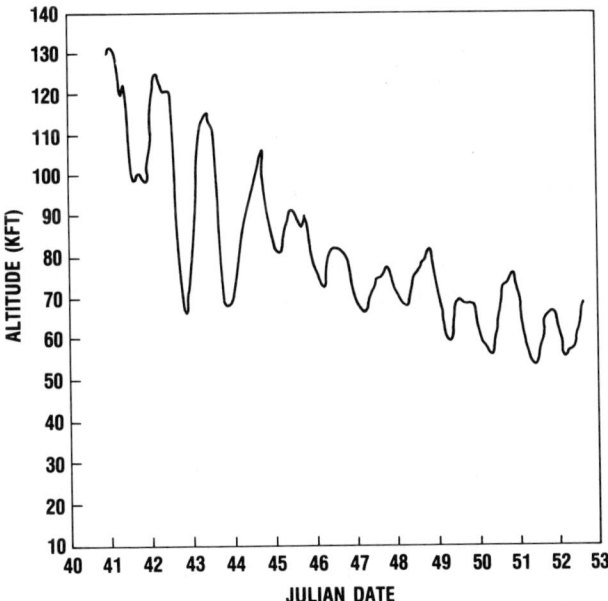

Fig. 3. Altitude profile of 1987 Australia-Brazil flight (courtesy NSBF).

of the long-duration payloads were recovered in South America. The flights without ballast relied on the so-called RACOON (Radiation Controlled Balloon) flight mode, whereby the balloon is allowed to descend naturally at night and re-ascend during the day.[5] This mode refers to the ability of stratospheric balloons to remain afloat for long periods (weeks or even months) with the help of terrestrial thermal radiation as the Earth's surface is cooling at night.

Figure 4 illustrates schematically the altitude dependence of the atmospheric temperature from the surface through the troposphere and stratosphere. The altitude of the mimimum temperature, which defines the tropopause, varies from around 30 kft for high latitude regions to around 50 kft at tropical latitudes. The tropopause temperature is typically about -50 °C at near-polar latitudes and about -70 °C at tropical latitudes.

A zero pressure balloon will remain aloft provided the change in temperature of the helium gas in the balloon is less than the change in temperature of the surrounding air.[5] Without appropriate ballasting, a zero-pressure balloon floating at full altitude during the day would typically decline in altitude during the transition from day to night as the atmosphere around the balloon cools. If the balloon descends below the tropopause it would quickly fall to the Earth's surface. On the other hand, the thermal radiation from the cooling Earth's surface can warm the balloon system so it remains above the tropopause for substantial periods of time. However, ballasting is required to maintain full daytime altitude. Note that zero-pressure balloons experience significant declines in altitude if substantial reductions in the thermal radiation from the Earth are encountered, e.g., if the balloon overflies cloud cover associated with an active thunderstorm.

A typical, unballasted RACOON flight might have a daytime float altitude of 125 kft (determined by the balloon volume, amount of helium, and suspended weight) and a nighttime altitude around 70 kft. The alternating day-night altitude pattern should continue, with some decay in the average altitude as a result of helium venting. Near the end of the flight the balloon could be expected to survive several day-night transitions by floating slightly above the local tropopause.

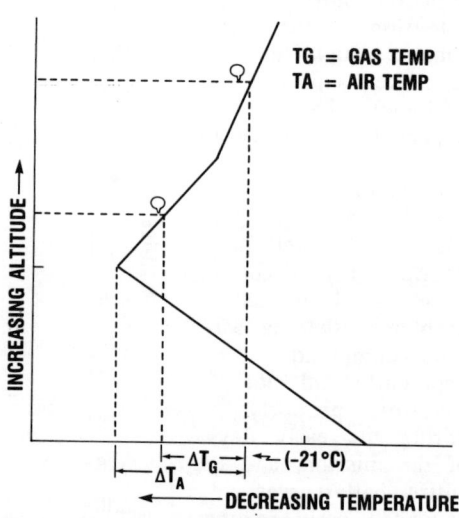

Fig. 4. Schematic temperature profile encountered by stratospheric balloon (courtesy WFF). RACOON requires $\Delta T_G < \Delta T_A$.

RECENT FLIGHTS IN ANTARCTICA

Concurrently with the 1988 NASA long-duration balloon flight campaign from Australia to Brazil, the Air Force Geophysical Laboratory (AFGL) and the National Science Foundation (NSF) conducted a one-flight campaign from McMurdo, Antarctica.[6] Investigation of gamma rays from the recent supernova SN1987A with the GRAD (Gamma Ray Advanced Detector) was a major scientific objective of that flight[7] which used an 11.8 MCF balloon. This is the largest balloon launched to date in Antarctica, and its success has greatly enhanced the interest in launches from Antarctica. Reference 6 gives an excellent description of the flight logistics, operations, launch, and recovery.

The balloon traveled in a circumpolar direction nearly along the launch latitude for about three days before its termination. In that time it covered almost 25% of the distance around the globe. Its altitude profile[8] is shown by the upper curve in Fig. 5. The initial float altitude was about 115 kft, and it experienced deviations of about 5 kft prior to termination. The only ballasting occurred during the first day of the flight, and the balloon altitude tended to generally track the Sun's declination angle, given by the lower curve in Fig. 5. Most of the decline in altitude could, in principle, be offset by adequate ballasting. Unfortunately from the standpoint of ballooning, the flight duration was too short to test our expectations that the balloon altitude profile should track the solar declination angle with with predictable altitude excursions for a full polar orbit.

In January 1989, a series of three "pathfinder" flights were carried out by AFGL-NSF, in order to collect meteorological data for future ballooning activities in Antarctica. None of those flights followed the 78th parallel, and they did not remain at float altitude as expected.[8] As illustrated in Fig. 6, these relatively small (0.25 MCF) balloons reached float altitudes of about 125 kft, but even with up to 20% ballast drops the altitude decayed to the preset termination altitude within a few days. The first flight lasted only 24 hours; the second fell to the tropopause within about 3 days and then, after the terminate mechanism failed to function, floated slightly above the tropopause for about 20 days; and the third was terminated after three days. It

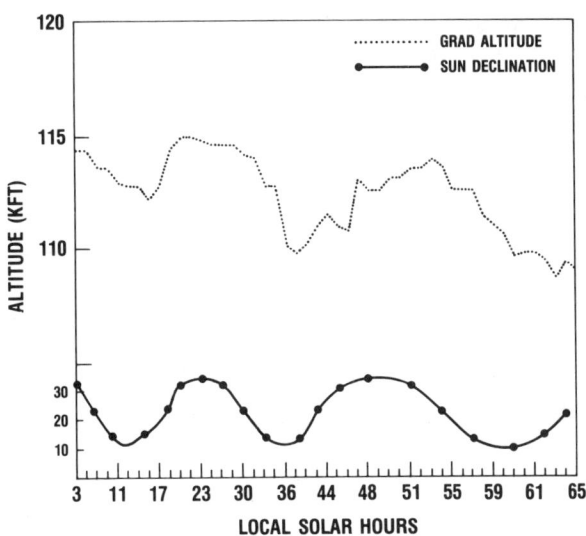

Fig. 5. Altitude profile of 1988 GRAD flight (upper) with solar angle (lower) in Antarctica (courtesy J. Ground).

is not known whether these balloons had unusual characteristics, whether they encountered atmospheric conditions that affected their performance, or whether turn-around effects may have caused them to experience random trajectories from the launch site.

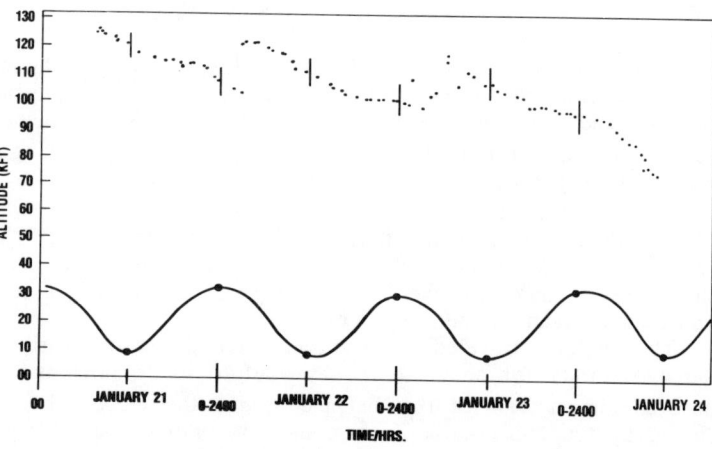

Fig. 6. Altitude of 1989 Antarctic test balloon flight (upper) with solar angle (courtesy J. Ground).

Evidence that the declining altitude profiles may not be unusual is given in Fig. 7, which shows the altitudes for five of the eight balloon flights carried out at the South Pole by the University of Houston during their 1985-86 austral summer study of the ionosphere.[8,9] Those flights were unballasted, and, with one exception, their altitudes decayed essentially monotonically within about three days to the value of the termination pressure-altitude setting. The one flight that flew at full float altitude in excess of 100 hr experienced an anomaly during launch that resulted in a very low free-lift; the protective shipping cover of the balloon was carried aloft, which resulted in a heavier effective payload and, therefore, slower rise to float altitude. There has been speculation that the slow ascent may have resulted in the improved balloon performance, but there has been no verification that this was indeed the case.

LONG-DURATION PLANS

A program is underway for developing global, long-

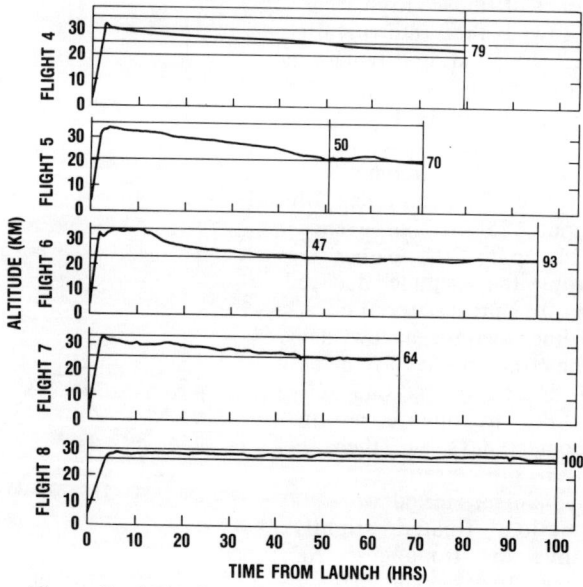

Fig. 7. Altitudes of University of Houston South Pole flights (courtesy J. Ground).

duration ballooning for both mid-latitudes and Antarctica; the latter will be a joint NASA-NSF effort. Both Antarctic and mid-latitude flights offer new research opportunities for all the disciplines served by the balloon program. The time-driver for long-duration development is the so-called Max '91 program, which will focus on studies of the Sun during the next period of maximum solar activity (1990-91). The current plan is to have an adequate Antarctic capability in place by December 1991 for the Max '91 investigations, while simultaneously preparing for mid-latitude flights. The latter are also in demand for the Max '91 program, especially in the northern hemisphere, since ground based data from observatories in the northern hemisphere could be coordinated with the balloon measurements.

The NASA objective is to support two long-duration campaigns each year, with at least 2 - 3 flights per campaign, initially. The need for stable winds requires that the campaigns be carried out during the local summer season, so the goal is to have one southern hemisphere campaign and one northern hemisphere campaign approximately six months apart.

The difficulties in maintaining altitude for two of the four flights from Australia to Brazil has emphasized the need for a focused effort to develop a reliable flight support system. The systems needed for Antarctic and mid-latitudes are similar, although there are some crucial differences, e.g., the satellites used for telecommunications would generally be different. For that reason, a modular system is being developed. As illustrated schematically in Fig. 8, the generic flight support system consists of the support instrumentation package (SIP) with electronics for controlling the flight, antennas for bi-directional telecommunications, solar panels with rechargeable batteries for power, and a crude (~ 5°) pointing capability to ensure that the solar panels will function adequately. Tests of the long-duration system components will begin in the summer of 1989 using continental U.S. (conus) flights. The first full-up conus flight test of the long-duration SIP is planned for the fall of 1990, with the first Antarctic test during the 1990-91 austral summer.

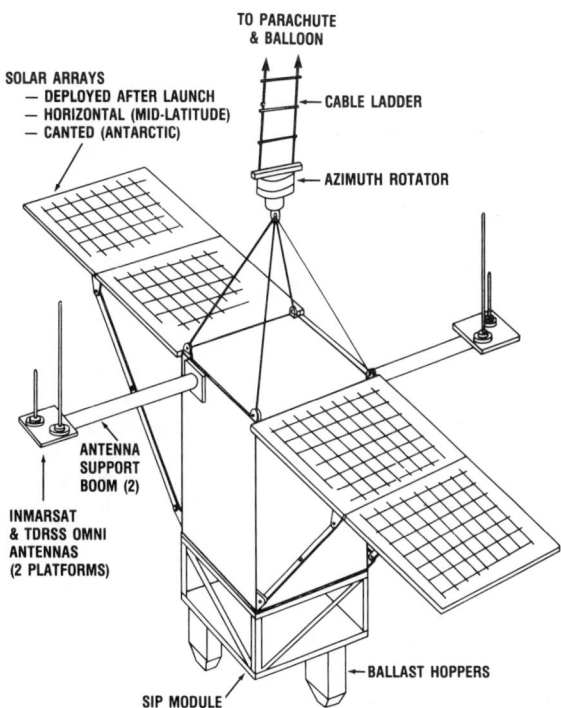

Fig. 8. Schematic diagram of long-duration balloon flight system (courtesy WFF).

The first NASA-NSF test campaign to Antarctica in December 1989-January 1990 will focus on tests of the long-duration subsystems and studies of the wind circulation patterns. The latter will be pursued with the launch of four small (~0.25 MCF) pathfinder balloons, in addition to the 28 MCF balloon used for the primary flight. The campaign will check the Antarctic launch technique and, hopefully, demonstrate the ability to circumnavigate the globe at nearly full alttitude. Based on the limited information from prior balloon flights in Antarctica and, especially, the erratic behavior of the January 1989 Pathfinder flights, it seems certain that a second test campaign will be required during the December 1990-January 1991 launch window, if an operational capability is to be in place for the Max '91 flights during the December 1991-January 1992 season.

The test campaigns to Antarctica in 1989-90 and 1990-91 will offer an opportunity for piggyback science payloads. Four separate payloads have been selected to fly on one gondola in 1989-90. These will carry out investigations of the supernova SN1987A, the Sun, iron isotopes in cosmic rays, and chip-upsets in the high particle density environment. Only after the successful completion of the subsystems tests will the fabrication of a second SIP be initiated, so it is likely that again in 1990-91 only one gondola can be flown. The payload(s) for that flight have not yet been selected.

The long-duration flights will employ standard zero-pressure balloons. Because of its superior altitude performance, the balloon-of-choice is the 28 MCF two-cap balloon which carries 3000 lb nominal suspended weight to a float altitude of 130 kft. For Antarctica, the nominal science weight allocation is 1500 lb, although payload weights up to about 1900 lb could probably be accommodated with sacrifice in altitude. These estimates are based on the as-yet-unproven expectation that less ballasting will be required for Antarctic flights, because of the continuous daylight during the launch window. Somewhat less science payload weights would be available for mid-latitude flights of similar durations and altitudes.

The agreement between NASA and NSF for carrying out long-duration flights from Antarctica calls for both agencies to do what they would normally do, i.e., launch and tracking of the balloons will be carried out by the NASA/NSBF, while logistics will be provided by the NSF. Williams Field near McMurdo will be used for the launch site, the NSF Delta-3 vehicle will serve as the launch vehicle, and a weighted skid for the mounting spool will be provided by the NSF.

Depending on availability and weather conditions, the NSF LC-130 aircraft will be used for both line-of-site telemetry during the flight and for payload recovery. The maximum payload cross-sectional area that can be accommodated by the LC-130 is 8 ft by 8 ft. The length of the payload is limited by the length of the cargo bay (TBD). It should be noted that payload recovery can not be guaranteed because the permissible recovery areas are very limited. Marie Byrd Land and the Ross Ice Shelf seem to offer the best chances of recovery, assuming the balloon circumnavigates the globe along the 78th parallel. The range of recovery operations is also limited, as is the time that can be spent on the ground during recovery operations. At least during the early flights it should be assumed that there is a high risk of no recovery, or perhaps a delay of up to a few years before recovery can be completed during a subsequent austral summer.

CONCLUDING REMARKS

The current balloon flight success is leading to increasing demands for longer and more sophisticated flights. Part of this demand is due to the anticipated shortage of space flight opportunities, and part is due to the recognition that high altitude balloons offer a viable substitute for certain types of space science investigations, e.g., those that need investigations on the order of days or weeks with payloads weighing a ton or more.

Interest in semi- or trans-global flights has grown in the wake of the demonstration of successful flights at both mid-latitudes and in Antarctica. Such flights are still in the experimental stage, but with adequate resources, and time, they should become routine. The potential scientific payoff is clearly high for these long-duration flights. However, the problems encountered with both Australia-Brazil and Antarctic flights show the need for a systematic approach to them.

Antarctica is especially appealing as a trans-global launch site, both from the standpoint of the unique science opportunities it offers and in terms of the low human/property casualty risk. For either of these reasons the development of ballooning in Antarctica is warranted. There is no doubt that the demands for scientific launches from either, or both, McMurdo and the South Pole will significantly exceed our near-term capacity to conduct such flights.

ACKNOWLEDGEMENTS

The author would like to take this opportunity to express his appreciation and the appreciation of the scientific community to the Balloon Project Office of the Wallops Flight Facility and to the balloon manufacturers, Raven Industries, Inc. and Winzen International, Inc. for the successful balloon recovery program. Likewise, we are extremely grateful to the staff of the National Scientific Balloon Facility for their yeomen efforts in conducting the flight program. Without the high level of success the program now enjoys, it would not be practical for us to be considering long-duration balloon flights in Antarctica and elsewhere.

REFERENCES

1. I. S. Smith, Adv. Space Research 5, No. 1, 9 (1985).
2. W. V. Jones, Journal of Spacecraft and Rockets (in press).
3. W. V. Jones, Y. Takahashi, B. Wosiek, and O. Miyamura, Ann. Rev. Nucl. Part. Sci., Vol. 37, p. 71 (1987).
4. R. P. Lin, et al., Solar Physics 113, 333 (1987).
5. V. Lally Proc. XXIV COSPAR Conf., Ottawa, Vol. I, p. 1.4 (1982)
6. J. Ground, in *Environmental Research Papers, No. 1015*, Preprint AFGL-TR-0265, Air Force Geophysical Laboratory, Hanscom AFB, MA 01731.
7. A. C. Rester, et al., Astrophys. J. Lett., in press; preprint (1988).
8. J. Ground, Private Communication at the Antarctic Project Initiation Conference, March 7, 1989, NASA Wallops Flight Facility.
9. J. Benbrook, University of Houston, Private Communication, 1989.

NEW DOORS OPENED BY THE POLAR PATROL BALLOON
--COSMIC QUARK MATTER, MICROGRAINS AND GAMMA RAYS--

Y.Fukada, Y.Hatano and T.Saito
Institute for Cosmic Ray Research, Univ. of Tokyo, Tanashi, Tokyo

R.Fujii
National Institute of Polar Research, Itabashi, Tokyo, Japan

H.Oda
Department of Physics, Kobe University, Kobe, Japan

T.Yanagita
Kansai Jyogakuin Junior College, Miki, Japan

ABSTRACT

Balloon Experiments in Antarctica, make possible to observe continuously the astrophysical objects in the southern polar sky during the flight of more than 20 days. New fields and instrument for the polar balloon experiment are proposed.

INTRODUCTION

The great advantage of balloon experiments in Antarctica is in the continuous watching for the astrophysical objects in the southern polar sky during the flight for more than 20 days. As an effective time available in the lower latitude balloon is at most 6 hours per one day, it needs more than 40 times flights to integrate the same exposure as the polar balloon flight. One polar balloon flight makes possible to get the collecting power more than [100 m^2 X hours] which is essential to open the doors of new scientific fields. The instrument for the polar balloon experiments is under development for studies involving the search for the quark matter from neutron stars, detection of high energy micrograins and atoms in the cosmic radiation, observations of gamma rays at energies of 1 to 1000 GeV.

OBJECTIVES AND EXPERIMENTAL CONSIDERATIONS

(1) COSMIC QUARK MATTER AND CENTAURO

The detections of high energy particles from Cyg X-3 have been reported in both surface and underground experiments which based on both their direction and their arrival time having the characteristic 4.8 h period of Cyg X-3. Several underground detectors, Sudan, NUSEX and recently IMB[1] have observed muons excess in the direction of Cyg X-3. On the other hand, the KAMIOKA, Frejus and recently GOTTHARD[2] gave negative results for the effect. At present, it is hard to understand the causes of these differences. Is it attributable to the differences in the observation period ? Is there some difficulties in the experiments ? If it is correct, the masses of several GeV/c^2 are still possible within the present experimental accuracy and the candidates for the quark matter have not ruled out, although the pulsar phase data would constrain the upper limit of masses to 10 MeV/c². On the other hand, some of anomaly phenomena observed in the past cosmic ray experiments are still surviving. A hypothesis[3] that Centauro[4] events found in the Mt. Chacaltaya emulsion chamber are caused by a primary glob of highly dense nuclear matter is able to be tested by the proposed direct observation.

If the anomalous particles are cosmic origin and if they are accelerated and propagated in the same way as cosmic rays, the anomalies will be found in the cosmic radiation at lower energies. Figure 1 shows the energy spectrum of cosmic rays and the fluxes of particles from CygX-3 and Centauro. In order to expected the fluxes of the anomalies from binary systems, the fluxes of the observed anomaly particles are extended to lower energy side. The numbers of anomalies at 10 GeV/n detectable by the experiments of [100 m² * hours] are shown at the left axis of Figure 1.

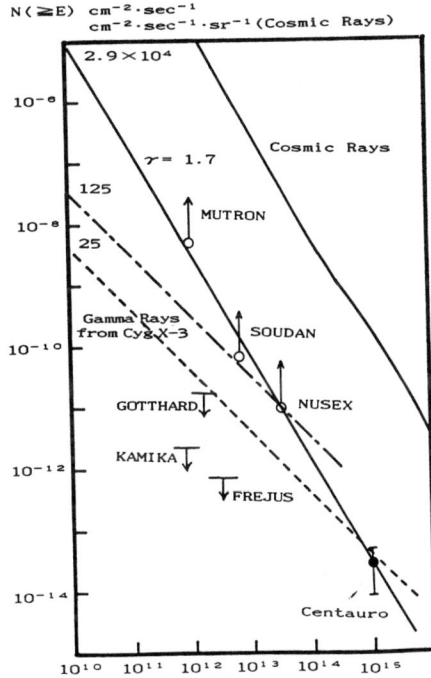

Fig. 1. Energy Spectra of cosmic rays, muons from Cyg X-3 and Centauro. Expected numbers at 10 GeV/n from the experiment of [100 m²*hr] are shown at left axis of Figure.

The incident direction of neutral particles is determined within an accuracy of 2 degree from the emission angle distribution of secondary charged particles from interactions of anomaly particles with detector materials. The backgrounds are mainly neutrons in fragmentation region being produced by collisions of cosmic ray protons with air nuclei. The numbers of neutrons is estimated as 5-10 particles within the cone of 4 degree at 10 GeV/nucleon in the observation of [100 m²*hr] at the atmospheric depth of 7 g/cm².

We would like to emphasize that the proposed experiment at the top of the atmosphere will make finally to confirm the existence of the anomalies particles from the point sources. If they exist it is possible to clear whether they are gamma rays, or quark matters like H particles, or another unknown particles by measuring the secondary products from the anomalies with the tracking detector.

(2) MICROGRAINS AND ATOMS

If cosmic rays are getting their energies by shock acceleration, neutral particles such as neutral atoms or micrograins will be found, in no small numbers of flux, in the cosmic radiation. Jets have been observed not only in extragalactic radio sources but also recently in the radio X-ray emissions from the galactic object SS433. If the observed gamma ray lines at 1.495 and 6.695 MeV with very narrow width (FWHM< 10 KeV) from SS433 can be interpreted as the blueshifted emissions from deexcitations of $^{24}Mg^*$ and $^{16}O^*$ in the jets[5], the micrograins will be found in the cosmic radiations.

We stress that the proposed polar balloon experiment make first possible to search for the particles at the level smaller than 10^{-8} of normal cosmic rays. Detection method is same as the tracking techniques used for the cosmic quark matter search.

(3) GAMMA RAYS AT 1 TO 1000 GeV

LMC, SMC

Before detecting the neutrino bursts, SN1987A, the LMC is one of the most interested object because it is the nearest galaxy outside of our Galaxy and it has many X-ray sources, LMC X-1, X-2, X-3 and X-4. It is important to observe the gamma ray sources at 1-1000 GeV in the regions with detector having high angular resolution.

There are theoretical difficulties for interpreting the unusual characteristics of a burst of gamma rays on 5 March 1979 which was identified in a young supernova remnant, N49, in the LMC.
If the 5 March 1979 event, as proposed by Alcock et al.[6], was originated from collision of a lump of strange matter with the rotating strange star, further gamma ray event may be expected from the same point. Similar gamma ray bursts may occur in other quark star binaries.

Extragalactic Radio Sources

Gamma ray observation from active galaxies provides important understanding on the activities of Radio galaxy, Seyfert galaxies and quasar. In order to explain these compact and supermassive objects, Kafetos proposed a model[7] which attributes the output within the active nucleus to Penrose process in the ergosphere of a massive Kerr black holes. The absence of a MeV break in the spectra of Cen A and 3C273 is insisted to be explained by the Penrose pair production. If true, a sharp break in the GeV region, around 2 GeV, is observed in the spectra of Cen A.
The proposed balloon experiment can check the model clearly as shown in Figure 2.

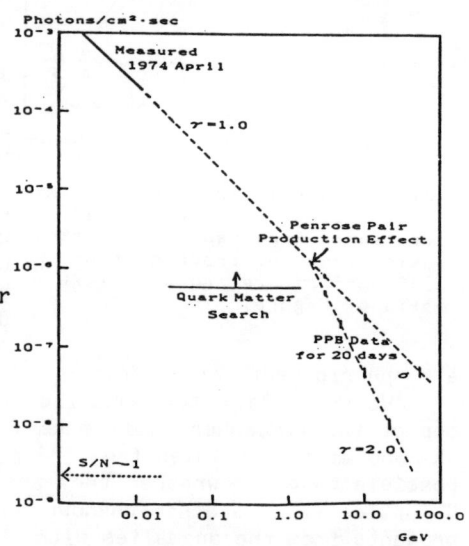

Fig. 2. Spectrum of gamma rays from Cen A.

Galactic Sources

More than 5 galactic sources emitting radio and/or X-rays are objects in the Southern Polar Sky. Vela is one of the highest luminous sources and its intensity is expected to be 1.54 X 10^{-6}

/cm^2*sec at 1 GeV for only pulsasing components, although the declination of Vela is 45 degree.

INSTRUMENTATION

Figure 3 shows a schematic diagram of instrument. It was designed within 250 Kg for maximum scientific pay load of the present Japanese polar balloon. The instrument is composed of (1) the converter of gamma rays, (2) detector of electro-magnetic cascades and (3) tracking detector for secondary charged particles. The converter consists of lead plates(3-10 mm thick), 2 xy arrays of scintillation fibers and 2 scintillation detectors. The detector of electro-magnetic cascades is 6 layers composing of 5 mm thick lead plate, 5 mm thick plastic scintillators and 2 xy arrays of scintillation fibers. The tracking detector is 6 stacks composing of 10 mm thick polyetylene plate, 5 mm thick scintillation counters and 2(x+y) of scintillation fibers.

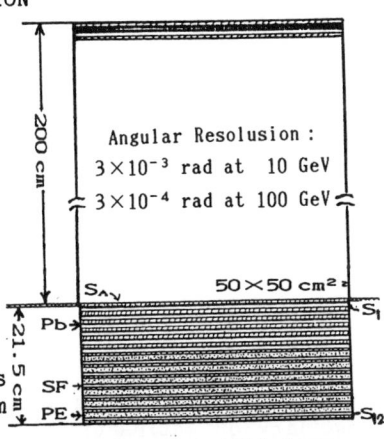

Fig. 3. Schematic diagram of instrument

The detector can keep watch on the regions to zenith angle of 60 degree in case of scanning for neutral particles and gamma rays. The incident direction of neutral particles is determined within 2 degree by measuring the emission angle of secondary particles with scintillation fiber. In case of observing the special gamma ray sources, detector is oriented to the source. At energy regions of 1 to 1000 GeV, the converted pair is not observed as the separated pair by using the present scintillation fiber photograph. As the multiple scattering angle dominates over the opening angle of the pair, the angular resolution of incident gamma rays depends on the multiple scattering depending on the relation of 0.03 $L^{1/2}$ X E^{-1}, where L is the radiation lengths of the converter and E the energy of gamma rays in GeV. Angular resolutions are shown in Figure 3. Details of the instrument will be presented elsewhere.

REFERENCES

1) G.Battistoni et al., Phys. Lett. 115 465 (1985).;
   M.L.Marshak et al., Phys. Rev. Lett. 54 2079 (1985);55 1965 (1985).
2) Y.Oyama et al., Phys. Rev. Lett. 56 991 (1986).
   Ch.Berger et al., Phys Lett. B174 118 1986).
3) J.B.Bjorken and L.D.McLerran, Phy. Rev. 20D 2353 (1979).
4) C.M.G.Lattes, Y.Fujimoto and S.Hasegawa, Phys. Pep. 65, 151 (1980)
5) R.Ramaty et al., Ap. J. 283 L13 (1985).
6) C.Alcock, E.Farhi and A.Olinto, Phy. Rev. Lett. 57 2088 (1986).
7) M.Kafetos, Ap. J. 236 99 (1980).

SEARCH FOR ANTIMATTER AT THE 10^{-7} LEVEL WITH THE POLAR PATROL BALLOON

Y.Fukada, Y.Hatano and T.Saito
Institute for Cosmic Ray Research, Univ. of Tokyo, Tanashi, Tokyo

R.Fujii
National Institute of Polar Research, Itabashi, Tokyo, Japan

H.Oda
Department of Physics, Kobe University, Kobe, Japan

I.Yamamoto
Faculty of Engineering, Okayama Univ. of Science, Okayama, Japan

ABSTRACT

A new annihilation technique for antimatter search is under development for the balloon experiments in the Antarctica. The capability of antiparticle-particle discrimination of the detector is about 10^{-9} by combining the Cherenkov technique, 10^{-5}, which was confirmed by using accelerator antiproton beams, and topological methods, 10^{-4}. The Polar balloon flight with the 1/4 m^2 detector will reach to the flux sensitivities around 10^{-7} level for antiprotons and antiheliums and the 10^{-6} level for heavier antinuclei.

INTRODUCTION

The discovery of unexpectedly high intensities of antiprotons observed by Buffington et al.[1] at energies below the kinematic threshold of antiproton production has forced us to reconsider our understanding on the origin of galactic cosmic rays and on the origin of the Universe. New experiments recently done by Ahlen et al.[2] and Mogomolov et al.[3] result the lower fluxes of antiprotons than that reported by Buffington et al.[1]. However, the flux sensitivities of their experiments are limited to at most 10^{-5}, and they remain to give only the upper limit from their negative results. It is essential to measure the energy spectrum of antiprotons in order to separate the proposed various hypotheses, involving the decay of primordial black holes[4] and the baryon symmetry of the Universe[5]. Moreover, the experiment with the 10^{-7} level is absolutely necessary in order to compare with new values based on the photino-antiphotino annihilation evaluated by J.Ellis et al.[6] which are lower than the value originally predicted by J.Silk and M.Sredniki[7].

For antihelium, although the probability of producing the antiheliums from collisions of cosmic rays with interstellar matter is negligible at any energy regions, the best limit of antihelium still remains to 2.2×10^{-5} obtained by Buffington et al.[1] in the energy range 130-370 MeV/nucleon. For antimatter heavier than helium, the fluxes also remain the upper limit of around 10^{-4} obtained by G.F.Smoot et al 15 years ago[8].

It is considered[9] that the contribution of the extragalactic cosmic rays to the local cosmic ray fluxes is considered as 10^{-5} to 10^{-4}. Assuming the baryon symmetric Universe as a whole, about a half of extragalactic cosmic rays would be antimatter. Flux of antimatter will be reduced by the fragmentation process in the original galaxy and by modulation in our Galaxy. So the experiments with a flux sensitivity of at least the 10^{-7} level are required in studying for "primordial" antimatter. We propose a new technique for discriminating antimatter from matter with the 10^{-9} level. The Polar rounding balloon experiment will reach to the flux sensitivities of 10^{-8} level for antiprotons, the 10^{-7} for antiheliums and the 10^{-6} level for heavier antinuclei.

DETECTING SYSTEM

1) Instrumentation

A schematic diagram and an outline of the instrument are shown in Figure 1 and Table 1, respectively.

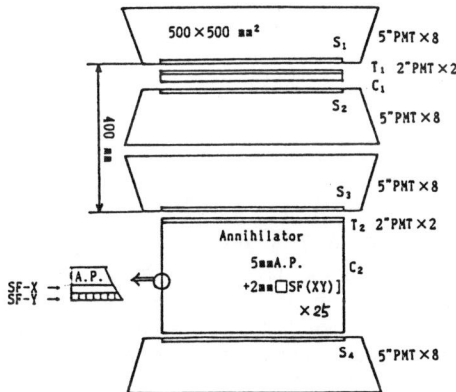

Fig. 1. Schematic diagram of the instrument

Table I Outline of instrumentation

EFFECTIVE SΩ	1300 cm²·sr [$S_1*S_2*S_3*(T_1*\overline{C_1})C_2$]
TRIGGER SΩ	2070 cm²·sr [$S_1*S_2*S_3*(T_1*C_1)$]
DIMENSIONS	50×50 cm²×100 cmH (Detectors)
	130×130 cm²×145 cmH (Packing)
WEIGHT	250 Kg
TRIGGER RATE (Photo Rates)	7.5/hr (Candidates & Backgrounds)
SENSITIVITIES (10 days)	≤5.6×10⁻⁸ for Antiprotons
	≤2.5×10⁻⁷ for Antiheliums
	≤6.7×10⁻⁶ for Heavier Antinuclei

Particles with velocities smaller than v/c=0.68 are selected with Cherenkov detector C_1.
The energies of particles are determined within an accuracy of 10 % by measuring the transition of dE/dX, S_1, S_2 and S_3. The incident direction of particles is determined with TOF detector of T_1 and T_2, and also with the measure of the variance of dE/dX.

The annihilator consists of 25 stacks of 5 mm thick Cherenkov radiators and two x-y cross layers of scintillation fibers of 2 × 2 mm² cross section.

The particles with energies below the kinematic limit (220 MeV for protons) do not make the Cherenkov signal in C_1 but the annihilation products of antimatter give a Cherenkov signal in C_2.
Under the trigger criterion of [$S_1*S_2*S_3*(T_1*C_1)*C_2$], the scintillating fibers are identified with the image-intensifier-CCD-camera system and the tracks of the events are reconstructed. If the triggered events are real antimatter, the track of incident particle, one vertex point, charged particles from annihilation point and development of electromagnetic cascades are seen in the CCD photograph. In order not to miss any annihilation event in which annihilation products travel backward through C_1, anticoincidence of C_1 in the (T_1*C_1) is set within 7

nanoseconds of T_1 pulse.

The scintillation fibers are connected to the image intensifier (I.I.) through the tapered light guide. It is estimated that the surviving photons are about 10 per one light guide at the entrance of the I.I. when a singly charged particle goes through the farthest point of one scintillation fiber. The photo-electron numbers per pixel in the CCD photograph is measured with the I.I. system of larger aperture (10 cm dia.) which will be available in a few year.

2) Background Estimation

The events satisfying the trigger criterion of $[S_1*S_2*S_3*(T_1*C_1)*C_2]$, that is the CCD photographic records of scintillating fibers, include the real antimatters and backgrounds. The main backgrounds originate in the case as follows. A low energy charged particle within the geometry satisfies the criterion of $[S_1*S_2*S_3*(T_1*C_1)]$ and, within the coincidence time, a relativistic charged particle out of geometry produces secondary particles in the annihilation boxes and gives a Cherenkov signal C_2. The rate of this kind of backgrounds is about 10^{-3}/sec for the present detector. As almost of backgrounds originating in the atmosphere above the detector, pion, neutron, electrons and gamma rays are eliminated by the counter logic, other sources do not make serious rate of backgrounds. The detector has no vessel and the wall thickness of the diffusion boxes is 0.1 g/cm^2. The annihilator does not contain lead plates in order to decrease the backgrounds producing in the lead plates. The capability for discriminating the antimatter from the background by the counter logic, 10^{-5}, was confirmed by using accelerator antiproton beams. The background events surviving from the counter logic are eliminated by scanning the TV-photographs in which more than two incident particles or plural vertex points are seen. The topological analysis up to 10^4 photographs is not serious. In the 20 days exposure of the present detector, the number of CCD-photographs for the topological analysis are at most 3000 pictures.

BALLOON EXPERIMENT AND EXPECTED RESULTS

Balloon flights in Antarctica are planned from 1992 to 1994 by The National Institute of Polar Research of Japan. The balloon launched in summer time from the Showa base in the Antarctica is expected to being back to near the launching site after the 20 days traveling. Expected flight curve is shown in Figure 2.

The Data are recorded during the flight and are returned by telemetry to the Showa base after the 20 days exposure. However, the balloon track from the Showa base is at the higher rigidity

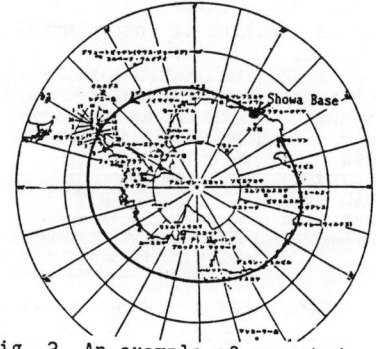

Fig. 2. An example of expected flight curve in case of launching in summer time from Showa base.

and about half of the 20 days
flight are outside of detection
threshold of the present plan.
The flight from the US base is
favourable for the antimatter
search.

The 20 days exposure of
1/4 m^2 detector (250 Kg)
will reach the total exposure of
2.25 X 10^9 cm^2*sr which
corresponds to the sensitivities
at 10^{-7} level for antiprotons
and antihelium, and the 10^{-6}
level for heavier antinuclei.

The ratios of antiproton to
proton fluxes evaluated by J.Ellis
et al.[7] basing on annihilation of different relic candidates
are quoted in Figure 3. As shown in
the Figure 3, the flux sensitivity
of 10^{-7} is required to study
the "primordial" antimatter. Regions from the proposed experiment
are shown by the solid line in Figure 3.

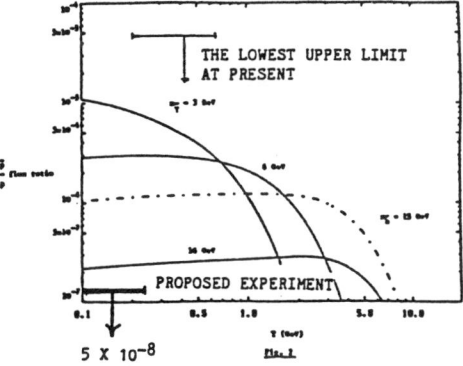

Fig. 3. The ratio of fluxes of antiprotons to protons expected from annihilations of different relic particles in the galaxy. The Fig. is taken from Ref.(6)

CONCLUSION

The annihilation techniques combining the Cherenkov method and
the topological method with scintillation fibers can separate
antimatter from matter with ability of 10^{-9}. The instrument having
1,300 cm^2*sr is under development. About 20 days exposure of the
instrument using the Polar rounding balloon will reach to flux
sensitivity of 10^{-7} which will make possible to detect "primordial"
antimatter before the experiments boarding on the Space Station.

We would like to acknowledge Professor A. Buffington who
taught us the sources of backgrounds in their annihilation experiment.

REFERENCES

1) A.Buffington, S.M.Syrovatskii and C.R.Pennypacker, Ap.J.248, 1179(1981).
2) S.P.Ahlen et al., Phys. Rev. Lett., 61 145 (1987).
3) E.A.Bogomolov et al., Proc. 20th ., 6 72 (1987).
4) P.Kiraly, J.Szabelski, J.Wdowczk and A.W.Wolfendale, Nature 285 386(1981). M.S.Turner, Nature 297 379 (1983).
5) F.W.Stecker, R.W.Orotheroe and D.Kasanas, Ap. and Sp.Sci.96 167(1983). F.W.Stecker and A.W.Wolfendal, Nature 309 37 (1984).
6) J.Ellis, R.A.Flores, K.Freese, S.Ritz, D.Seckel and J.Silk, CERN-TH.5062/88 (1988).
7) J.Silk and M.Srednicki, Phys. Rev. Lett., 53 642 (1984).
8) G.F.Smoot, A.Buffington and C.D.Orth, Phys.Rev.Lett.35 258(1975).
9) Y.L.Ginzburg and S.I.Syrovatskii, Origin of Cosmic rays (1964). S.Hayakawa, Cosmic Ray Physics (1969).

A Measurement of the Isotopic Composition of Iron-group Elements in the Galactic Cosmic Rays using Balloon-borne Track-recording Detectors in Antarctica

Andrew J. Westphal, P. Buford Price and Daniel P. Snowden-Ifft
Department of Physics
University of California, Berkeley, Ca. 94720

ABSTRACT

Using a new track-recording glass detector which we have recently developed, we will measure the isotopic composition of iron-group elements in the galactic cosmic radiation. This measurement has lately become a high priority in the field of cosmic ray physics because of its important role in the elucidation of stellar evolution and nucleosynthesis in highly evolved stars, of the acceleration mechanisms of cosmic rays, and of the structure of the interstellar medium. The detectors will be carried to high altitude for exposure to cosmic rays by a balloon to be launched from Antarctica in the austral summer of 1989-90.

INTRODUCTION

An extraordinary amount of astrophysical information can be extracted from isotopic abundances of iron-group nuclei (isotopes of iron, nickel and neighboring elements) in the cosmic rays. Unfortunately, past experimental efforts to study these isotopes have been limited by inadequate statistics and poor mass resolution. We are constructing and will fly a balloon-borne instrument that has sufficient mass resolution and sufficient collecting power to study these astrophysically important isotopes. Our design is based on the remarkable properties of a newly discovered phosphate glass detector called BP-1.

For many years our group has pioneered in the development and application of track-recording solids to fundamental research in a variety of fields.[1] Such solids include plastic films, glasses, and many nonconducting crystalline solids. On traversing a solid-state track detector a sufficiently highly ionizing particle produces a continuous, submicroscopic, chemically reactive trail of radiation damage. Because of an enhanced etching rate along the track, this can be developed into conical pits at the points of entry and exit of the particle from the solid when etched in a suitable corrosive solution, and into a test-tube-shaped pit at the end of the particle's range (see figure 1). The size and shape of the etchpit depend on the choice of etching conditions and on the charge and velocity of the particle; we exploit the differential slowing of particles of different isotopes to measure the mass of each particle. We use a fully automated microscope system to scan and measure etchpit diameters and lengths in large areas of detectors without human supervision. This makes it possible, with detectors that have been carefully calibrated with beams of heavy ions, to identify nuclear fragments of high-energy interactions[2,3] and to measure the charge and energy distribution of heavy cosmic ray nuclei[4]. Solid-state track detectors have tremendous advantages for use in cosmic-ray observations: they are light, inexpensive, rugged and simple, and a detector can be made easily of virtually any size and collecting power. A disadvantage is that they require recovery for analysis in the laboratory after exposure.

Recently we discovered the remarkable track-recording properties of a phosphate glass denoted VG-13[5], a laser filter glass. Although it is in some ways similar to plastic track detectors, we found that it is capable of unprecedented charge resolution, far better than is attainable with the best plastic detectors and better than any electronic detectors for the identification of extremely heavy nuclei. VG-13 has a number of serious practical problems; a glass which we developed and christened BP-1[6] exhibits slightly better sensitivity, and has none of the practical difficulties of VG-13. We will fly detectors made with BP-1 in our balloon experiment in Antarctica.

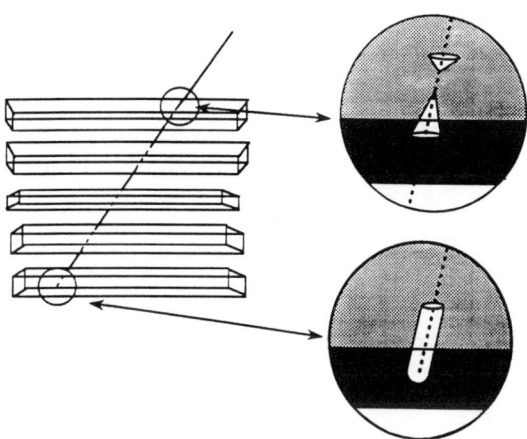

Figure 1: *The passage of a highly ionizing particle (such as a cosmic-ray iron nucleus) through a stack of track-detecting sheets produces a trail of damage which can be exposed by etching in a suitable acid. Conical pits are formed at the entry and exit points of the particle; a test-tube-shaped pit is formed when the etchant reaches the particle's end of range.*

SCIENTIFIC OBJECTIVES

Knowledge of isotopic abundances of the iron group elements tells us about the origin and acceleration of cosmic rays, conditions in the interstellar medium, and processes of nucleosynthesis in the late stages of stellar evolution. The current state of detector technology is such that cosmic ray physicists have published data with moderate statistics and resolution only for isotopes of the relatively abundant even-Z elements up through calcium ($Z = 20$). Figure 2 (from ref. 7) compares the rather skimpy existing data with predictions of a model in which cosmic rays originate in regions of the Galaxy that are metal-rich compared with the solar system and a model in which a fraction of the cosmic rays originate in material expelled by Wolf-Rayet stars. Beyond silicon the error bars are too large to be useful. To go to higher atomic number has been made difficult both by the decreasing fractional mass difference between adjacent isotopes and by the decreasing fluxes of high-Z nuclei.

The synthesis of iron-group elements is thought to take place in high-temperature, high-density regions of massive stars under the conditions of nuclear equilibrium. The abundances are found to be most sensitive to neutron excess in the evolving material; neutron excess, in turn, is a sensitive function of the evolutionary history of these massive stars. Hainebach et al.[8] found that three zones of different neutron excess were needed to explain solar system abundances of the Fe and Ni isotopes. Rare isotopes such as ^{57}Fe, ^{58}Fe, ^{58}Ni and ^{61,62}Fe can be used to check the predictions of the three-zone e-process model. A statistically significant measurement of rare iron group

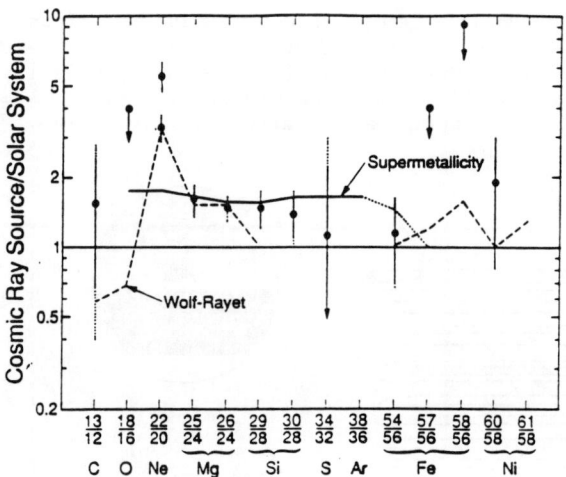

Figure 2: Abundance of rare isotopes in the cosmic-ray source compared to abundance in the solar system. Lines indicate results of calculations using the supermetallicity or the Wolf-Rayet model

isotopes would permit a direct comparison with nucleosynthetic predictions of a variety of models which have been extensively discussed in the literature.[8-9] The determination of source abundances is greatly simplified by the fact that precise corrections for nuclear spallation in the interstellar medium can be made since relatively few partial cross-sections need to be considered.

The time delay between nucleosynthesis and acceleration to cosmic ray energies, $\Delta t_{syn \to acc}$, is largely unknown. The small ratio of elemental abundances of nickel to iron has been used to argue that $\Delta t_{syn \to acc} \gtrsim$ few weeks. Abundance measurements of such isotopes as ^{56,57}Ni and ^{57}Co which decay only by electron capture would put severe constraints on $\Delta t_{syn \to acc}$, thus determining whether supernova ejecta are a significant source of galactic cosmic-rays. Also, ^{60}Fe can act as a clock for the total time-of-flight of the cosmic rays, since it is suffers beta-decays without electron capture.

A careful measurement of the abundances of ^{54}Fe, ^{55}Fe and ^{55}Mn would permit a self-consistent determination of the amount of material traversed by the cosmic rays *and* the relative abundances of hydrogen to helium in the interstellar medium (assuming the relevant partial cross-sections are known to the necessary accuracy). Radioactively decaying isotopes of manganese, such as ^{53}Mn and ^{54}Mn, can be used to determine the distribution of matter traversed by the cosmic rays.

Lastly, a comprehensive measure of iron group isotopic abundances can be combined with isotopic spectra obtained from lighter masses as a way to test various cosmic ray acceleration hypotheses.[10]

INSTRUMENTATION

The basic unit of our detector is a stack of 10 identical glass plates, each plate measuring 5 cm × 5 cm × 1.25 mm; there will be approximately 100 such stacks in our package. This modular feature enables construction of a detector of arbitrary size and lay-out. After a balloon exposure, the plates are separated and etched for several weeks in a strong acid. Particles that stop between the second and tenth sheet in a stack can be identified; this corresponds to an energy window of 260 MeV nucleon^{-1} to 370 MeV nucleon^{-1} as measured at the top of the atmosphere. We will fly $\sim$ 10 kg of glass; the total collecting power of the instrument will be $\sim$ 3000 cm^2 sr in each payload; the exposure time will be on the order of ten days. This total collecting power surpasses by a factor of twenty to forty the collecting power of any other detector. With this advantage we can study very rare isotopes of iron-group elements as well as already rare trans-iron elements.

After exposure in Antarctica, the glass sheets will be exposed to monoenergetic calibration beams of iron, lanthanum and gold, using the Bevalac at LBL. We will then have a standard reference with which we will remove the effects of any variations in glass properties or etching conditions.

The glass will be etched in 49% fluoboric acid (HBF$_4$) at elevated temperature ($\sim 50°C$) for several weeks. Several etching sessions will take place; between each session, the glass will be scanned (with the automated track measurement discussed below) so as to obtain as much information as possible from each track. Tracks will then be reconstructed and isotopic bands displayed on scatter plots of s as a function of range.

Because the distance to the end of range has to be known very accurately for each measurement, the local thickness of each piece of glass must be gauged. For the purpose of thickness mapping, we have developed an automated system which uses an electro-mechanical thickness gauge and a computer controllable x-y stage. Using this system, we can map thousands of glass detectors quickly and to an accuracy of 1μ over the entire surface.

In order to measure etchpit dimensions for the thousands of tracks resulting from accelerator or cosmic-ray exposures, we have developed two automated track measurement systems. Our second generation instrument, developed by one of us (DPS-I), is based on a SUN 3/140 computer and a Datacube image processing system, with a CCD camera, microscope and computer-driven stage. This system is capable of making complete three-dimensional measurements of etchpits, which makes possible the study of stopping particles and gives better resolution than the first system. The system is sufficiently fast that we will be able to etch and scan each sheet of glass repeatedly to glean as much information from it as possible.

We have accumulated extensive experience with VG-13 and BP-1 as experimental tools. In charge pickup studies of gold at 900 MeV nucleon^{-1} at the LBL Bevalac, we obtained a charge resolution of .15e at a single surface; this resolution was improved in the usual way by measurements at several surfaces in an exposed stack of glass. An exposure to Ne and Mg with $\beta = 0.15$ at the LBL Hilac gave a resolution of 0.06e. A measurement of the *isotopic* resolution at Fe using VG-13 gave $\sigma_A = 0.35$; we anticipate doing much better with BP-1 because of its greater sensitivity and cleaner properties. We are in the process of measuring the mass resolution of BP-1 at iron, but are being slowed by some difficulties with its etching; we are making rapid progress in finding an etching solution in which the etching rate is insensitive to reaction products.

WHY ANTARCTICA?

A very long exposure is necessary to observe very rare isotopes such as those of Ni and to achieve sufficiently high statistics for some of the more abundant isotopes to be able to correct for secondary production in the interstellar medium and in the atmosphere. Long-duration circum-global flights are possible at mid-latitudes, but they are not suitable for low-rigidity cosmic ray studies because of the high cut-off rigidity.

On this continent, only northern Canada has both the requisite high geomagnetic latitude and suitable terrain for payload recovery, but both the atmospheric circulation and political constraints make long-duration ballooning in the northern hemisphere impossible. In the southern hemisphere, the geomagnetic cut-off even as far south as New Zealand is far too high. A space exposure on a satellite or free-flying platform in a high-inclination orbit would only spend a small fraction of the time in regions of low cutoff, and such an exposure is only possible in the distant future. In contrast, the entire continent of Antarctica lies within the zone of low geomagnetic cutoff required for our experiment. In the austral summer a stable high-pressure cell sits on Antarctica, inhibiting mixing with the rest of the atmosphere, thus discouraging balloons from drifting north. The virtually complete lack of diurnal variation gives balloons a constant environment, relaxing requirements for temperature control of the payload, and enabling the balloon to stay at nearly constant altitude, thereby conserving helium which is usually lost in losing and gaining altitude due to day/night cycles at mid-latitudes. Except for the logistical difficulties in launching balloons from Antarctica, it is easily the best place in near-earth space for an iron-isotope experiment.

LIST OF REFERENCES

1. R. L. Fleischer, P. B. Price and R. M. Walker, **Nuclear Tracks in Solids: Principles and Applications** (California Press, Berkeley, 1975).

2. P. B. Price, M. L. Tincknell, G. Tarlé, S.P. Ahlen, K. Frankel and S. Perlmutter, Phys. Rev. Letters $\underline{50}$, 566 (1983).

3. G. Gerbier, W. T. Williams, P.B.Price, Ren Guoxiao and G.-R. Vanderhaeghe, Phys. Rev. Lett. $\underline{30}$, 2535 (1987).

4. J. Drach, P.B.Price, M.H. Salamon, G. Tarlé and S.P.Ahlen, Proc. 19th Int. Cosmic Ray Conf., La Jolla, $\underline{2}$, 131-135, (1985).

5. P.B. Price, H.-S. Park, J. Drach and M.H. Salamon, Nucl. Instr. and Meth. $\underline{B21}$, 60-67 (1987).

6. Shicheng Wang, S.W. Barwick, D.P. Ifft, P.B. Price, A.J. Westphal and D.E. Day, Nucl. Instr. and Meth. $\underline{B35}$, 43-49 (1988).

7. R.A. Mewaldt, in Proc. 13th Texas Symposium on Relativistic Astrophysics, ed. M.P. Ulmer (World Scientific, Singapore, 1987),p. 573

8. K.L. Hainebach, et al., Ap.J. $\underline{193}$, 157 (1974).

9. S.E. Woosley, Ap. Space Sci. $\underline{39}$, 103 (1976).

10. G.M.Raisbeck, et al., Proc. 14th Inter. Cosmic Ray Conf., Munich, $\underline{2}$, 560 (1975).

Table 1: Expected events by isotope for a thirty-day balloon flight at solar maximum with 6000 cm² sr collecting area and an energy interval of 264 MeV nucleon⁻¹ to 372 MeV nucleon⁻¹ at the top of the atmosphere; sources are: SS – Solar system abundances for stable isotopes, Hainebach's 3-zone e-process model for unstable isotopes; Sp – spallation (spallation cross-sections from ref. 16) – assumes 5 g cm⁻².

Isotope	Expected counts	Source	Significance
^{56}Fe	30 000	SS	Primary iron
^{55}Fe	1 800	SS,Sp	Decays by pure EC (2.6y). Absence would imply spallation only at the source, with delay between synthesis and acceleration $\Delta t_{syn \to acc} > 5y$
^{54}Fe	2 800	SS,Sp	If spallation cross-sections are well-measured, gives pathlength in ISM to 6% assuming solar-system composition at source and no contribution from ^{54}Mn β^- decay. Factor of 2 enhancement predicted from supermetallicity model.
^{57}Fe	720	SS	Entirely primary origin in low-η zone from ^{57}Ni.
^{58}Fe	~120	SS	Entirely primary origin in high-η zone from ^{58}Ni. An excess would be a direct probe of mixing of highly neutronized material from r-process in supernovae
^{60}Fe	?	SS	Produced at ~1% of ^{56}Fe in high-η zones. Acts as a clock with $t_{\frac{1}{2}} \sim 10^6$ yr.
^{56}Ni	?	SS	Decays by EC (6d).
^{57}Ni	?	SS	Decays by EC (36h)
^{58}Ni	?	SS	Primary nickel component; produced in low-η zones.
^{59}Ni	30-120	SS	Decays by EC (8×10^4y)
^{60}Ni	480	SS	Produced in high-η zones.
^{61}Ni	480	SS	Predicted from Hainebach's 3-zone e-process model. Solar system abundances would suggest factor of 10 less ^{61}Ni.
^{62}Ni	36	SS	Unambiguous detection of only a few events would confirm e-process predictions.
^{64}Ni	12	SS	Unambiguous detection of only a few events would confirm e-process predictions.
^{57}Co	90	SS	Decays by EC (273d)
^{59}Co	60	SS	e-process underproduces solar-system value; is this true cosmic rays?
^{53}Mn	1 000	Sp	Decays by EC (3.6×10^6y) - absence implies $\Delta t_{syn \to acc} \gtrsim 10^7$ y
^{54}Mn	720	Sp	Decays by EC (314d) - absence implies $\Delta t_{syn \to acc} \gtrsim 2y$
^{55}Mn	720	Sp	Stable, produced by spallation.

ZERO PRESSURE BALLOON BEHAVIOR in ANTARCTICA

John R. Ground

Geophysics Laboratory
Aerospace Engineering Division, Hanscom AFB, MA 01731

In January 1988 an 11,820,000 cubic foot zero pressure single cell capped polyethylene balloon, The GRAD Experiment, was launched from McMurdo Station, Antarctica. This balloon floated for three days between 110,000 and 115,000 feet MSL (Standard Pressure Altitude), before we terminated the flight about 1000 miles west of McMurdo Station. During the float time, the balloon responded to the high and low sun angles; sinking as the sun angle decreased and rising as the sun angle increased. The altitude excursions were on the order of 5000 feet between high and low sun. No ballast was dropped during the float period to maintain altitude. The GRAD float profile is shown in figure 1 below.

H88-01 GRAD Float Profile
William Field, Antarctica   January 8, 1988

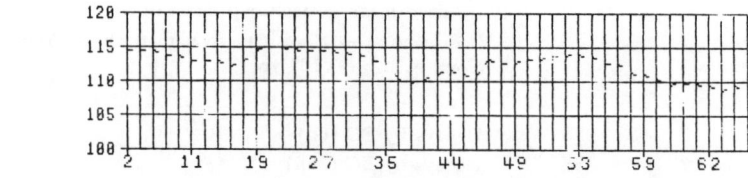

Time (hours)

Figure 1. Float profile for the GRAD flight, January 8, 1988. Altitude excursions of 5000 feet were noted between high and low sun elevations.

The following year, January 1989, we launched three 250,000 cubic foot zero pressure single cell uncapped polyethylene balloons from McMurdo Station, Antarctica. These balloons reached their design float altitude in two to three hours, then immediately began a slow continuous descent. The first two balloons had 20 percent free lift and each was rigged with one 10 percent ballast drop which was activated by an aneroid at 105,000 feet MSL. At each drop the balloons rose rather rapidly to their new float altitude and then again began the slow descent. The third balloon had 15 percent free lift and was rigged with two 5 percent ballast drops set to be dropped by aneroid at 105,000 and 100,000 feet MSL. At each of these drops the balloon rose to its new float altitude and again began the slow descent. During the descent there appears to be a response to the high sun, in that the descent rate slows slightly. These float profiles are shown in figures 2, 3, and 4.

H89-01 Pathfinder Float Profile (ARGOS 2292)
Williams Field Antarctica January 12, 1989 (0414Z)

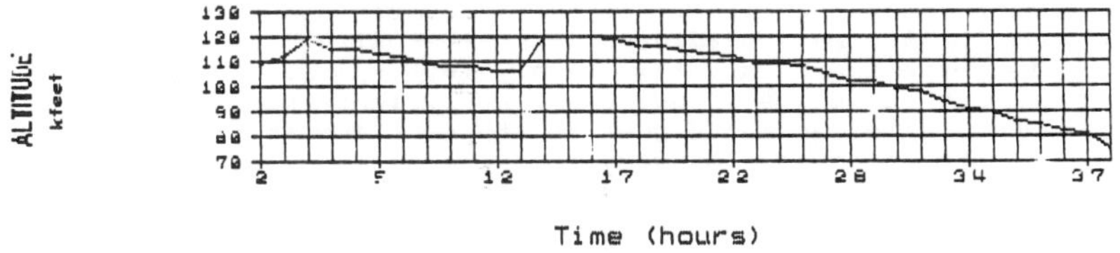

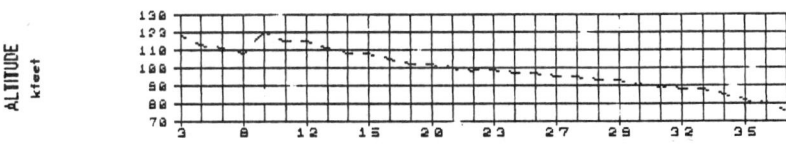

Figure 2. Pathfinder float profile for January 12, 1989. One 10% ballast drop. 20% free lift.

H89-04 Pathfinder float profile (ARGOS 2291)
Williams Field Antarctica January 19, 1989 (2207Z)

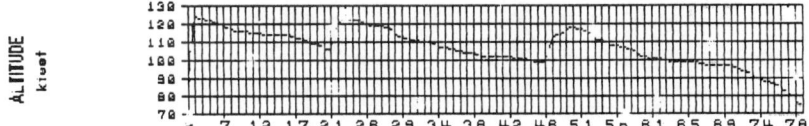

Figure 3. Pathfinder float profile for January 19, 1989. One 10% ballast drop 20% free lift.

H89-05 Pathfinder float profile (ARGOS 1158)
Williams Field, Antarctica January 21, 1989 (0355Z)

Figure 4. Pathfinder flight profile for January 21, 1989. Two 5% ballast drops. 15% free lift.

The source of the altitude data is from an onboard, Computer Instruments Corporation, pressure transducer. Altitude data above 75,000 feet pressure altitude were acquired real time at McMurdo through the ARGOS system. An extremely fine data mesh is not available, but sufficient data are available to provide a useful altitude curve.

The decay of the float altitudes of the small balloons came as a surprise as we had expected a float profile similar to the GRAD flight. Also, these 250,000 cubic foot balloons had exhibited good short term flight characteristics at 30 degrees latitude. After this experience the flights conducted by Bering at Amundsen-Scott Station in 1986 were examined for similar flight characteristics. In this experiment balloons of a similar size, 180,000 cubic feet, were used. All but the last flight had a float decay signature, similar but slower to those flown from McMurdo. The last flight which had a very low free lift floated steady for almost 100 hours. Interestingly, these flights were flown from the south pole and should not have been appreciably affected by the changing sun angle, yet their float profiles decayed similar to those of lower latitudes.

Modeling of the GRAD and Pathfinder flight profiles by George Conrad, Physical Science Laboratory, New Mexico State University produced interesting results. Good agreement was obtained between the model and the actual flight profile for the 11,820,000 cubic foot GRAD balloon. However, the model predicted a more stable float profile for the smaller 250,000 cubic foot balloons. The inability of the model to correctly simulate their profile is not understood at this time. The modeled GRAD flight is shown as figure 5.

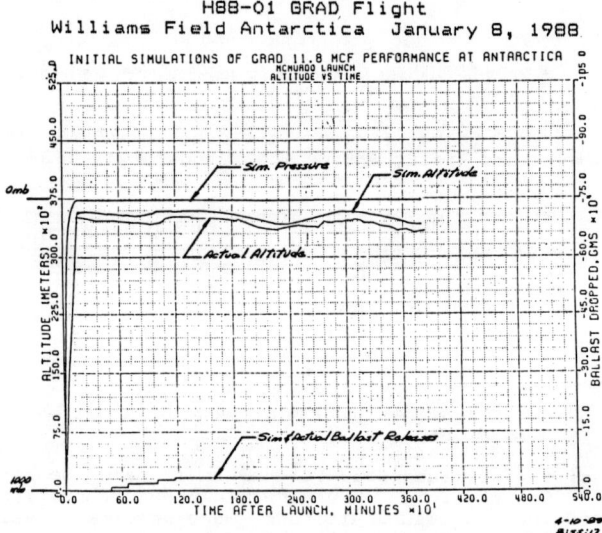

Figure 5. Modeled GRAD flight profile. The predicted and real altitudes track nicely. Conrad

An interesting aside to these flights was that the second flight failed to terminate at 50,000 feet as programmed. The balloons had an aneroid set to terminate the flight at 50,000 feet. When the balloon reached this altitude a relay should close and fire the termination device. This relay closure was monitored via a telemetry link through the ARGOS system. We know that the relay closed and did not reopen indicating the balloon stayed below 50,000 feet for the next 17 days. During this period the ARGOS positions showed the balloon eject from a low pressure vortex over the Ross Ice Shelf, cross the Antarctic continent - almost directly over Amundsen-Scott Station - and enter and eject from a low pressure vortex in the vicinity of Berkner Island. The time to cross the continent was about five days. The signal was lost at the end of the nineteenth day.

These flights have raised questions about using small floating balloons in the Antarctic and prompted a reexamination of their past use in other areas. This research has shown that the gradual decay of the float altitude appears to be a characteristic of small lightly loaded balloons in general. The decay signature bears marked similarities with varying sizes, latitudes, and times of day. There is some indication that free lift plays a role in that a lower free lift may produce increased float stability. At what point a balloon ceases to be small or lightly loaded is not clear. Caution must be used when trying to equate the behavior patterns of these small balloons to large balloons such as used on the GRAD flight. New studies are being planned to gain additional knowledge about this puzzling phenomenon.

REFERENCES

Bering, E. A. III, *The 1985-1986 South Pole balloon campaign,* Antarctica Journal of the United States, 1986 Review, Volume XXXI-No. 5

Byrne G. J., *Summertime stratospheric wind measurements above the south pole,* Paper November 1, 1987. Physics Department, University of Houston, Houston Texas.

Conrad, G., Private correspondence. Physical Science Laboratory. New Mexico State University, NASA Wallops, Wallops Island, VA.

Bering E. A. III, Private correspondence. Physics Department, University of Houston, Houston, Texas.

ROBFIT: A GENERAL PURPOSE SPECTRAL ANALYSIS PACKAGE

G. J. Bamford, R. L. Coldwell, A. C. Rester
Institute for Astrophysics and Planetary Exploration
University of Florida, Gainesville, FL 32611

ABSTRACT

We show how the spectral analysis code ROBFIT has been used in analyzing gamma-rays output from Supernova 1987A. The observations were performed using the GRAD detector which was on-board a balloon launched from McMurdo Station, Antarctica on January 8 1986. A spectral analysis code, ROBFIT, capable of distinguishing peaks at the noise level was used to analyze the data. ROBFIT is a general purpose spectral analysis package which has been developed at the University of Florida.

ROBFIT represents gamma-ray peak shapes and background continuum as combinations of spline functions. The background is fitted over the entire spectrum using a robust fitting technique. This removes the need to independently fit small regions of a spectrum which can lead to mis-representation of the background under certain conditions. The code has the ability to model any peak shape as a combination of spline functions. Both background and peaks are adjusted simultaneously until χ^2 is minimized. Then the background can be modified and new peaks searched for and a new minimum found. This iterative procedure is carried out until a user supplied limit is reached. ROBFIT then outputs peak positions, widths strengths and their associated errors.

INTRODUCTION

This paper describes the computerized spectral analysis package ROBFIT. ROBFIT has been developed to aid the analysis of spectra containing numerous peaks, many overlapping, all of which reside on a complex background. ROBFIT runs on a number of computer systems, to date we have codes that operate on IBM and Macintosh II personal computers and VAX and IBM mainframe computers. Here we describe the operation of the code and show how it can be used to study complex spectra.

ROBFIT has been used in the analysis of gamma-rays output from Supernova 1987A[1]. The observations were performed using the GRAD detector[2], which was on-board a balloon launched from McMurdo Station, Antarctica on January 8 1986. Analysis of this data has been made harder by the failure of a power supply in the veto shield. The subsequent limited coverage required an analysis method capable of distinguishing peaks at the noise level. ROBFIT was chosen as the analysis routine because of its ability to accommodate many peak shapes and to deal in an accurate manner with compex backgrounds and overlapping peaks.

In its operation ROBFIT separates spectra into two functions, background and foreground. The background contains slowly varying features of the spectra while the foreground contains the high frequency content. Accurate separation of these functions is what allows the code to detect small peaks on the background and decompose multiplet peak structures. ROBFIT iterates on background and foreground fitting to successively move smaller peaks from the background to the foreground. An earlier version of the code has been described previously[3], significant changes have since been made to the code.

BACKGROUND FITTING

The background is fitted over the entire spectrum as a set of cubic splines with adjustable knots. The user specifies the number of knots to add and can thus "stiffen" or "slacken" the background fit as required. Fitting over the whole spectral range allows the background features to be continuously fitted with fewer constants. This results in a more accurate representation than is possible when fitted in short sections along with the peaks. Two algorithms make this possible. The first, data compression, uses a robust averaging technique to reduce the peak contribution to the background. This involves averaging over a sixteen channel region and reducing the contribution of any high lying points. The second, a minimization algorithm SMSQ minimizes χ^2 with respect to the constants of the background and foreground. Having represented the background as a smoothly varying function, peaks can be identified as regions of the spectra which lie above this background curve.

FOREGROUND FITTING

The foreground contains the peak content of the spectra and consists of a small number of peak types of various shapes, each of which may be repeated many times. These shapes, or standards, are each fitted as a collection of back-to-back cubic splines. ROBFIT allows the user to create a standard from the raw data, thus accommodating complex peak types. The standards are then, in turn, fitted to the spectrum to determine where peaks of that shape reside. Clumps of overlapping peaks are fitted simultaneously. A fast residual locater for finding peak regions, a routine for systematically controlling the allowed peak widths and SMSQ, mentioned above, are the principle algorithms used in peak determination.

CONTROLLING ROBFIT

To control ROBFIT the user inputs the number of constants to be fitted to the background and selects the peak shapes to be found within the data. Additionally the level at which the code will stop looking for peaks needs to be set. This determines how large a peak must be with respect to the background fluctuations, before it is picked up by the code. In practice this level is slowly reduced so that successively smaller peaks can be found. During this cycling the user can monitor the fitted background function and decide upon the optimum number of background constants. The peak standards can also be changed, new ones added or old ones subtracted until the optimum set of standards have been selected.

USING THE CODE ON A REAL SPECTRUM

The following is a brief explanation of how ROBFIT has been used on a real spectrum. The figure below shows a region of the gamma-ray spectrum of Supernova 1987A taken by the GRAD experiment. Analysis of this data by ROBFIT[1] has been performed and all peaks have been identified.

We wish to point out the power of ROBFIT in analyzing this portion of the spectrum. Though the figure only shows a small section of the spectrum the background fit shown has been determined using the whole 4000 channels of data. This provides an extremely accurate representation of the background over the

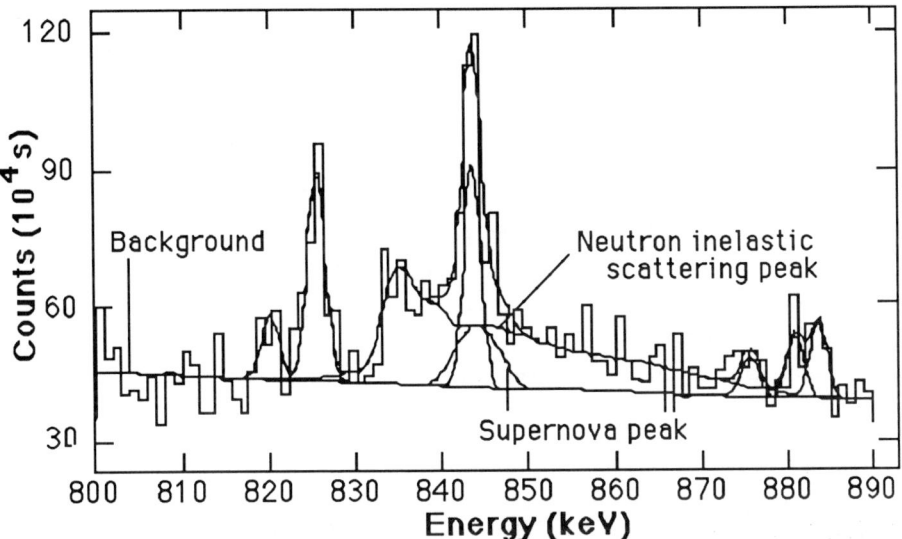

region shown. We can now search for small peaks and deconvolve multiple peaks with greater accuracy. The second feature shown is the codes ability to represent complex peak shapes. Within the region shown a neutron inelastic scattering peak has been found. This is as expected, but required the generation of an odd shaped standard. The standard was created using a neutron peak in a cleaner region of the spectrum. ROBFIT's ability to remove these complex shapes and accurately determine the background means that an estimate of the Supernova signal could be made.

CONCLUSIONS

We have described the operation of our spectral analysis code ROBFIT. Though the code has been mainly used in analyzing gamma-ray data we hope other areas of spectral analysis will find uses for the code. Its ability to represent complex shaped peaks and accurately determine complicated backgrounds makes the code a valuable tool in these fields.

REFERENCES

1. A. C. Rester et al, Ap. J. (Letters) in print.
2. A. C. Rester et al, IEEE Trans. on Nucl. Science Vol. 33 732(1986).
3. R. L. Coldwell, Nucl. Instr. and Methods A242, 455 (1986).

JACEE Long Duration Balloon Flights

The JACEE Collaboration:
T. Burnett[i], S. Dake[b], J. Derrickson[h], W. Fountain[h], M. Fuki[d], J. Gregory[f],
T. Hayashi[f], R. Holynski[j], J. Iwai[i], W. V. Jones[e,g], A. Jurak[j], J. J. Lord[i],
O. Miyamura[c], H. Oda[b], T. Ogata[a], T. A. Parnell[h], E. Roberts[h],
S. Strausz[i], T. Tabuki[h], Y. Takahashi[f], Y. Tominaga[g],
J. W. Watts[h], J. P. Wefel[g], B. Wilczynska[j], H. Wilczynski[j],
R. J. Wilkes[i], W. Wolter[j], B. Wosiek[j]

[a]ICR, Tokyo; [b]Kobe Univ.; [c]Osaka Univ.; [d]Matsusho Gakuen Junior College, Matsumoto; [e]NASA, Washington, DC; [f]Univ. of Alabama at Huntsville [g]Louisiana State Univ.; [h]NASA / Marshall Space Flt. Ctr.; [i]Univ. of Washington; [j]INP, Krakow.

ABSTRACT: JACEE balloon-borne emulsion chamber detectors are used to observe the spectra and interactions of cosmic ray protons and nuclei in the energy range 1-100A TeV. Experience with long duration mid-latitude balloon flights and characteristics of the detector system that make it ideal for planned Antarctic balloon flights are discussed.

I. INTRODUCTION

Experience with JACEE[1] clearly shows that passive, large-acceptance emulsion chambers represent a detector system that can reliably achieve large exposure factors via balloon flight. The key to the success of this project has been the simplicity and self-contained nature of the detectors, allowing experimenters to take advantage of flight opportunities denied to heavier, more complex, or more costly detector systems. In particular, the relatively low pre-analysis investment in the detectors allows us to take flight success risks unacceptable to other groups.

Recent improvements in long duration flight capability as evidenced by several successful transcontinental flights greatly increase the collecting power available to balloon experiments. For example, the JACEE-7 balloon flight from Australia to Paraguay in January, 1987 logged over 140 hours at float altitude, and this performance was repeated the following year with a successful 120-hour flight from Australia to Brazil. Altitude profiles for these flights are shown in Figure 1. Following the successful long-duration flight and recovery of the GRAD experiment[2], planning is underway for expeditions to Antarctica where flight durations of 10-20 days will be possible[3].

II. EXPERIMENTAL TECHNIQUES

A typical emulsion chamber (EC), shown in Figure 2, consists of double-coated nuclear emulsion plates, X-ray films, and Pb sheets 0.5-2.5 mm thick. Each chamber unit, measuring approximately 40x50x20 cm and with mass ~100 kg, is sealed in an airtight, waterproof bag, to eliminate plate motion during pressure changes and for environmental protection following landing. Four such units can be accomodated on a gondola of total weight 1500 lbs, well within the scientific payload limit for a standard 28 MCF balloon[3].

The EC satisfies several basic requirements: (1) large geometrical factor, (2) accurate charge determination, with charge resolution essentially independent of energy, (3) reliable energy measurement, with energy resolution independent of (and to some extent, improving with) energy, and (4) simplicity and reliability. The latter factor has been especially important in allowing long cumulative exposure times via balloon flight. Since the emulsion chamber is primarily used as a vertex detector (although external atmospheric interactions can also be observed), the initiating particle is identified.

The detectors were assembled in flight-ready condition in the US, and then shipped by commercial air freight to Alice Springs, Australia, where they were interfaced to the NSBF flight equipment. Local liaison, logistics, and communications in Alice Springs were quite satisfactory. Although preflight preparation facilities were inadequately air conditioned (with outside ambient temperatures approaching 40°C), we encountered no significant problems.

All JACEE data were recorded onboard, including temperature and pressure altitude readings, using a data recorder employing high-density memory chips. An aneroid control system was used to shift a topside film layer out of registration with the main emulsion chambers when pressure altitude dropped below a preset level; shifter status was included in the data transmitted to NSBF by the ARGOS satellite tracking system. On these flights, 28 MCF balloons of the same type proposed for planned Antarctic flights were used. Total ballast carried was approximately equal to the 600 kg scientific payload, in order to maintain altitude above approximately 6 mbar for 5-6 nights. Due to weight and cost limitations, it was not possible to provide adequately for sea recovery in case of balloon failure. In the case of JACEE-7, unanticipated sudden altitude loss resulted in automatic flight termination over northern Paraguay. Despite logistical and diplomatic difficulties, the payload was returned to the US after several weeks' storage at local ambient temperatures. Although background fog levels in the emulsions and x-ray films were higher than usual, the payload is fully analyzable, and data reduction is well under way. For the JACEE-8 flight, termination and recovery proceeded as planned in southern Brazil. Despite advance preparation and energetic local liaison, the return of the apparatus was delayed several weeks due to bureaucratic delays, although in this case the emulsion chambers were stored in a controlled environment with no tangible degradation of data.

The Southern hemisphere balloon flight expeditions of 1987-88 approximately doubled the statistical weight of the JACEE data on primary composition, and we have reached the point where additonal factors of 2~10 in exposure will provide significant information on composition and spectra in the critical 10^{15} ev knee region[4].

III. SUITABILITY FOR ANTARCTIC FLIGHTS

JACEE detectors are in many ways very well matched to the requirements for Antarctic flights. Antarctica presents no population hazard to limit flight paths and durations, as is the case for northern hemisphere routes. Since Antarctica is an international zone, there are no national boundaries to cross, and the equipment remains under the control of US agencies throughout flight and recovery. While some diurnal variation of altitude will occur, ballasting requirements are greatly reduced by the absence of sunset, and the weight available for scientific payload is correspondingly increased.

The flight path is overland at all times, and it is in principle possible to recover the package from a much larger fraction of the underlying terrain than in the case of mid-latitude flights, where the package is over open ocean most of the time. JACEE payloads are modular, with a basic unit of dimensions roughly 40x50x20 cm and mass 100 kg. Thus the gondola can be broken down into components that are easily handled by a recovery crew working at the high altitudes which characterize the East Antarctic plateau, and essential components of limited weight can be removed if it is necessary to abandon the main gondola. In case of delayed recovery, the ambient environment in Antarctica (characterized by extremely low temperature and humidity) is much less hostile to photographic media than in Australia or South America, where high temperatures promote latent image fading and background fog. Although southern mid-latitude flights have the advantage of relatively high geomagnetic cutoff (hence reduced soft background in the emulsion and reduced fog level in the x-ray films), emulsion chamber detectors are inherently self-shielding, and to the extent that background levels will be higher than previous flights, they will provide us with an opportunity to confirm our ability to analyze plates with enhanced background in preparation for planned Space Station exposures[5].

Recovery is essential for emulsion chamber flights, but this also means that the package has negligible telemetry requirements, reducing the burden on flight operations facilities. For example, the typical JACEE detector is an ideal "guinea pig" for flight systems tests, because it is self contained and passive, can be prepared on relatively short notice, and loss of a given package can be tolerated since the main investment of effort by experimenters occurs following successful recovery.

REFERENCES
1. W. V. Jones, et al, Ann. Rev. Nucl. Part. Sci. 37:71 (1987); T. H. Burnett, et al, Nucl. Instr. Meth. A251:583 (1986).
2. C. Rester, this conference; J. Grounds *et al*, Air Force Geophysical Laboratory Report AFGL-TR-88-0265 (1988).
3. W. V. Jones, this conference.
4. T. H. Burnett, *et al, Particle Astrophysics Workshop,* (Berkeley, 12/88), AIP Conference Proceedings (to be published 1989).
5. T. Parnell, *et al, SCINATT Proposal,* (unpublished) NASA/MSFC, 1988.

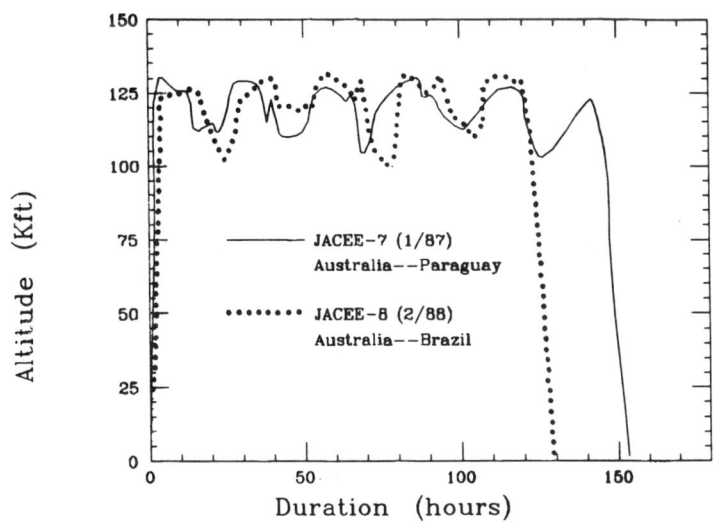

Fig. 1. Altitude profiles for JACEE-7 and -8, from ARGOS satellite data.

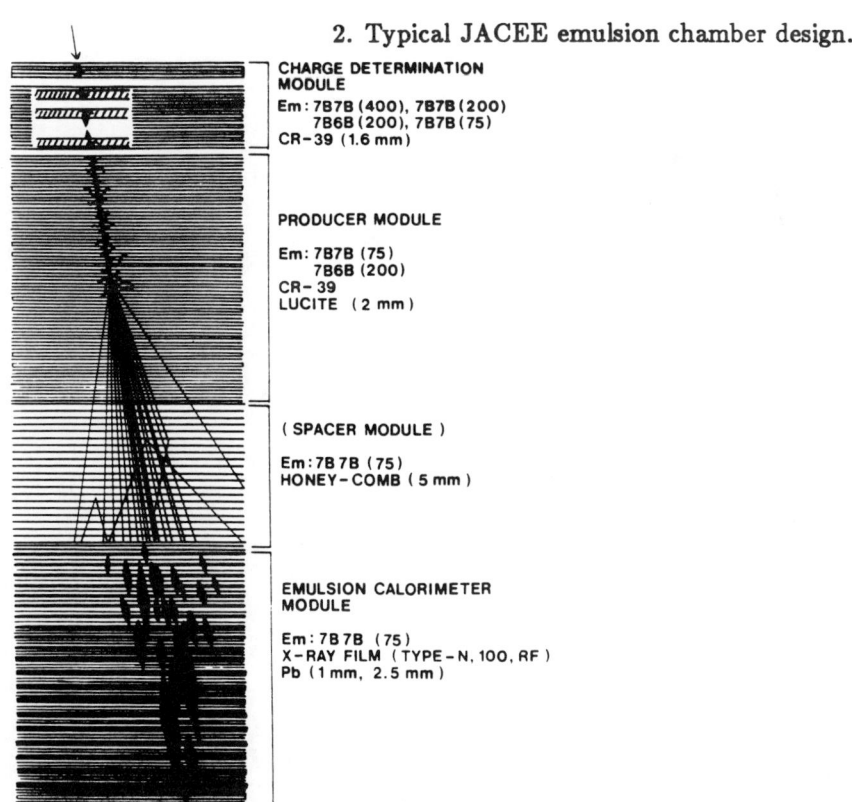

2. Typical JACEE emulsion chamber design.

Solar and Stellar Astronomy

SOLAR AND STELLAR OBSERVATIONS FROM THE SOUTH POLE

Jeffrey L. Linsky[*]
Joint Institute for Laboratory Astrophysics
National Institute of Standards and Technology
and University of Colorado, Boulder, CO 80309-0440

ABSTRACT

An astronomical observatory located at the geographic South Pole could provide important new insights into the physical bases of stellar variability by monitoring stars for long periods of time with minimal interruptions by the day-night cycle. I summarize here three broad topics that could be studied with monitoring techniques -- magnetic phenomena on stars, helioseismology, and asteroseismology.

INTRODUCTION

Like most astronomers, I have never been to the South Pole nor do I have any direct experience with research programs underway in Antarctica. Since I can bring no direct experience to this meeting, perhaps I may offer some fresh perspective based on my experience of studying solar-like active phenomena on stars using high-resolution spectroscopy in the optical and ultraviolet, in conjunction with radio and x-ray observations.

Establishing an astronomical observatory at the South Pole, or elsewhere on the continent, should not be attempted unless there are unique advantages to the site, given the harsh environment, communication difficulties, and logistical support problems. On the other hand, economic considerations for obtaining observations from the South Pole are compelling when the only alternative site is in space.

I see two major advantages of the South Pole as an astronomical observing site. One is to exploit the extended atmospheric windows through which ground-based observers study cosmic sources. The very low water vapor column density above this high and very dry site opens useful windows for infrared, submillimeter, and millimeter wave astronomy as described by other speakers at this meeting. In addition, the unwanted ozone hole may extend the usable near-ultraviolet region; for example, the Be II resonance lines[1] at 3130.4 and 3131.1 Å are useful for studying the abundance of this light element which was formed in the Big Bang and subsequently in the cores of stars. Also, there are many Fe II lines in the 3100-3300 Å region that appear in emission in cool supergiants[2] and may be used to study the chromospheres and winds of these interesting stars. Observations of a few of these lines are barely feasible from Mauna Kea, but observations of many additional lines may become feasible from the South Pole when the ozone column density is low. I am not advocating the destruction of atmospheric ozone for the benefit of astronomical research, but near-ultraviolet observations from the South Pole may become a growth area.

A second advantage of the South Pole site lies in the time domain. Solar and stellar astronomers now recognize the importance of monitoring; that is, observing targets for long periods of time with minimal interruptions in order to understand the physical basis for their variability. In particular, those phenomena with time scales close to one terrestrial day cannot be studied from an observatory at moderate latitudes, and phenomena with time scales of say one-third to three days cannot be studied well from low latitudes because of time domain aliasing. I will emphasize this point later, but astronomers, being very clever people, are beginning to circumvent even the day-night cycle by establishing networks of telescopes around the world to observe the same phenomena (or star) continuously for days at a time. A premier example is the GONG network to study solar oscillations from 6 sites, but stellar observers are establishing a global network of Automatic Photometric Telescopes[3] and they are beginning to plan for a similar network of spectroscopic telescopes. Thus the particular merits of the South Pole as the site for an astronomical monitoring observatory must be compared with the merits and costs of a network of telescopes at low-latitudes that are properly distributed in longitude.

MAGNETIC PHENOMENA ON STARS: THE SOLAR-STELLAR CONNECTION

The fields of solar physics and the study of solar-like phenomena on stars, often called the solar-stellar connection, have been particularly active in the last few years with a number of new discoveries and increased understanding of previous data. Two major topics of particular relevance to potential observations from the South Pole are magnetic phenomena on stars and solar/stellar oscillations. I summarize in Table 1 the different magnetic phenomena already detected in different types of stars. In this table stars are divided into those with radiative envelopes (spectral types O, B, and A) or with convective envelopes (spectral types F, G, K, and M), because it is generally assumed that convective envelopes are required for the dynamo amplification of magnetic fields. The presence of magnetic fields in stars with radiative envelopes thus indicates that the fields are either remnant or that the fields can be amplified by processes other than conventional dynamos.

In the category of pre-main sequence stars I include the T Tauri stars (TTS), which exhibit infrared and ultraviolet excesses indicative of circumstellar disks, and the so-called naked T Tauri stars (NTTS), which show little or no evidence of disks but otherwise appear identical to the TTS. The main sequence (MS) stars include those stars in clusters like the Hyades and Pleiades which have only recently reached the main sequence. The post-main sequence stars include the late-type giants and supergiants, as

Table 1. MAGNETIC PHENOMENA ON STARS

	RADIATIVE ENVELOPES	CONVECTIVE ENVELOPES
PRE-MS		Magnetic fields on NTTS Starspots on NTTS, TTS Gyrosynchrotron emission on TTS Coronae on NTTS (sat), TTS (acc) Chromospheres on NTTS (sat), TTS (acc) Flares on NTTS, TTS
MS	Magnetic fields on Bp, Ap Gyrosyn. emission on Bp, Ap, O Coronae on Bp, Ap, O	Magnetic fields on F8-M5 V Starspots on young G-M, dMe Gyrosynchrotron emission on dMe Coronae on A7-M8, dMe (sat) Chromospheres on A7-M8, dMe Flares on dMe
POST-MS		Starspots on RS CVn, Algols, W UMa Gyrosyn. emission on RS CVn, Algols Coronae on RS CVn, Algols, W UMa MS Chromo. on RS CVn, Algols, W UMa Flares on RS CVn, Algols, W UMa, M supergiants (?)

well as evolved stars in tidally-synchronous binary systems like the RS Canum Venaticorum (RS CVn), Algol, and W Ursae Majoris (W UMa) systems. Magnetic fields have been directly measured in stars using both the Robinson line-broadening technique for late-type stars[4] and polarization techniques for the early-type stars.[5] Indirect indicators of magnetic fields include dark photospheric structures (starspots), gyrosynchrotron radio emission, ultraviolet emission lines formed in stellar chromospheres, and x-ray emission formed in stellar coronae. By analogy with the Sun, bright ultraviolet and x-ray emission are likely formed in active regions consisting of closed magnetic loop structures. Finally, rapid increases in the optical, uv, x-ray and radio emission during flares probably indicate the rapid conversion of magnetic energy into heat and relativistic electrons.

Table 1 provides an overview of a major field of research, and I direct interested readers to important recent reviews on magnetic fields,[6] starspots,[7] gyrosynchrotron radio emission,[8] coronae,[9] chromospheres,[10] and flares.[11] Those stars for which the x-ray or ultraviolet emission lies at the empirically largest values for the convective energy available in the subphotosphere[12] are indicated as saturated (sat), while those stars for which the accretion of material from the disk may be the energy source for the observed bright emission are indicated by (acc). The different types of binary systems are distinguished by the spectral types of the component stars and whether or not they are detached (both stars do not fill their Roche lobes), semidetached (one star fills its Roche lobe), or contact systems (both stars fill their Roche lobes). RS CVn systems[13] are generally detached, Algols[14] are generally semidetached with a disk around the more massive star, and W UMa systems[15] are generally contact systems.

TIME SCALES FOR MAGNETIC PHENOMENA ON STARS

Figure 1 summarizes some of the important time scales over which stellar activity is known to vary. The rotation of a star establishes the time scale with which long-lived structures on or above the stellar surface will appear and disappear from view. The stellar emission will vary in phase in a repeatable manner when the features do not vary appreciably in brightness over a rotational period, when only a few such features are located on the surface, and when these features are not circumpolar (i.e., not always visible). This type of variability, termed "rotational modulation", can be caused either by dark structures such as starspots or bright structures like active regions that are observed with high contrast especially in the ultraviolet and x-ray regions. Low brightness contrast with the quiet atmosphere and the presence of many structures well-distributed in longitude will lead to no measurable rotational modulation signal even when bright or dark structures are present on the stellar surface. Also, intrinsic variations in the brightness of an active region on time scales comparable with or shorter than the rotation period can mask the rotation modulation signal. It is essential to monitor active stars for a number of sequential rotational periods in order to disentangle spatial from temporal variations.

I include in Fig. 1 the rotational period of the Sun (25 days), typical periods for TTS (2-8 days),[16] dMe stars (0.5-5 days),[17] RS CVn systems (1-100 days),[13] W UMa systems (0.2-1.6 days),[15] and Algol systems (0.6-38 days).[14] These periods are generally derived from variations in the optical light curves due to the presence of starspots or reflection effects, or they can be derived from variations in the orbital radial velocity for synchronous binary systems. Typical values for the rotational periods of K giants (200-1000 days) and M supergiants (>1000 days) are based on estimates of the rotation velocities and measured radii. In addition, a number of well-studied low mass x-ray binaries (LMXB) have orbital and rotational periods of 0.2-2

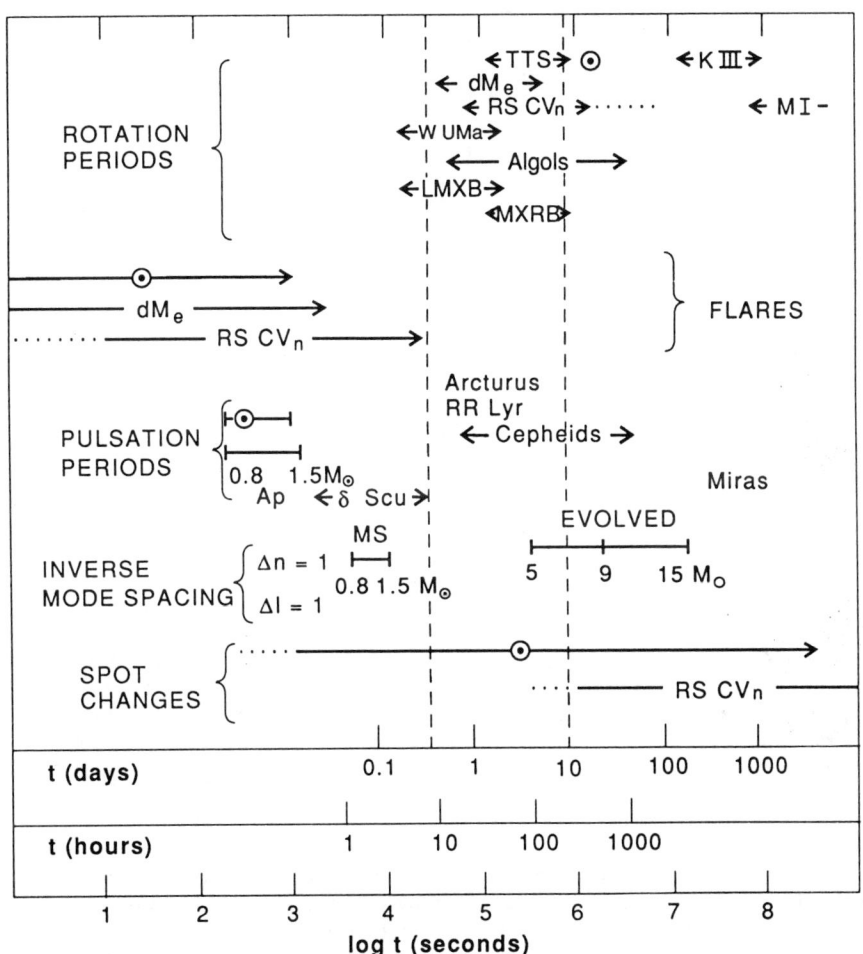

Fig. 1. Typical timescales for stellar rotation, flares, pulsations, and spot changes. See text for the sources of the data and the meanings of the symbols. The dashed vertical lines for 0.3 and 10 days indicate the timescales for which observations from the South Pole are most useful.

days,[18] and the well-studied massive x-ray binaries (MXRB) have periods in the range 1.4-9 days.[18] Thus many interesting stars and binary systems have rotational periods in the range 0.3-10 days where an observatory at the South Pole would be particularly useful for nearly continuous monitoring. Some important stars observable from the South Pole with rotational periods in this range are the dMe binary AB Dor (12.3 hours), the dMe star AU Mic (4.8 days), and the RS CVn system TY Pyx (3.2 days).

Also included in Fig. 1 are periods of different types of intrinsic variability. Flares on the Sun and dMe stars can have very short rise times for their impulsive phases (less than 0.1 seconds for the Sun) and longer decay times (as much as 2 hours for the uv and x-ray emission). The decay times for the uv and x-ray emission during flares on RS CVn systems can be roughly 6 hours, but the enhanced radio emission in major flare events can persist for up to 10 days. Sunspots and spot groups change in size and structure on a wide range of timescales including the famous 11 year cycle. Spots on dMe and RS CVn systems also change on many time scales and there is evidence for spot cycles in the range 5-100 years. Chromospheric emission from a number of solar-type dwarf stars varies cyclically both on rotational timescales and periods of 3 to >20 years that may indicate analogs[19] of the sunspot cycle.

This cursory examination of stellar activity indicates that the interesting activity phenomena that an observatory at the South Pole is particularly well-suited for monitoring include rotational modulation of surface structures on TTS, dMe, and close binary systems, and the intrinsic variability of starspots. There may also be other phenomena with periods of 0.3-10 days that are unknown because of the difficulty of their detection from low latitude sites.

PHOTOMETRIC STUDIES OF MAGNETIC PHENOMENA ON STARS

One of the simplest types of astronomical observations can be made with a telescope of less than one meter aperture and a simple photomultiplier detector. This arrangement is suitable for broadband optical photometry of stars brighter than about 15th magnitude. This type of telescope can be fully automated and remotely controlled as are the Automatic Photometric Telescopes[13] now being built at many locations. Such telescopes can readily study spots on dMe and close binary systems. Analysis of very precise optical light curves can determine the sizes, locations, and brightness contrast of up to three large spots on one star, although there is a uniqueness problem for single stars where it is difficult to determine the angle of the rotation axis with respect to the line of sight. Figure 2 demonstrates the magnitude of this uncertainty by comparing the distribution of two or three large spots that are inferred from the analysis of the same light curve of the single-lined RS CVn system II Peg[20] when the inclination of the rotation axis with respect to the line of sight is assumed to be 35, 60, or 90 degrees, respectively.

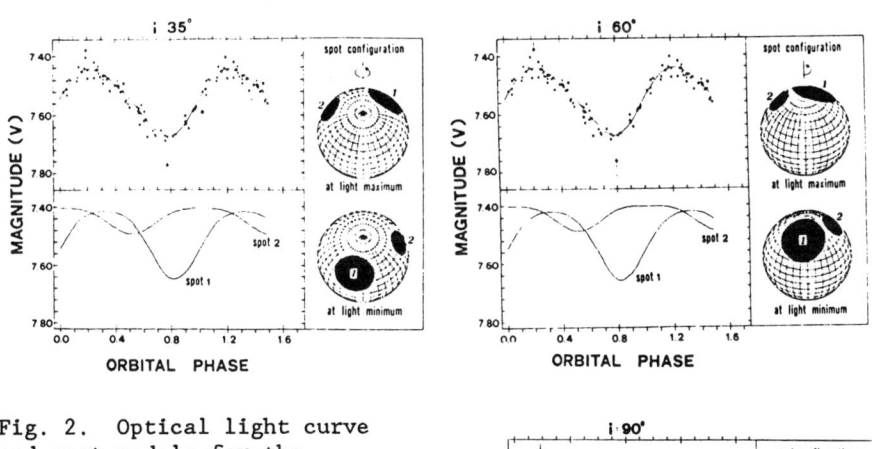

Fig. 2. Optical light curve and spot models for the single-lined RS CVn system II Peg[20] for three assumed values of the angle (i) of the rotation axis with respect to the line of sight.

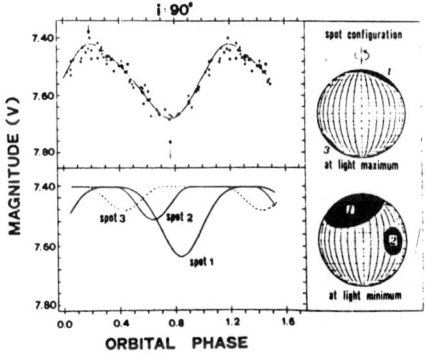

Optical light curves are typically constructed from observations obtained over many rotational periods with few observations within any given period. Figure 3 is an example of a light curve of this sort for the pre-main sequence star HDE 283572,[21] which has a rotation period of 1.548 days. Continuous optical monitoring of such stars is needed to separate rotation from possible temporal changes. An APT at the South Pole could monitor a star continuously for several days to obtain a very high precision light curve that will separate time-independent horizontal brightness contrasts across the stellar surface from temporal changes in the starspots. This experiment can be done with a telescope at the South Pole for stars with rotational periods of 3-48 hours. A similar study of the active regions in the chromospheres of such stars using a narrow-band filter centered on the Ca II resonance lines (3933 and 3968 Å) is needed to separate spatial from temporal variability. With a similar narrow band filter and a two-dimensional detector like a CCD, one can study chromospheric variability in a cluster of stars simultaneously.

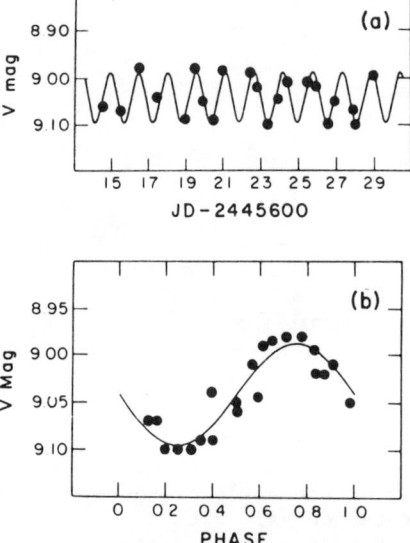

Fig. 3 Optical light curve[21] for HDE 283572 as a function of time (panel a) and phase (panel b), with a derived period of 1.548 days. Note the sparse sampling of data.

SPECTROSCOPIC STUDIES OF MAGNETIC PHENOMENA ON STARS

Spectroscopy provides a number of useful tools for studying magnetic phenomena on stars. These tools become more powerful when a star is observed systematically at many rotational phases to separate temporal from phase-dependent variations and to map the location of specific features onto the stellar surface. Magnetic field strengths and fractional filling factors are now routinely measured for late-type dwarfs from the analysis of high-resolution line profiles in unpolarized light.[6] These techniques have been used to measure the magnetic parameters of the active star ξ Boo A (G8 V) at many phases in order to construct a magnetic image of this star.[22]

A second application of monitoring spectroscopy is to obtain Doppler images of a star. One may now infer the sizes, shapes, brightness contrasts, and locations of both bright and dark features on the surface of a rapidly-rotating star by using maximum entropy algorithms to reconstruct the stellar image from spectral line profiles obtained at many phases. Figure 4 demonstrates the capability of this technique to reconstruct the image of a simulated Ap star.[23] Analogous techniques have been used to infer the locations and sizes of bright active regions on the surfaces of both stars in the RS CVn system AR Lac[24] and the dMe system AB Dor.[25] The latter system is particularly interesting because it is in the southern sky and has a rotation period of 12.3 hours. A sizable number of southern hemisphere stars could be studied spectroscopically with a telescope located at the South Pole.

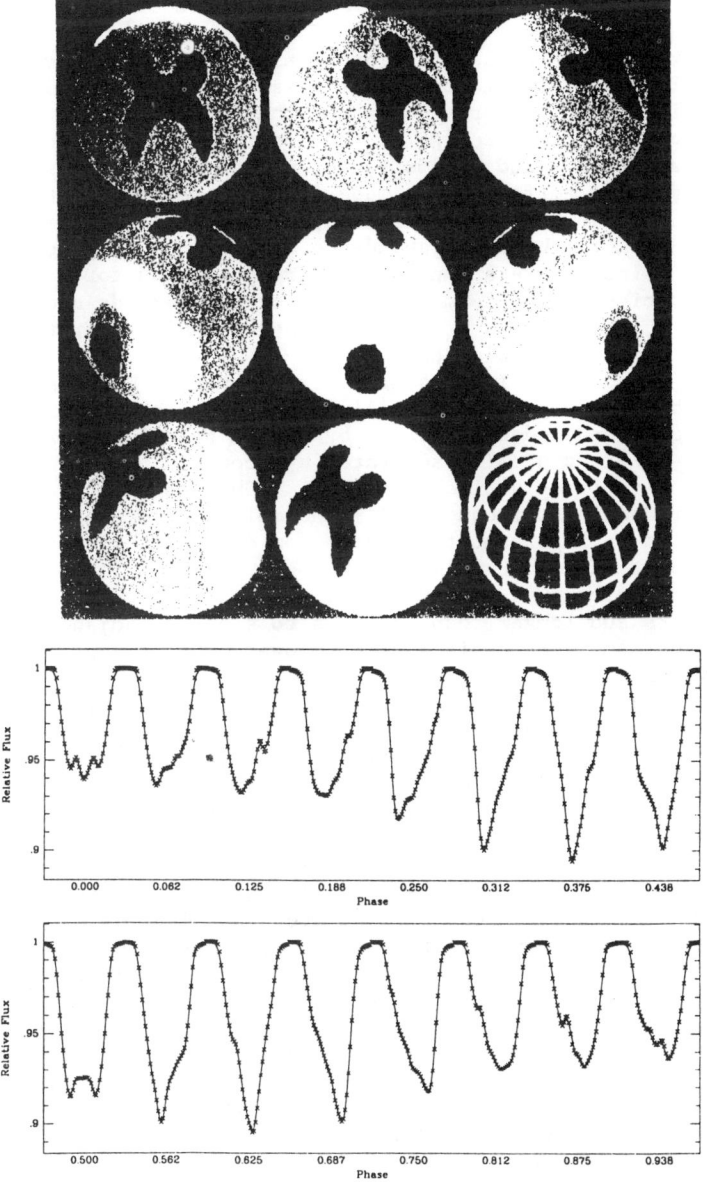

Fig. 4. An example of the Doppler imaging technique.[34] (Bottom) Absorption line profiles at 16 evenly spaced rotational phases for a simulated Ap star with a pattern of dark cross-shaped spots, a dark spot and a bright ring. (Top) The distribution of dark and bright features across the surface of the star reconstructed from the 16 absorption line profiles.

HELIOSEISMOLOGY AND ASTEROSEISMOLOGY

The study of solar oscillations (helioseismology) and the newly developing field of stellar oscillations (asteroseismology) are providing unique information about the interiors of stars, since the periods and spacings of the observed pressure modes (p-modes) measure the sound speeds (and thus the temperatures) far below the stellar surface. The mode spacings may therefore be used to measure such fundamental stellar properties as mass, radius, and central hydrogen abundance, which is a measure of evolution. Since there are no other direct methods for measuring these quantities, accurate measurements of the periods and amplitudes of these modes are of enormous value in understanding the constitution and evolution of stars.

Solar p-mode oscillations have periods near 5 minutes, but modes which differ by one in radial order ($\Delta n=1$) are separated by about 136 μHz, and modes differing by one in spherical harmonic degree ($\Delta \ell=1$) are typically separated by about 10 μHz.[26] Since observations covering a time interval T induce spurious power in the Fourier domain at frequency T^{-1} and multiples of this frequency, it is important to observe for durations longer than the inverse of the expected mode spacings. For the Sun, this requires observing runs longer than 2 hours to separate adjacent radial orders and longer than 28 hours to separate adjacent spherical harmonics to avoid spurious power or sidelobes with 10 μHz spacings (see Fig. 5).

Included in Fig. 1 are the p-mode oscillation periods for the Sun (5 minutes) and the predicted increase in period for one solar mass stars from age zero (4 minutes) to 12 billion years (15 minutes).[27] The periods for main sequence stars between 0.8 and 1.5 solar masses are predicted to increase with mass.[27,28] The predicted inverse mode spacings corresponding to $\Delta n=1$ are included for main sequence stars and evolved stars as a function of mass.[27] Also included in Fig. 1 are the observed fundamental mode pulsation periods for the typical K giant Arcturus (1.82 days),[33] and the range of observed values for RR Lyrae, Cepheid, Mira, Ap, and δ Scuti variables.[29] An observatory at the South Pole capable of observing pulsations either by variations in intensity or radial velocity should be very useful in studying the periods of RR Lyrae, Cepheids and late-type giants, as well as studying the mode spacings ($\Delta n=1$ or $\Delta \ell=1$) for both main sequence and giant stars.

High-order, low-degree p-mode oscillations of the Sun viewed as a star have been studied from the South Pole both by measuring intensity fluctuations[26] and Doppler shift variations.[30] The more difficult stellar observations are beginning to be made first from low latitude sites. For example, intensity fluctuations in the chromospheric Ca II K line indicate global p-mode oscillations in the K2 V star ϵ Eridani[31] with 172 μHz spacing. Also, there is now a report[32] of 79.4 μHz p-mode spacings for Procyon (F5 IV-V) and 165.4 μHz spacing for α Centauri A (G2 V). The extension of this work to evolved stars and to the measurement of spherical

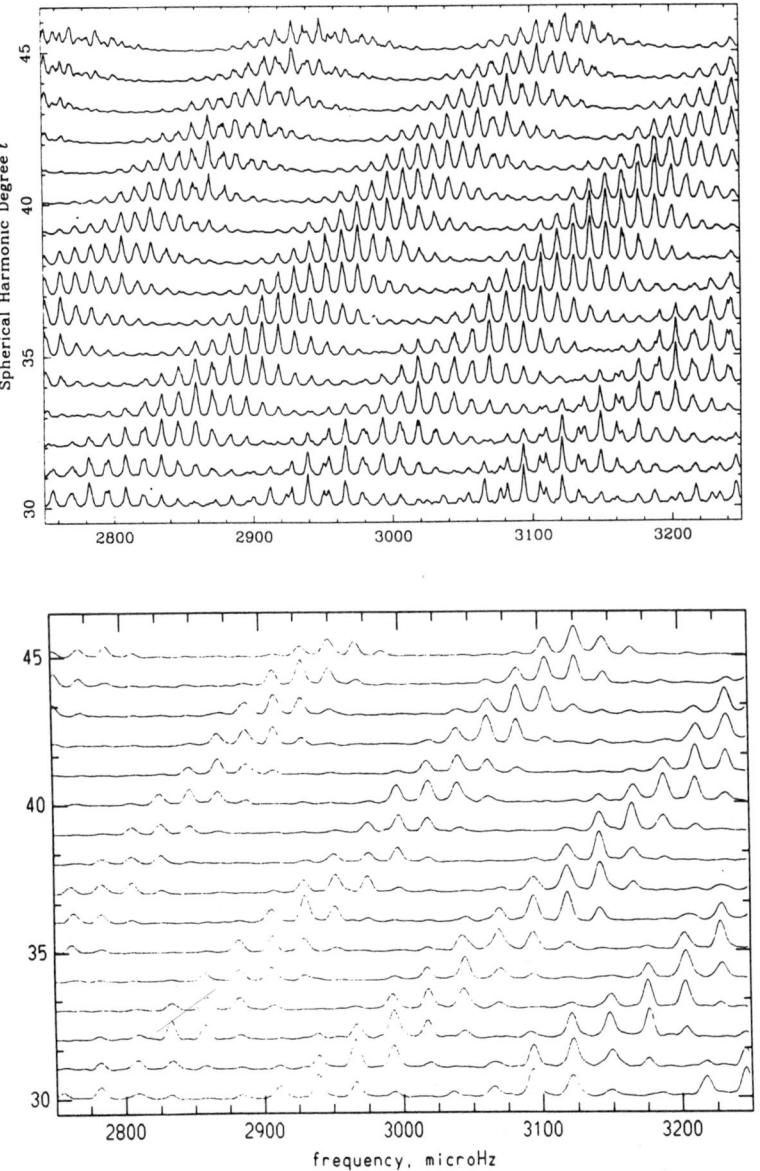

Fig. 5. (Top) Segments of averaged power spectra[26] obtained from a 12 day series of solar Doppler shift observations from Big Bear Solar Observatory with night-time interruptions. (Bottom) A continuous 2 day set of intensity measurements from the South Pole. Note that the 12 day Big Bear data set shows narrower peaks than the 2 day South Pole data set, and the South Pole data set is free from the temporal sidelobes contaminating the Big Bear data.

harmonic mode spacing ($\Delta \ell=1$) for both dwarfs and giants should be feasible from an observatory at the South Pole. Bright stars that can be studied from the South Pole include α Cen A (G2 V) and B (K1 V), Canopus (F0 I), β Hyi (G2 IV), and ϵ Indi (K4 V).

This work was supported by NASA grants NGL 06-003-057 to the University of Colorado and W-15130 to the National Institute for Standards and Technology. I would like to thank Dr. A. Brown for his comments on the text, and Drs. J. Harvey and R. Gilliland for their suggestions. I also thank the Bartol Institute for its hospitality and a most interesting meeting.

REFERENCES

1. A.M. Boesgaard, W.D. Heacox, and P.S. Conti, Ap. J. 214, 124 (1977).
2. A.M. Boesgaard and H. Boesgaard, Ap. J. 205, 448 (1976).
3. R.M. Genet, L.J. Boyd, and D.S. Hall, in Cool Stars, Stellar Systems, and the Sun, ed. J.L. Linsky and R.E. Stencel (Springer-Verlag, Berlin, 1987), p. 473.
4. R.D. Robinson, Ap. J. 239 (1980).
5. E.F. Borra, J.D. Landstreet, and I. Thompson, Ap. J. Supp. 53, 151 (1983).
6. S.H. Saar, in Cool Stars, Stellar Systems, and the Sun, ed. J.L. Linsky and R.E. Stencel (Springer-Verlag, Berlin, 1987), p. 10.
7. M. Rodono, in Highlights of Astronomy, ed. J.P. Swings, p. 429.
8. G.A. Dulk, Ann. Rev. Astron. Astrophys. 23, 169 (1985).
9. J.L. Linsky, in From Einstein to AXAF, to appear.
10. J.L. Linsky, Ann. Rev. Astron. Astrophys. 18, 439 (1980).
11. P.B. Byrne, Solar Physics, to appear.
12. O. Vilhu, in Cool Stars, Stellar Systems, and the Sun, ed. J.L. Linsky and R.E. Stencel (Springer-Verlag, Berlin, 1987), p. 110.
13. K.S. Strassmeier, D.S. Hall, M. Zeilik, E. Nelson, Z. Eker, and F.C. Fekel, Astron. Astrophys. Suppl. 72, 291 (1988).
14. G. Giuricin, F. Mardirossian, and M. Mezzetti, Ap. J. Suppl. 52, 35 (1983).
15. S.W. Mochnacki, Ap. J. 245, 650 (1981).
16. J. Bouvier and C. Bertout, Astron. Astrophys. 211, 99 (1989).
17. B.R. Pettersen, in Activity in Red-Dwarf Stars, ed. P.B. Byrne and M. Rodono (D. Riedel, Dordrecht, 1983), p. 17.
18. F.A. Cordova and I.D. Howarth, in Scientific Accomplishments of the IUE, ed. Y. Kondo et al. (D. Reidel, Dordrecht, 1987), p. 395.
19. S.L. Baliunas et al., Ap. J. 275, 752 (1983).
20. M. Rodono et al., Astron. Astrophys. 165, 135 (1986).
21. F.W. Walter, A. Brown, J.L. Linsky, A.E. Rydgren, F. Vrba, M. Roth, L. Carrasco, P.F. Chugainov, N.I. Shakovskaya, Ap. J. 314, 297 (1987).

22. S.H. Saar, J. Huovelin, M.S. Giampapa, J.L. Linsky, and C. Jordan, in Midnight Sun Workshop on Activity in Cool Star Envelopes, ed. O. Havnes (D. Reidel: Dordrecht, 1988).
23. A.P. Hatzes, G.D. Penrod, and S.S. Vogt, Ap. J. 341, 456 (1989).
24. J.E. Neff, F.M. Walter, M. Rodono, and J.L. Linsky, Astron. Astrophys. 215, 79 (1989).
25. R.D. Robinson and A. Collier-Cameron, Proc. Astr. Soc. Australia, 6, 319 (1986).
26. T.L. Duvall, J.W. Harvey, K.G. Libbrecht, B.D. Popp, and M. Pomerantz, Ap. J. 324, 1158 (1988).
27. J. Christensen-Dalsgaard and S. Frandsen, Solar Phys. 82, 469 (1983).
28. J. Christensen-Dalsgaard, in Advances in Helio- and Asteroseismology, ed. J. Christensen-Dalsgaard and S. Frandsen (D. Riedel: Dordrecht, 1988), p. 295.
29. W. Dappen, W.A. Dziembowski, and R. Sienkiewicz, in Advances in Helio- and Asteroseismology, ed. J. Christensen-Dalsgaard and S. Frandsen (D. Riedel: Dordrecht, 1988), p. 233.
30. G. Grec, E. Fossat, and M.A. Pomerantz, Solar Phys. 82, 55 (1983).
31. R.W. Noyes, S.L. Baliunas, E. Belserene, D.K. Duncan, J. Horne, and L. Widrow, Ap. J. 285, L23 (1984).
32. B. Gelly, G. Grec, and E. Fossat, Astron. Astrophys. 164, 383 (1986).
33. P.H. Smith, R.S. McMillan, and W.J. Merline, Ap. J. 317, L79 (1987).

DRIVING OF THE SOLAR P-MODES BY RADIATIVE PUMPING IN THE UPPER PHOTOSPHERE

Juan M. Fontenla *, A. Gordon Emslie
The University of Alabama in Huntsville, Huntsville, AL 35899

Ronald L. Moore
NASA/Marshall Space Flight Center, Huntsville, AL 35812

Two possibilities have been advanced for driving the Sun's observed global p-mode oscillations. These are radiative pumping below the photosphere[1] and stochastic excitation by turbulent convection in and below the deep photosphere[2,3]. The degree to which these drivers actually power the p-modes remains uncertain[4]. Our purpose in this note is to point out that another viable driver of the p-modes is radiative pumping in the upper photosphere where the opacity is dominated by the negative hydrogen ion. This new possibility is suggested by the similar magnitudes of two energy flows that have been evaluated by independent empirical means: (1) the power to sustain all of the p-modes (found by Libbrecht[4] to be of order 10^8 erg/cm^2/s) and (2) the excess radiative heating of the upper photosphere (which, as shown in Figure 1, is also of order 10^8 erg/cm^2/s in empirical model solar atmospheres of Avrett[5]). This coincidence suggests that the p-modes are radiatively pumped in the upper photosphere and thereby provide the needed nonradiative cooling.

Further, we have found[6] that radiative pumping should occur in the upper photosphere because the radiative heating increases as the gas is compressed as in a p-mode oscillation. In Figure 2, we show that in Avrett's model solar atmospheres, the radiative relaxation time for temperature perturbations in the upper photosphere is short relative to the 3-10 minute periods of the observed p-modes. This indicates that the compressions and expansions in the p-mode oscillations in the upper photosphere take place roughly isothermally. For isothermal perturbations, we find the results shown in Figure 3: in Avrett's model solar atmospheres the perturbation in radiative heating in the upper photosphere is positive for a positive perturbation in density. This indicates that the p-modes should be radiatively pumped in the upper photosphere.

Finally, from the computed rate of increase of radiative heating with density increase (Figure 3) together with the observed amplitude of the ensemble of p-mode oscillations in the upper photosphere, we have estimated[6] the magnitude of the radiative pumping and have found it to be of the order needed to sustain the p-modes. The magnitudes of the net radiative heating density q_R and the net radiative heating energy flux F_R absorbed by the radiative pumping are given by multiplying $dq_R/dln\rho$ and $dF_R/dln\rho$ by the magnitude of the oscillations in $ln\rho$, $\Delta\rho/\rho$, due to the p-modes in the pumping layer. From the observed velocity

---

* Also Member of the Carrera del Investigador, CONICET, Argentina.

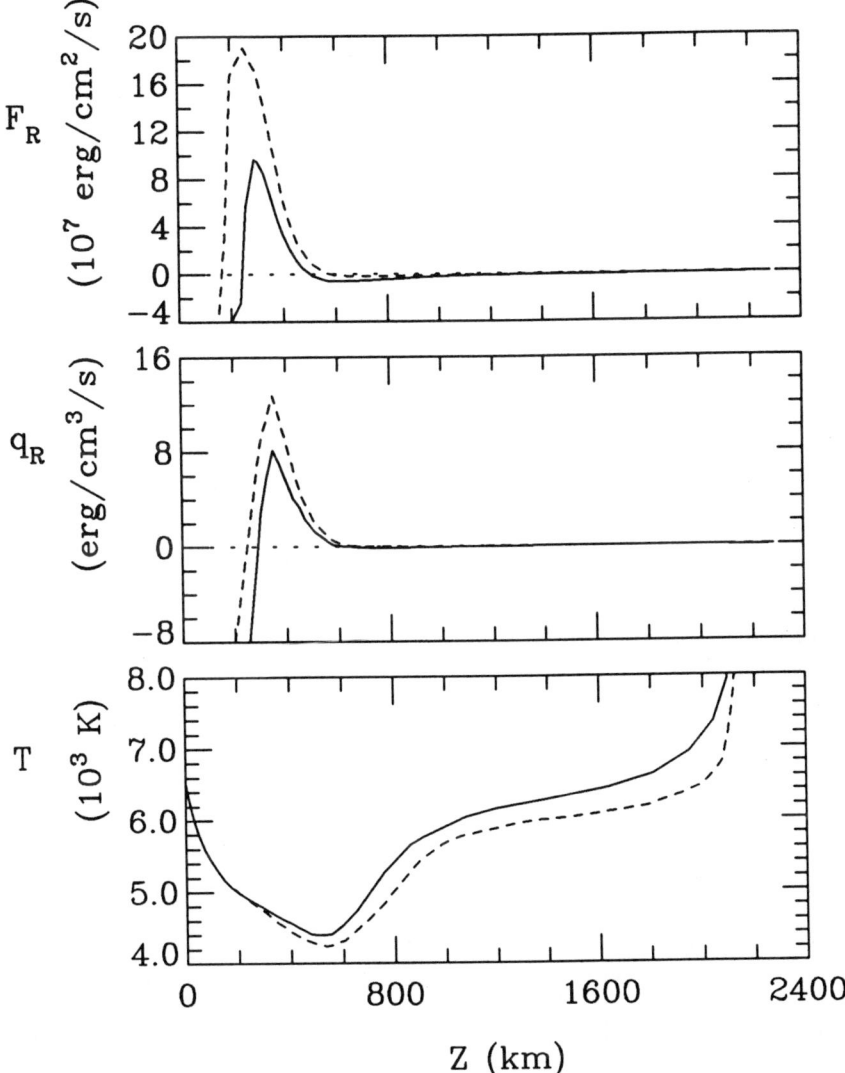

Fig. 1. Radiative heating and cooling in two empirical models of the quiet solar photosphere and chromosphere computed by Avrett[5]. The temperature T, net radiative heating density q_R, and the net radiative heating energy flux $F_R = \int_Z^\infty q_R dz$ are plotted against height Z above the level in the photosphere where optical depth $\tau_{5000} = 1$. In each panel, the solid curve is for the model derived from the observed radiation averaged over the bright lanes and the dim cells of the chromospheric network; the dashed curve is for the cells alone.

amplitude of about 0.5 km/s for the oscillations in the upper photosphere, we estimate[6] $\Delta\rho/\rho \simeq 0.5$, and hence that the radiative energy flux absorbed by radiative pumping is of order 10^8 erg/cm^2/s. This indicates that radiative pumping in the upper photosphere may indeed be a major driver of the solar p-mode oscillations.

Because the inferred radiative pumping is in the upper photosphere, it is open to observational confirmation or denial.

This work was supported by NASA through its Solar Physics Branch and by NSF through its Atmospheric Science Division.

REFERENCES

[1] Ando, H. and Osaki, Y. Publ. Astron. Soc. Jpn. **27**, 581-603 (1976).

[2] Goldreich, P. and Keeley, D.A. Astrophys. J. **211**, 934-942 (1977).

[3] Goldreich, P. and Kumar, P. Astrophys. J. **326**, 462-478 (1988).

[4] Libbrecht, K.G. Astrophys. J. **334**, 510-516 (1988).

[5] Avrett, E.H. in Chromospheric Diagnostics and Modelling (ed. B.W. Lites) 542 (National Solar Observatory, Sunspot NM, 1985).

[6] Fontenla, J.M., Emslie, A.G., and Moore, R.L. MSFC Space Science Laboratory Preprint No. 89-113 (1989).

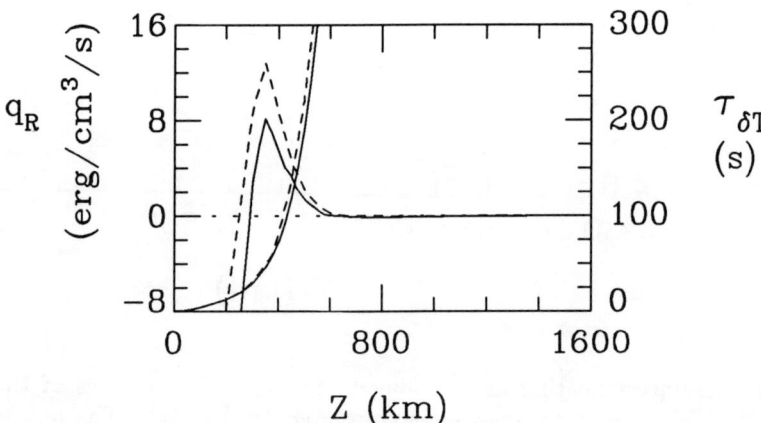

Fig. 2. Radiative relaxation time $\tau_{\delta T}$ for temperature perturbations in the two model atmospheres of Figure 1. The curves from Figure 1 for the net radiative heating density q_R are also plotted for comparison with the $\tau_{\delta T}$ curves. It is seen that $\tau_{\delta T}$ decreases rapidly with depth to less than 50 s at the peak in net radiative heating.

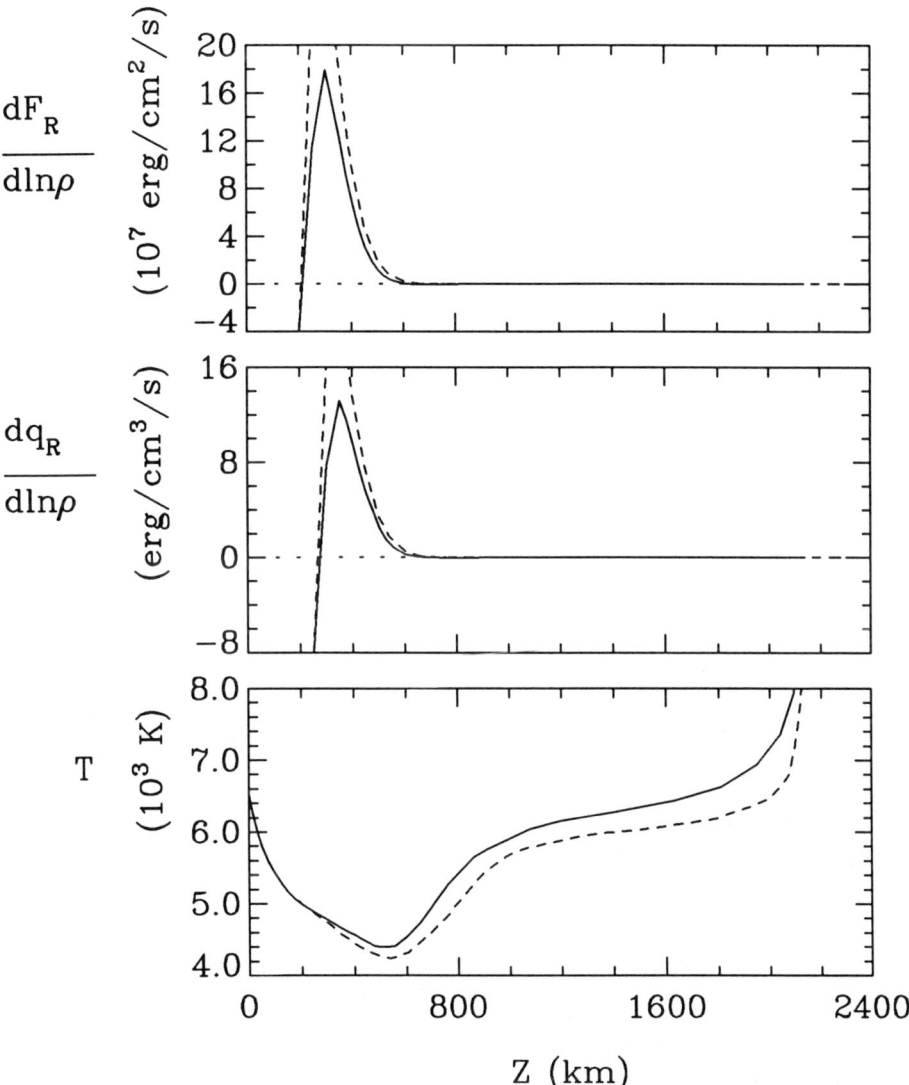

Fig. 3. Radiative pumping in the two model solar atmospheres. P-mode oscillations should be radiatively pumped in the layer of the upper photosphere where $dq_R/dln\rho > 0$. Analogously to F_R in Figure 1, $dF_R/dln\rho = \int_Z^\infty (dq_R/dln\rho)dz$. The curves for $dq_R/dln\rho$ and $dF_R/dln\rho$ are similar, in position, shape and magnitude, to those for q_R and F_R in Figure 1, as they should be if the net radiative heating in the upper photosphere is balanced by cooling due to radiative pumping of the p-modes.

A SOLAR TRACKING PLATFORM FOR USE AT THE SOUTH POLE

S.M.Jefferies, R.Pfeiffer, M.A.Pomerantz, L.Schulman
Bartol Research Institute, University of Delaware, Newark, DE 19716

W.Ball
National Optical Astronomy Observatories[†], Tucson, AZ 85726

ABSTRACT

This paper describes the design and performance of a solar tracking platform specifically built to capitalize on the outstanding observing conditions available at the geographic South Pole.

INTRODUCTION

During the Austral summer, the South Pole offers some unique advantages as a site for solar observations: a) The sun is continuously above the horizon. This allows for observations which are interrupted only by bad weather (i.e. clouds or ice-crystal precipitation in the first few hundred meters above the surface). Experience has shown that one can expect clear sky for periods of up to a week. b)The sun changes zenith angle very slowly with time. c)The altitude of the site (10,000 ft.) coupled with the extremely low water vapor content of the atmosphere and essentially zero pollution, provide observing conditions which can not be matched elsewhere on Earth.

Any experiment which requires the sun to remain stationary at the center of the field of view in the sky, is plagued with two problems; a) the sun changes in azimuth rapidly and b) atmospheric turbulence ("seeing") causes the sun to have an apparent motion. In order to help realize the potential of such a site, any solar tracking device must correct for these problems without perturbing the spatial and temporal information to be measured.

DESCRIPTION OF APPARATUS

The solar tracking platform facility shown in fig.1 and schematically in fig.2, is one of two such identical devices available for experiments to be conducted at the South Pole. The platform area for mounting an experiment is 50.8 cm wide and 71.1 cm long and is at a convenient working height. The detector for the photoguidance system (see below) is located on the underside of the platform and counter-weights can be added to the rear of the platform as necessary. The right ascension (RA) drive consists of a large worm gear driven by a stepper motor via a secondary worm gear reduction. A tensioning device provides antibacklash force. The declination (DEC) drive consists of a fine pitch, threaded rod which is pivoted at one end and driven by a large nut mounted on a worm wheel which is driven by a stepper motor. In order to alleviate cable "wind-up" problems, provision has been made for all the cables from the platform area to pass through the center of the RA drive shaft into the platform support structure/gear housing unit. Use of nylon clad cables sheathed in a low temperature neoprene jacket means that the cables only have to be unwound every two days or so. Sufficient cable is provided to enable the platform to be

---

[†] Operated by the Association of Universities for Research in Astronomy,Inc. under contract with the National Science Foundation.

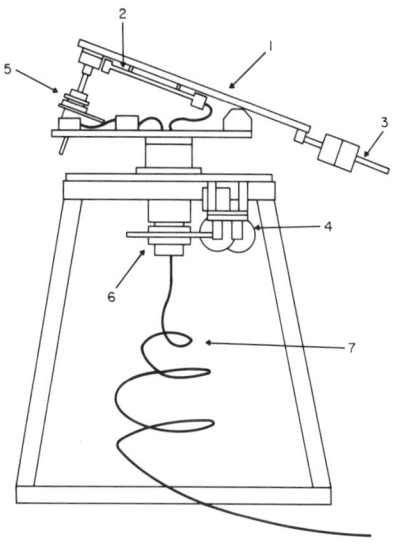

Fig.1 The solar tracking platform in operation at South Pole in 1987 with an experimental package mounted on top (Jefferies et al [2]). The fencing in the background is designed to reduce wind near the telescope[6].

Fig.2 A schematic view of the solar tracking platform. 1. Platform for mounting an experiment 2. Detector for guidance system 3. Counter balance weights 4. Stepper motor drive and secondary gear reduction for RA 5. Stepper motor drive, threaded rod and worm wheel for DEC 6. Main worm gear for RA 7. Cable "wind-up".

sited up to one hundred feet from the main experimental area. This helps reduce any atmospheric disturbances from "local" heat sources (see section on performance).

PHOTODETECTOR GUIDANCE

The principles of operation are shown schematically in fig.3. A small Galilean telescope is used to form an image of the sun on the surface of a quadrant photodiode detector. The diameter of the image is slightly smaller than the active area of the diode, most of which is covered by an opaque circular disk. The detector is aligned so that the signals between opposite quadrants correspond to image motion in RA and DEC. It is important that this alignment is accurate so as to minimize any "cross-talk" between adjacent quadrants. For each of the directions, RA and DEC, the quadrant signals are fed into a differencing amplifier. The output signal is then integrated and amplified before being processed to produce a pulse train whose frequency is proportional to the magnitude of the difference signal. This pulse train is used to drive a stepper

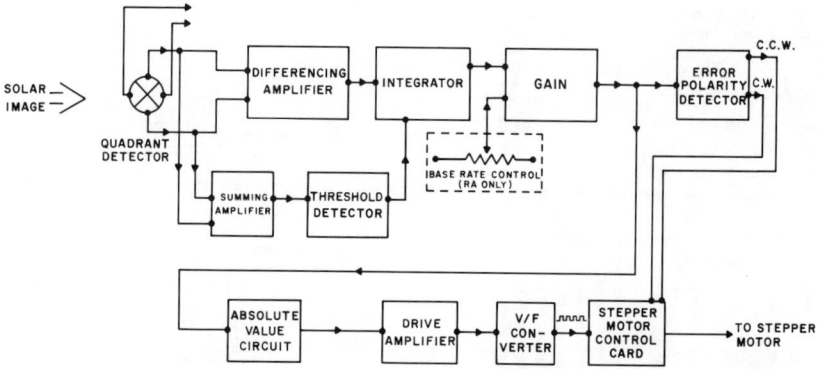

Fig.3 A schematic view of the photodetector guidance electronics.

motor on the guidance platform in the direction required to recenter the solar image on the detector. In RA, an additional dc level superimposed on the error voltage provides a base speed equal to the diurnal rate. The base speed is set in the laboratory and is verified in the field by turning off the error signals and studying the image offset as a function of time.

One disadvantage with any scheme based on detecting changes in intensity is that a cloud or other obstruction passing in front of the sun will cause the platform to drift in the direction of the obstruction. An attempt to overcome this problem is provided in the form of a threshold detector which turns the integrator off when the integrated intensity in opposite quadrants falls beneath some preset level. A better solution, which is under developement, is the use of a detection scheme based on locating the position of the solar limb.

PERFORMANCE

The platform was designed and built in 1982 and has been successfully used during four[1,2,3,4] of the subsequent Austral summers. The results reported here were obtained using the data of Jefferies et al[2]. Their experiment used a CCD camera with 260*260 pixels to collect full disk solar images which were integrated over 75 seconds before being recorded. A determination of the position of the center of the recorded solar image on the CCD array therefore gives a measure of the RA and DEC errors integrated over 75 seconds. The camera control computer was used to produce a real time estimate of the center coordinates of the recorded image and to output (via a digital-to-analogue converter) an error voltage which was fed into the input of the guidance electronics. This additional "computer guiding" was used to eliminate low frequency drifts of the image on the CCD.

Figs.4(a) and 4(d) show the image displacement in RA and DEC, relative to the center of the CCD array, over a period of 55 hours. The large increase in the r.m.s. of the error signals after 8 hours and 46 hours, is due to thin cirrus passing in front of the sun (the threshold circuitry had not been implemented when this data was taken), where as the slight increase at about 16 hours and 40 hours is due to the platform pointing over the observing hut. Although the observing hut was buried under several feet of ice, sufficient heat was escaping through the double door to provide a localized region of degraded "seeing". Figs.4(b) and 4(e) show the distribution of the RA and DEC errors

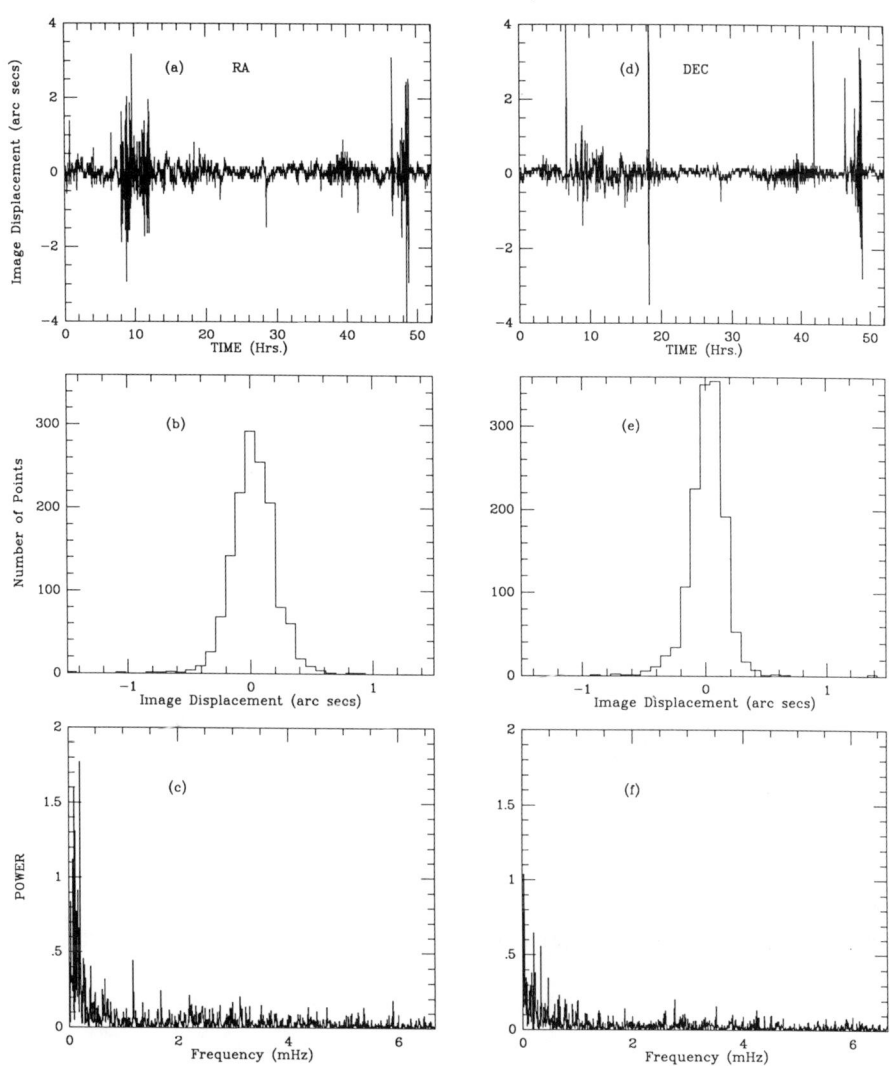

Fig.4 Plots demonstrating the performance characteristics (see text).

measured during the interval between appearances of cirrus. The r.m.s. errors in RA and DEC are 0.15 and 0.11 arc seconds respectively. Since the total response time of the guidance system is approximately 0.5 seconds, an estimate of the uncorrected high frequency guiding error was obtained from a video made of an oscilloscope display of the RA and DEC error signal before entering the integrator. This estimate gave an r.m.s. of order 1 arc second in both axes (for a "clear" sky) and is in good accord with the report of Harvey et al [5] for "seeing" conditions at South Pole.

Estimates of the frequency spectra due to the guidance system are shown in figs.4(c) and 4(f) and were obtained by taking the Fourier transform of the error signals shown in figs.4(a) and 4(d) over the same period of time as used for the distribution plots in figs.4(b) and 4(e). It is interesting to note that there is no significant power at frequencies corresponding to the RA gear tooth frequency (2.777 mHz) and its harmonics. This is in good agreement with the results of Jefferies et al [2] from an analysis of their solar oscillation power spectra. It is suspected that the increase in power at low frequencies may be due to an inadequacy in the real time "computer guiding" algorithm which caused a systematic "pulling" of the image center coordinates due to changing brightness patterns near the limbs used to compute the error signal[7].

ACKNOWLEDGEMENTS

The authors would like to thank Jack Harvey for many interesting discussions and suggestions. This work was supported by NSF grant DP819627. S.M.Jefferies was supported by the NSF grant DP8715791 as a visiting research fellow at the National Solar Observatory.

REFERENCES

1. B.Gelly, E.Fossat, G.Grec and M.Pomerantz, in "Advances in Helio- and Astero- seismology", (eds. J.Christensen-Dalsgaard and S.Frandsen, IAU symposium 123, D.Reidel), p.21, 1988.
2. S.M.Jefferies, M.A.Pomerantz, T.L.Duvall,Jr., J.W.Harvey and D.B.Jaksha, in "Seismology of the Sun and Sun-Like Stars", (ed. E.Rolfe, ESA SP-286, Noordwijk), p. 279, 1988.
3. S.M.Jefferies, M.A.Pomerantz, T.L.Duvall,Jr., J.W.Harvey and R.S.Aikens, to be reported.
4. M.A.Pomerantz, Antartic Journalof the USA, 18, no.5, p.266, 1983.
5. J.Harvey, M.Pomerantz and T.Duvall,Jr., Sky and Tele., 64, p.520, 1982.
6. R.H.Hammerschlag and C.Zwaan, Publ. Astron Soc Pacific, 85, p.468, 1973.
7. S.L.Keil and S.P.Worden, in "Nonradial Nonlinear Stellar Pulsation", (eds. H.A.Hill and W.A.Dzeimbowski, Lecture Notes in Physics 125, Springer), p.219, 1980.

SOLAR OBSERVING CONDITIONS AT THE SOUTH POLE

J. Harvey
National Solar Observatory,
National Optical Astronomy Observatories*, Tucson, AZ 85726

ABSTRACT

Austral summer weather conditions, atmospheric seeing and transparency of importance to solar observing at the South Pole are described.

INTRODUCTION

While setting up a cosmic ray detector at the South Pole in 1964, Martin Pomerantz of the Bartol Research Foundation realized the potential of that site for unique solar observations. Preliminary, year long, seeing tests were conducted in 1968-9 by the Bartol cosmic ray observer using a Questar and a 5-inch telescope with encouraging results. In 1970, Arne Wyller, then at the Bartol Research Foundation, contributed a chapter on polar astronomy to "Polar Research - A Survey", a study conducted by the National Academy of Sciences.[1] This study includes monthly averages of clear days, wind velocity and temperatures at the South Pole and other Antarctic stations. It was clear that there was good potential for astronomical observations from the South Pole and it was recommended that seeing tests be undertaken. In 1975-6 further seeing tests were made by Bernard Jackson. He and W. Smythe also measured precipital water vapor by means of balloons and infrared photometry.[2] They found a yearly mean of 0.45 mm with a standard deviation of 0.25 mm and pointed out that the South Pole appeared to be an outstanding site for infrared work, particularly in the winter. They also presented a table of cloudiness from 1957 through 1974.

The first serious South Pole solar observations were done in January 1979 by Ulf Kusoffsky and Martin Pomerantz. An 8-cm refractor fed by a 15-cm heliostat produced an image of the sun which was photographed through an H-alpha filter for a period of about 120 hours.[3,4] Image quality was described as limited by the optics rather than seeing conditions. Following these pioneering observations, solar observations have been done at the South Pole nearly every summer.

CLOUDS

The most unique advantage of the South Pole for solar observing is the opportunity to obtain continuous observations for a long duration. How long? Visual estimates of cloud cover have been made daily since 1957. These estimates are of some value but do not include details about whether or not the sun's disk is covered. An atlas of global cloud cover[5] includes 11 years of cloud data from the South Pole. From this we learn that the frequency of completely clear sky ranges from about 50 to 20 per cent during winter and summer respectively. The most frequent type of cloud is cirrus with altostratus and stratus also significant (however, reporting policies have mixed the latter two types). Personal experience from three summers confirms that cirrus and low-lying stratus are the most frequent types of clouds.

Fortunately, solar radiation monitoring equipment has been operated at the South Pole regularly since as early as 1976. This provides information about how frequently the sun's

---
*Operated by the Association of Universities for Research in Astronomy, Inc., under contract with the National Science Foundation.

disk is clear. These data have been analyzed separately by Newkirk[6] and Stebbins and Wilson[7] who find similar results included in figure 1.

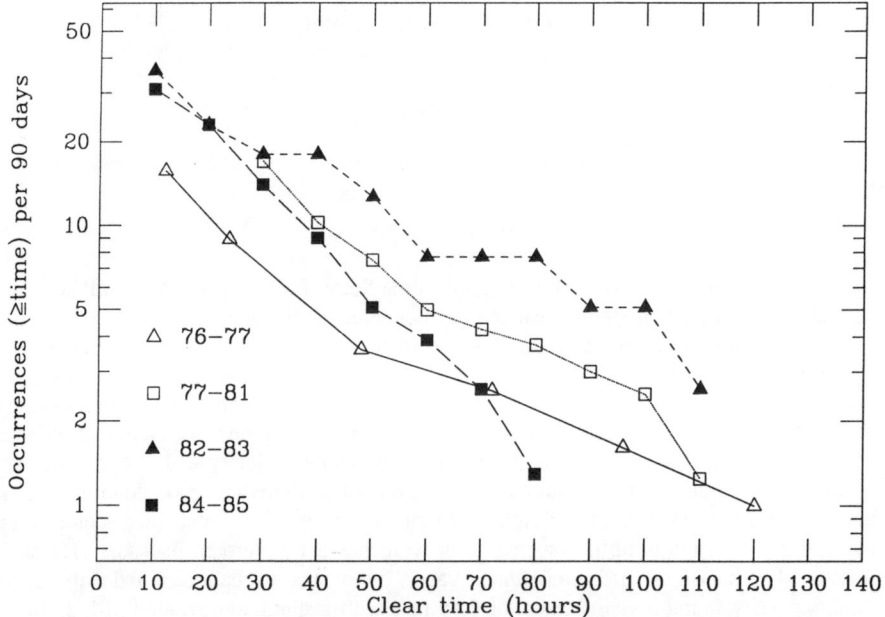

Figure 1. The number of occurrences of unbroken (or almost unbroken) runs of clear solar disk of duration t hours or longer. The data have been normalized to a season of 90 days duration. Data from 76-77[6] and 77-81[7] are from a radiation monitoring device while the 82-83[8] and 84-85[9] data are based on solar observing runs.

Some conclusions can be drawn from this and unpublished experience. First, the probability of obtaining an unbroken run in excess of say 160 hours during one summer is quite low, while there is a reasonable chance of obtaining one run of about 110 hours duration. Reports of 120 hour runs are fairly frequent. Second, the cumulative amount of clear solar disk time is about 50% from figure 1, considerably in excess of the reported frequency of totally clear sky. This leads to a high probability of high-duty-cycle observations for fairly long periods. Thus, we find reports of 600 clear hours at 34% duty cycle[8], 932 hours at 55%[9], 704 hours at 45%[10], 253 hours at 96%[11], 333 hours at 79%[12] and 343 hours at 84%.[13] Third, the best chance for clear weather is early and late in the observing season (late October - early November and late January). However, the literature shows that this is not a strong tendency and several day clear runs can occur any time. Fourth, the weather is highly variable from year to year. Fifth, aircraft operations produce contrails which sometimes interfere with solar observations.

SEEING

Good, quantitative measures of seeing at the South Pole do not exist at present. By seeing is meant the sharpness of the solar image rather than the clarity of the sky. In the literature, these two characteristics are sometimes confused. One might expect good seeing based on the lack of relief and the constant albedo of the surface for hundreds of km in all directions. However, the strong temperature gradient in the first few hundred meters above the surface combined with wind mixing act to degrade seeing quality. Furthermore, local heat sources must be rigorously controlled to avoid producing bad seeing.

Two seeing reports appeared in 1981. The first of these was based on observations with a 20-cm tower telescope with which it was reported that the seeing was within a factor of two of the telescope diffraction limit.[14] This implies about 1 arc sec. The second report is based on a one-time measurement with an elevated 10-cm telescope.[11] Here the seeing was estimated to be better than 2.5 arc sec. During the next observing season, photographs were made in H-alpha with the tower telescope.[15] These indicated seeing of about 1.5 arc sec. I made a one-time test with this telescope by putting a knife edge adjacent to a small piece of the solar limb and projecting an image of the pupil onto a card. What was seen was a familiar pattern of light and dark blotches moving rapidly across the pupil in the wind direction. The size of the blobs was perhaps a third of the pupil diameter which suggested seeing in the 1-2 arc sec range. This was consistent with the appearance of the solar image itself although the optics were not of perfect quality. It was not clear whether or not the seeing was degraded by heat from the observing room or if the seeing was intrinsically limited by conditions in the atmosphere.

The next seeing measurements seem to have been made in 1987.[12, 16] As part of an experiment to record the solar disk every minute with 8 arc second pixels, we used a quadrant photodiode to guide the solar image. The signals from the 1.5-cm guiding telescope were fed to an oscilloscope arranged to provide an x-y display of the instantaneous pointing error. We found typical motions of an arc sec or so with episodes of larger errors. These were found to coincide with pointing the telescope over heated areas. During the quietest times, the error signal suggested seeing much better than 1 arc sec. This inspired me to use the 8-inch Celestron telescope available at the South Pole station. I placed the telescope on the surface several km from the station, away from all local heat sources except myself (located downwind). I looked at the moon, Venus and the distant horizon with the highest magnification available. This was done on two different occasions with similar results. There was no detectable sign of seeing effects; the image quality seemed excellent but it would have been difficult to detect seeing between 0.5 and 1 arc sec.

A year later, in November 1988, conditions were unusual. The wind was frequently very light and blew erratically from abnormal directions. Our guider error signal showed periods of very stable conditions as well as bad conditions of several arc second excursions. We had taken additional precautions to avoid local heat sources which proved to be effective. The periods of bad seeing did not seem to correlate with the telescope pointing in any particular direction. Solar observations in 1988 consisted of one-minute integrations of the solar disk at 394 nm using 4 arc sec pixels. These images indicate that the seeing is certainly better than 4 arc sec for the vast majority of the time. Our experience indicated that doubling the resolution to 2 arc sec was warranted. Whether experiments which require 1 arc sec or better seeing are possible at the South Pole remains an open question.

SKY TRANSPARENCY

The air at the South Pole is the cleanest on earth. Accordingly, there is little or no problem from unnatural aerosols and man-made pollution. Typically, the sky is blue right up to the solar limb. Two measurements of the brightness of the sky near the limb appear in the literature. Both were made using an Evans sky photometer. One[4] is a remarkable average value of 8 millionths of the solar disk intensity and the other,[11] during a period of bad weather is an unremarkable 20 millionths. My own experience, based on three summers is that the sky can be outstanding. However, there are frequent occasions of faint, featureless aureoles near the sun. More spectacular are ice crystal precipitation storms. These produce a sparkling sky with halo phenomena of striking beauty. While it is often possible to observe the sun during minor storms, the quality of the observations is certainly degraded.

Solar oscillation measurements allow us to quantify the variability of sky transmission. A frequency spectrum of full-disk intensity measurements at the South Pole in the 0.1 to 10 mHz range indicates a power spectral density which varies with the -2 power of frequency rather than the -5/3 power one might expect.[17] At 3 mHz, the level is about 10^{-6} in units of $(\delta I/I)^2$ per Hz. This means that it has been possible to detect the degree zero oscillations of the sun in intensity observations, a feat not readily accomplished elsewhere on earth.

CONCLUSION

Aside from the details and references in this note, there are two semi-popular articles containing useful information.[18, 19] The South Pole has already demonstrated its unique characteristics for serious solar observations. As the second decade of such observations begins, there is every expectation that it will continue as an outstanding site for solar research. However, more sophisticated observations depend critically on observing conditions and more needs to be done to specify what the conditions really are like.

REFERENCES

1. A. Wyller, *Polar Research - A Survey* (National Academy of Sciences, Washington, D.C.), Ch. 8, pp. 170-177 (1970).
2. W. D. Smythe and B. V. Jackson, *Appl. Opt.* **16**, 2041 (1978).
3. U. Kusoffsky and M. A. Pomerantz, *Highlights of Astro.* **5**, 89 (1980).
4. M. A. Pomerantz, A. A. Wyller and U. Kusoffsky, *Solar Instrumentation - What's Next?* (National Solar Observatory, Sunspot, NM), pp. 379-384 (1981).
5. S. G. Warren, C. J. Hahn, J. London, R. M. Chervin and R. L. Jenne, *Global Distribution of Total Cloud Cover and Cloud Type Amounts Over Land* (National Center for Atmospheric Research, Boulder CO), NCAR/TN-273 (1986).
6. G. Newkirk, *Solar Cycle and Dynamics Mission Final Report* (Goddard Space Flight Center, Greenbelt MD), Appendix C, pp. 103-107 (1980).
7. R. Stebbins and C. Wilson, *Solar Phys.* **82**, 43 (1983).
8. R. Stebbins and R. Mann, *Antarctic J. U. S.* **18** (5), 268 (1983).
9. R. Stebbins, R. Ronan and M. Arrambide, *Antarctic J. U. S.* **20** (5), 219 (1981).
10. B. Gelly, E. Fossat, G. Grec and M. Pomerantz, *Advances in Helio- and Asteroseismology* (D. Reidel, Dordrecht), 21 (1988).
11. R. Stebbins and R. Mann, *Antarctic J. U. S.* **16** (5), 223 (1981).
12. S. Jefferies, M. A. Pomerantz, T. L. Duvall, Jr., J. W. Harvey and D. Jaksha, *Antarctic J. U. S.* **23** (5), in press (1988).
13. S. Jefferies, M. Pomerantz, J. Harvey and T. Duvall, Jr., *Antarctic J. U. S.* **24** (5), submitted (1989).
14. M. A. Pomerantz, A. A. Wyller and U. Kusoffsky, *Antarctic J. U. S.* **16** (5), 221 (1981).
15. M. A. Pomerantz, J. W. Harvey and T. Duvall, Jr., *Antarctic J. U. S.* **17** (5), 232 (1982).
16. S. Jefferies, R. Pfieffer, M. Pomerantz, L. Shulman and W. Ball, these proceedings.
17. J. W. Harvey, *Advances in Helio- and Asteroseismology* (D. Reidel, Dordrecht), 497 (1988).
18. J. Harvey, M. Pomerantz and T. Duvall, Jr., *Sky and Telescope* **64**, 520 (1982).
19. D. H. Smith, *Sky and Telescope* **77**, 598 (1989).

FULL-DISK HELIOSEISMOLOGY IN THE ANTARCTIC

Eric Fossat, Bernard Gelly, Gerard Grec
and Francois-Xavier Schmider
Universite de Nice, France

ABSTRACT

The South Pole has been an important site for unimaged helioseismology since 1979. With the rapid development of helioseismology, observations have evolved toward international network and space projects. Future observations from the South Pole will have an important role in validating data from low-latitude networks, obtaining certain measurements comparable only to those made in space and for imaged disk data. The utility of the South Pole could be enhanced by combining measurements from the pole and at another Antarctic station such as Dome C.

HISTORICAL INTRODUCTION

In December 1979, the first helioseismological campaign was organized at the geographic South Pole by Martin Pomerantz from Bartol, Eric Fossat and Gerard Grec from Nice. The instrument was an optical resonance spectrophotometer used without imaging, and thus measuring the radial velocity spatially integrated over all the visible solar surface. Such a full-disk measurement gives access to the lowest degree solar eigenmodes, namely the radial (degree zero) and slightly non-radial (degrees 1 to 3) modes. These are the eigenmodes which penetrate deeply inside the solar core, to the very center in the case of the radial ones. After the first evidence for a discrete frequency spectrum of the data obtained with such an observation[1], this campaign has since been regarded as a milestone in the history of helioseismology because it provided the first unambiguous identification of individual solar eigenmodes in the five-minute p-mode range[2,3].

The initial scientific goal was twofold. It was necessary to obtain uninterrupted data time series much longer than 12 hours in order to achieve the frequency resolution required for the detection of individual eigenmodes. On the other hand, the optimal atmospheric transparency and the constancy of the angular elevation of the sun in the sky were used to guarantee the long-term stability of the velocity signal necessary for the investigation of the long-period range of the solar g-mode oscillations.

Five days of totally uninterrupted data were obtained during the first week of January 1980, with one additional day after a cloudy break of a few hours. This success was such that many other helioseismological campaigns have been organized again at the South Pole, most of them being quite successful too.

PRESENT MOTIVATION

Now, a few years later, the situation has significantly changed. The pioneering campaigns of the early 80's have been followed by a tremendously rapid development of helioseismology. It has now become clear that the few days of continuous clear sky, at most between 10 and 15, allowed by the South Pole cloud statistics, will not be long enough for most observing programs to be developed. Also, the total duration of the summer accessible season has proven to be too short (at most 3 months), both for the sharper and sharper required frequency resolution and for following continuously the variation of helioseismic parameters along the solar magnetic cycle.

The consequence is that all important international projects in helioseismology, now under development in the second phase of this young science, are either worldwide networks or space projects.

However, the situation of the South Pole in the context of helioseismology remains unique. No other site on the earth can claim a comparably good sky transparency, and of course, no other site can offer the peculiarity of a constant angular elevation.

At the Nice University, we are now developing a network of 7 stations named IRIS, which makes use of a more reliable version of the same instrument successfully operated at the South Pole in 1979/80. In parallel, we still propose to organize campaigns in the Antarctic for the four main reasons listed in the next section.

PROPOSED FUTURE PROGRAMS

1. The analysis of the network data will face a difficult problem with the data merging. In general two, sometimes three or even more stations will be operated at the same time. Moreover, the data, in most cases, will merge when the sun is relatively low, in the afternoon on one side and in the morning on the other side, i.e. when it is not the best. The quality of the data used in the low frequency range will depend very strongly upon the quality of the data merging software. Certainly, the best solution for testing such a software will be to have a reference data set without interruption or merging, obtained with a single instrument. This reference data set, as described, is the one that can be obtained at the South Pole. It can contain, during one season, several time series of several days, just what is needed.

2. In any case, the sky quality and the constant angular elevation of the sun in the Antarctic will make the quality of the data collected there far superior to the network data, even with optimized data merging software. To date, nobody knows what is the amplitude of the g-modes, but everyone knows that the detection of g-modes would be of extreme interest for the physics of the solar core, where the p-modes alone cannot

eliminate all ambiguities. The top quality of the data obtained at the South Pole can probably be compared with what can be done in space, only the time of integration not being limited by clouds favors observation in space.

3. This limitation can be made less severe, or in other words the duty cycle can certainly be improved by the use of a limited "Antarctic network". The French government plans to operate in a not too distant future a permanent station at the Dome C site (S 74°, E 123°). Operating two full-disk instruments during one complete summer season, at the Dome C and at the Amundsen-Scott South Pole station, is one of our future plans. A quick look at the limited meteorological information available seems to indicate that a duty cycle of the order of 75 percent could be achieved, with the data merging characteristic times being much longer than the expected g-mode periods.

4. Imaged disk helioseismology has been conducted successfully at the South Pole by scientists from NSO and NASA collaborating with Bartol. We also propose to run a very simple instrument doing white light imaging. It has been shown by a Japanese group that such a simple technique can be used even in the extremely polluted sky of downtown Tokyo[4]. The reason is simply that for modes of higher and higher degree, the atmospheric transparency is less and less a limitation, and the reason for that is the large angular coherence of atmospheric transparency fluctuations compared to the apparent angular "wavelength" of the medium and high-degree solar eigenmodes. Certainly, the sensitivity of such a white-light measurement will not compete in the low-degree range with the South Pole imaging measurements done earlier with narrowband filters. However, the instrument to be operated is so simple that it seems worth doing it systematically.

CALENDAR

The IRIS network is planned to be fully deployed by the end of 1990, and then to be operated during one complete solar cycle. Our next Antarctic campaign could then be planned for 1991/92.

REFERENCES

1. A. Claverie, G. R. Isaak, C. P. McLeod, H. B. van der Raay and T. Roca Cortes, *Nature* **282**, 591 (1979).
2. G. Grec, E. Fossat and M. Pomerantz, *Nature* **288**, 541 (1980).
3. G. Grec, E. Fossat and M. Pomerantz, *Solar Phys.* **82**, 55 (1983).
4. J. Nishikawa, S. Hamana, K. Mizugaki and T. Hirayama, *Publ. Astron. Soc. Japan* **38**, 277 (1986).

Infrared Solar Physics from the South Pole

Drake Deming
Planetary Systems Branch, Code 693
Goddard Space Flight Center

Abstract: Infrared (IR) observations of the Sun could greatly benefit from the quality of the South Pole as an IR site, and the potential for multi-day sequences of uninterrupted observations. A nearly continuous picture of the evolution of the magnetic field in solar active regions could be obtained using vector magnetographs, especially vector magnetographs which incorporate IR array detectors. Observations of the Sun over a range of wavelengths in the IR continuum could also be used to study the vertical propagation characteristics of the solar p-mode oscillations.

I. INTRODUCTION

To date, most solar research from the south pole has been in the area of helioseismology (e.g. Duvall and Harvey 1983, 1984). Helioseismology requires long uninterrupted time series data, in order to obtain p-mode oscillation frequencies and frequency splittings without interference from sidelobes due to the day-night cycle. The quality of the south polar helioseismology results should prompt us to ask if other solar observations could benefit from the properties of the south polar site. Two examples of such observations are briefly discussed here.

II. EVOLUTION OF SOLAR ACTIVE REGION MAGNETIC FIELDS

The emergence of magnetic flux from the solar interior to form active regions on the solar surface, and the subsequent evolution and decay of these active regions, is a largely mysterious process. One promising method of shedding light on this question is to study the vector magnetic field in active regions, and its time evolution. With the recent introduction of sensitive IR array detectors (Wynn-Williams and Becklin 1987), it becomes possible to measure the vector field using IR lines. The ratio of Zeeman splitting to line width increases linearly with wavelength, so IR lines are more sensitive to magnetic fields than are visible lines. Very Zeeman sensitive lines occur near 1.6 μm (Stenflo, Solanki and Harvey 1987), and in the thermal IR near 12.3 μm (Brault and Noyes 1983, Deming et al. 1988). Consequently, it seems likely that work on solar vector magnetic fields will make increasing use of observations at IR wavelengths. While the quality of the South Pole as an IR site will be of benefit to such observations, the main reason to study solar active regions from the south pole is the possibility of following the evolution of the magnetic field, uninterrupted by the day-night cycle. An example of the need for such observations is illustrated in Figure 1. This figure shows the magnitude and orientation of the transverse vector field at selected points in a decaying sunspot. These observations were made using the 12 micron lines (Hewagama et al. 1989), and a single detector. The technology for vector field measurements at 12 microns is advancing rapidly, and it is expected that array detectors will soon make it possible to obtain complete vector field maps using these lines. However, another limitation on such measurements is already apparent. Significant changes in the field can occur during the night, preventing a continuous record of the magnetic field evolution. For example, the field in Figure 1 was largely radial on 5/10/88, but on the following day the spot had shrunk in area, and the field had developed a more noticeable azimuthal twist. It would be desirable to have data in the intervening period, as would be possible from an antarctic solar observatory. Ideally, the telescope for such a solar observatory would be of moderately large aperture ($\sim$ 3 meters), in order to obtain the best diffraction limited spatial resolution at 12 microns. However, very useful work could be done even with a 1-meter telescope.

III. VERTICAL PROPAGATION CHARACTERISTICS OF THE P-MODE OSCILLATIONS

Helioseismology investigations have established that the solar 5 minute oscillation consists of a rich spectrum of trapped modes, some of which have very long lifetimes, and very narrow linewidths. In the solar atmosphere the modes appear largely as evanescent standing waves, since most of the mode energy is trapped. However, there are deviations from standing wave behavior which are interesting, and not well understood. Such deviations appear in the phase of the wave velocity or temperature perturbation versus altitude, and can be derived from simultaneous observations of lines formed at different altitudes (e.g. Staiger 1987). The phase difference between the velocity and intensity in a single spectral line also gives information on the propagation characteristics of the modes. It is known that the mode linewidths increase rapidly above 4 mHz (Libbrecht and Zirin 1986), suggesting that the higher frequency modes are damped by propagation to the upper solar atmosphere, where they may play a significant role in the energy balance of the chromosphere, i.e. may contribute to chromospheric heating. Recently, Fontenla et al. (1989) have suggested that the solar p-modes are excited by radiative pumping in the upper photosphere. Such an excitation mechanism would apparently lead to net upward propagation above the pumping region, with net downward propagation in the mid- to low photosphere.

In order to investigate questions such as the role of the p-modes in chromospheric heating, and their possible excitation by radiative pumping, it is necessary to determine the net energy propagation versus altitude and frequency throughout the observable solar atmosphere. Although most observations of wave phase relations have shown evidence for some upward propagation, the upward energy flux is very uncertain, and there are even indications of a significant downward flux (Deming 1987). IR continuum observations could shed considerable light on these problems, especially if made simultaneously with velocity observations. The IR continuum is formed over a large range in altitude in the solar atmosphere, from the deepest observable layers at 1.6 μm, to chromospheric altitudes in the far-IR. The opacity source over this wavelength range (H⁻ free-free) is simple, well understood, and in LTE. This makes IR continuum observations relatively simple to interpret in terms of the wave temperature fluctuations, and the altitude of formation versus wavelength can be reliably calculated. Lindsey and Kaminski (1984) demonstrated that the far-IR continuum responds significantly to p-mode temperature oscillations, and phase delays versus wavelength were observed by Lindsey and Roellig (1987). Such studies require a very dry site such as the antarctic, due to extensive water vapor absorption in the far-IR. Observations in the far-IR benefit from relatively large telescope apertures, which improve the diffraction-limited spatial resolution. However, significant work could be done with a 1-meter telescope, especially at the shorter IR wavelengths ($\lesssim 20$ μm).

REFERENCES

Brault, J.W. and Noyes, R.W. 1983, Ap.J. (Letters) 269, L61.
Deming, D. 1987, in Cool Stars, Stellar Systems, and the Sun, proceedings of the Fifth Cambridge Workshop, eds. J.L. Linsky and R.E. Stencel, Springer Verlag, New York, p. 361.
Deming, D., Boyle, R.J., Jennings, D.E. and Wiedemann, G. 1988, Ap.J. 333, 978.
Duvall, T.L.Jr. and Harvey, J.W. 1983, Nature, 302, 24.
Duvall, T.L.Jr. and Harvey, J.W. 1984, Nature 310, 19.
Fontenla, J.M., Emslie, A.G. and Moore, R.L. 1989, preprint.
Hewagama, T., Jennings, D.E., Deming, D. and Zipoy, D. 1989, B.A.A.S. 21, 839.
Libbrecht, K.G. and Zirin, H. 1986, Ap.J. 308, 413.
Lindsey, C. and Kaminski, C. 1984, Ap.J.(Letters) 282, L103.
Lindsey, C. and Roellig, T. 1987, Ap.J. 313, 877.
Staiger, J. 1987, Astr.Ap. 175, 263.
Stenflo, J.O., Solanki, S.K. and Harvey, J.W. 1987, Astr.Ap. 173, 167.
Wynn-Williams, C.G. and E.E. Becklin, eds. Infrared Astronomy with Arrays, proceedings of the Workshop on Ground-based Astronomical Observations with Infrared Array Detectors, 1987, Univ. of Hawaii, Institute for Astronomy, Honolulu.

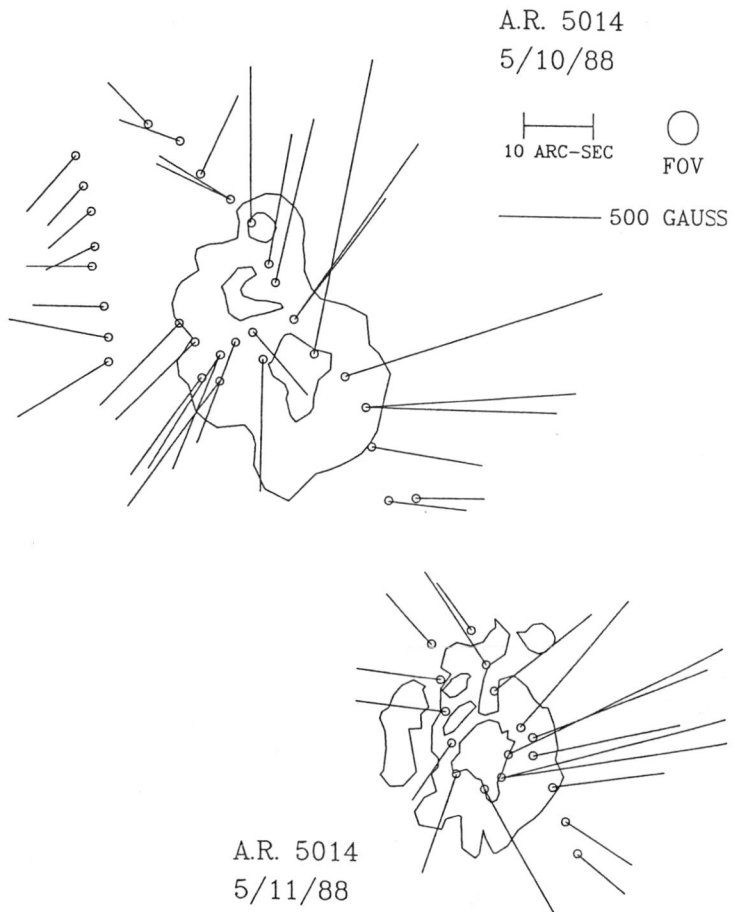

Figure 1. Transverse magnetic field vectors at selected locations near a decaying sunspot, obtained using observations of the 12.3 μm MgI emission line (Hewagama et al. 1989). The sunspot is illustrated in line-drawing form, having three umbrae in a single penumbra. In the period between the two observations the sunspot shrank in area, and the magnetic field developed an overall azimuthal twist. Antarctic observations could obtain a continuous record of such changes.

PHOTOMETRY OF SELECTED VARIABLE STARS AT
THE SOUTH POLE

Kwan-Yu Chen
University of Florida, Gainesville, Fl. 32611

ABSTRACT

Light variations of the many stars of different kinds show multiple periodicities. These are shown in the light curves of both the intrinsic and the close binary stars. The spectra of one group of close binary systems have H and K emission lines. Some of their variations of light curves are described as distorted wave or migrating wave. This phenomenon is interpreted as starspots, or spot region, which are observed at different orbital phases. The South Pole is the unique location on earth for the continuous stellar observation. Not only the cycle-to-cycle light variation, but also the eruptive feature can be investigated. In the starspot model, the activity cycle of the stars can be studied in detail.

INTRODUCTION

The systematic studies of the variations of stellar brightness began with Goodricke's observation of Algol[1]. Since then about 28,450 stars have been known to vary, as listed in the <u>Fourth Edition of the General Catalogue of Variable Stars</u>[2] which contains various types of eruptive variable, pulsating variable, rotating variable, cataclysmic variable, close binary eclipsing systems, and optically variable close binary sources of strong variable x-ray radiation. One common phenomenon in many important studies of these variable stars is the multiplicity in their periodic variations. A major discussion of this topic was presented in the 1975 International Astronomical Union Colloquium No. 29, "Multiple Periodic Variable Stars,"[3]. Continuous deviations from mean light curves are not uncommon for many stars of different types. In 1982, the 71st Colloquium of the International Astronomical Union, "Activity in Red-Dwarf Stars,"[4] presented many theoretical and observational researches on activity cycles which have been related to stellar oscillations and/or migration of starspots. In the discussion of that Colloquium summary, Dr. J. L. Linsky remarked, "... a great deal of what has been said at this meeting tells us that we need to monitor objects for long period of time. The allocation of telescope time, both in space and on the ground, is not usually made by people who recognize that elementary fact.... So the moral is that there is a great virtue in long-term monitoring and this should be recognized."

A polar observatory would allow continuous observations of stellar objects to be made at constant altitudes above the horizon. These observations, made with a single telescope-photometer system would reduce or eliminate many of the errors associated with conventional observations such as the altitude of the observed object, and the uncertainty of combining the observations made at different sites with different telescope photometer systems.

We are building a 40-cm reflecting telescope with a photoelectric photometer for use at the Amundsen-Scott South Pole station. This

© 1989 American Institute of Physics

close binaries (south of decl. = $16°S$) with H and K emission[7,8,9,10,11,12]

Name	V_{max}	Period (d)
CF Tuc	8.3	1.84
TY Pyx	6.9	3.20
IL Hya	7.5	12.87
CP-41°3888	9.0	54.95
TT Hya	7.6	6.95
RY Lib	9.8	10.72
V824 Ara	6.8	1.68
NSV 08931	9.1	31.96
AD Cap	9.3	2.99

chromospherically active binary stars (south of decl. = $20°S$)[12]

V_{max}	Period(d)	HD no.	V_{max}	Period(d)
8.80	0.66	118238	9.9	22.62
8.1	32.28	119285	7.86	12.00
8.19	18.24	127535	8.63	5.97
8.76	1.56	137164	7.0	46.19
7.99	0.96	165141	7.11	34.60
8.67	2.53	319139	10.4	2.45
7.69	20.38	175190	4.99	———
7.0	28.22	177716	3.32	———
9.05	———	181809	6.76	61.0
7.9	2.44	182776	8.50	45.18
7.1	43.76	188088	6.16	46.82
7.61	13.07	195040	8.98	23.21
7.0	11.47	202134	7.72	61.73
6.36	19.34	204128	8.7	22.35
7.95	11.66	214479	9.05	4.39
5.17	56.03	217344	8.42	1.65

The construction of the 40-cm optical telescope is a joint project of the scientists in the University of Florida, Beijing Astronomical Observatory, and Yunnan Observatory. It is a fork-mount telescope with an effective f-ratio of 15 at the Nasmyth focus where the photometer is to be mounted. Based on the experience of operating the 8-cm telescope, this new telescope is designed particularly for functioning at the South Pole during the long wintry night. Temperature monitoring of the filters, detectors, telescope section, instrumentation section, entrance section, and the outside air will provide the computer information for temperature control. This environmental control is implemented for the prevention of window frosting and for proper function of the electronic and optical parts.

A compact single-board version of the IBM PC is chosen for the basic control systems — telescope motion, stellar observation, data acquisition. The DOS and the control programs will be stored on 3.5 inch "micro-floppy" diskettes. The computer will be housed in a sealed heated enclosure.

automated system woul(...)
austral winter and would n(...)

PHOTOMETRIC ST(...)

Since the few close binari(...)
by Struve[5] the number of stars(...)
photometric and/or spectroscopi(...)
radio wave. It is natural, then,(...)
caused by starspots and chromosp(...)
scale as compared to the solar acti(...)
events, for example, the excellent re(...)
Venaticorum binary systems by Linsky(...)
light curves of some relatively short-p(...)
show migration-wave distortion. Howev(...)
covering a complete cycle. It is impos(...)
variation from cycle to cycle, orbital and/(...)
Optical telescope will be in the unique pos(...)
activities, cyclical and/or eruptive, in thes(...)
this class of star systems, close binaries wi(...)
photoelectric observations during the first obs(...)

The stars in Table I are selected from th(...)
table. Dr. Bopp's list gives the CaII surface fl(...)
gives the high energy fluxes. Additional stars a(...)
are taken from the list of Strassmeir et al. Inforn(...)
as celestial coordinates and brightness, will be sto(...)
detailed computer programs will be written for autor(...)
best practical efficient order.

The observing programs may be altered in light of(...)
made before observations are actually initiated. A pro(...)
made by the winter-over scientist who will be in re(...)
communication with the scientists at the University of F(...)
mind also the idea that, after a successful initial year(...)
inevitable startup problems are overcome, the limited bin(...)
will be expanded to include the other fields of astronomy t(...)
from polar observation. Observations, in general, will be(...)
"standard" U, B, V, R and I wavelength regions. Interferei(...)
wavelengths centered at the CaII H and K lines and Hα lines(...)
considered. Continuous light curves covering complete cycl(...)
variation of a close binary with H and K emission are inv(...)
understanding the nature of stars of this kind, in particular, in the(...)
the starspot model.

INSTRUMENTATION

In January 1986, the Department of Astronomy, University of Flo(...)
installed a small telescope at the Amundsen-Scott South Pole Station. T(...)
optical telescope is a two-mirror siderostat with a 8-cm lens[13]. T(...)
automated operation of the telescope system is computer controlled.
Valuable observational data have been obtained on the Wolf-Rayet binary γ^2(...)
Velorum[14].

Table I. C(...)

HD no.
5305
77137
81410
83442
97528
12817(...)
15555(...)
15839(...)
2060(...)

Table II. C(...)

HD no.
8435
10909
14643
16157
17084
26354
3480(...)
3784(...)
3957(...)
399(...)
425(...)
466(...)
61(...)
72(...)

Normally the computer will be operated from a remote terminal located in the telescope control room or in the South Pole Station computer room. The observational data will be recorded and transmitted to the Station main computer, and then transmitted via the communication satellite to Malabar, Florida, then to the Department of Astronomy in Gainesville. Thus preliminary study of the data can be made soon after the observation. This allows changes of observational programs, if needed, by sending messages from Florida to South Pole. The installation of the telescope will be a major step toward the investigation of southern variable stars, in particular, their starspots and asteroseismology.

REFERENCES

1. J. Goodricke, Phil. Trans. **73**, 474 (1783).
2. P.N. Kholopov, N.N. Samus, M.S. Frolov, V.P. Goranskij, N.A. Gorynya, E.A. Karitskaya, E.V. Kazarovets, N.N. Kireeva, N.P. Kukarkina, G.I. Medvedeva, E.N. Pastukhova, N.B. Perova, and S. Yu Shugarov, General Catalogue of Variable Stars, Fourth Edition. (Nauka, Moscow, 1987).
3. W.S. Fitch, ed., Multiple Periodic Variable Stars, Proceedings of the International Astronomical Union Colloquium No. 29. (Academiai Kiado, Budapest, 1976).
4. P.B. Byrne and M. Rodono, eds., Activity in Red-Dwarf Stars, Proceedings of the 71st Colloquim of the International Astronomical Union (D. Reidel Publ. Co., Dordrecht: Holland, 1983), pp. 655, 659, 660.
5. O. Struve, Ann. d'Astrophys. **9**, 1 (1946).
6. J.L. Linksy, in Cool Stars, Stellar Systems and the Sun, eds. S.L. Baliunas and L. Hartmann (Springer-Verlag, Berlin, 1984), p. 244.
7. D.M. Popper, Information Bull. on Var. Stars, No. 1083, (1976).
8. O.J. Eggen, Information Bull. on Var. Stars. No. 1426, (1978).
9. D.S. Hall, in Solar Phenomena in Stars and Stellar Systems, eds. R.M. Bonnet and A.K. Dupree (D. Reidel Publ. Co., Drodrecht: Holland, 1981), p. 431.
10. B.W. Bopp, in Activities in Red-Dwarf Stars, eds. P.B. Byrne and M. Rodono (D. Reidel Publ. Col., Dordrecht: Holland, 1983), p. 363.
11. P.A. Charles, ibid. p. 415.
12. K.G. Strassmeier, D.S. Hall, M. Zeilik, E. Nelson, Z. Eker and F.C. Fekel. Astron. Astrophys. Suppl. Ser. **72**, 291 (1988).
13. K-Y. Chen, J.P. Oliver, and F.B. Wood, Antarctic J. of the United States, 1986 Review, Vol. XXI, No. 5, 281 (1988).
14. MJ. Taylor, K-Y. Chen, J. McNeill, J.E. Merrill, J.P. Oliver, and F.B. Wood, Publ. Astron. Soc. Pacific **100**, 1544 (1988).

SOLAR-STELLAR ASTRONOMY
WORKING GROUP SUMMARY

J. Harvey
National Solar Observatory,
National Optical Astronomy Observatories*, Tucson, AZ 85726

ABSTRACT

The South Pole and other Antarctic stations offer unique possibilities for observations of the sun, stars and planets. Past, current and future use and options for such observations are considered. There is a clear need for more detailed information about observing conditions at the South Pole. A meter-class telescope installed at the South Pole and equipped with specialized instruments promises to provide unique and valuable solar, stellar and planetary observations.

INTRODUCTION

On the afternoon of June 9, 1989 a working group of about 20 persons convened to exchange information about making solar and stellar observations in Antarctica and to consider future possibilities. This is a report of that meeting. The unique advantages of polar astronomy were recognized long ago. Serious solar observations were made from Thule, Greenland during July 1966.[1,2] A committee of the National Academy of Sciences reported on the possibilities of polar astronomy in 1970.[3] The emphasis was on the Antarctic venue and, happily, several of the suggestions and possibilities mentioned by the committee have been achieved. Serious solar observing from the South Pole started in 1979 and has continued nearly every summer since then.[4,5] The infrared solar spectrum was observed from the South Pole in 1980 with the resulting discovery of 12-μm emission lines of great astrophysical interest.[6] Stellar photometry was first attempted in 1984 with an improved instrument installed in 1986.[7]

Early polar astronomy was done on a modest scale, often 'bootlegged' as part of other projects. The science is now recognized on its own merit as a valuable part of a broad Antarctic research program. The most recent National Academy long-range study of research in Antarctica[8] specifically notes as a major research objective 'detailed astronomical studies of the sun, stars, and galaxies'. The study also indicates that 'large mirror, multiuse telescope facilities' will be needed.

The morning plenary meeting of this conference heard a review by J. Linsky of solar and stellar astrophysics which provided a good basis for our afternoon discussion. Other working groups were generally technique oriented and often included discussions of solar and stellar research opportunities. For example, there are at least three balloon experiments under development to study the sun from above Antarctica. Such projects are not included in this summary; reference is made to other reports in these proceedings. Our working group exchanged information, affirmed the unique characteristics of Antarctic astronomy, heard about some projects for the near future, considered future research possibilities and reached some conclusions and a recommendation.

---

*Operated by the Association of Universities for Research in Astronomy, Inc., under contract with the National Science Foundation.

INFORMATION EXCHANGE

S. Jefferies described a solar tracking platform which has been in use at the South Pole since 1983. This facility can accommodate a wide range of instrument packages. K.- Y. Chen recalled the rationale for stellar photometry at the South Pole, gave some results from the second generation, 7.5 cm instrument and showed plans of the 40-cm, third generation stellar photometer. This instrument should be capable of reaching 11-12 magnitude to allow observation of nearly all of the known close binary systems with periods of special interest to the South Pole site. It was also reported that past experience indicates that South Pole extinction in the blue and yellow is comparable to that of Kitt Peak and Cerro Tololo.

L. Kay gave the perspective of an astronomer who had wintered over at the South Pole in 1985. She emphasized problem areas which must be recognized in future operations of astronomical equipment during the winter. These included: power outages, snow blowing into nominally closed volumes, bearing lubrication, twilight, moonlight and auroral light and the general difficulty of working at -90° F. She showed records of temperature and cloudiness which supported the notion that clear nighttime periods are correlated with low temperatures. She also mentioned that recent work indicates that the winter is cloudier than previously thought. New estimates now suggest that summer and winter seasons are equally cloudy. In this connection, polar stratospheric clouds may be significant during the winter.

Additional advice was offered concerning equipment. Frosting and ice crystal precipitation on optics can be a problem. Many experiments are underengineered to cope with the frigid Antarctic environment; developers should be prepared to stay for an adequate amount of time to get their equipment to operate properly. Finally, it was stated that the library at South Pole is inadequate as a research facility useful to a winterover scientist who might be writing a doctoral thesis. During group discussion, it was agreed that information on actual observing conditions is inadequate and that both day and night optical experiments need protection against frost veiling and seeing-degrading local heat sources.

NEAR TERM RESEARCH POSSIBILITIES

Reasons for using the South Pole as an observing site were discussed. Three principal characteristics are evident: The ability to observe time scales from about 1/2 to several days either uninterrupted or at least with a very high duty cycle. The ability to observe at infrared (and possibly near ultraviolet) wavelengths inaccessible to other ground-based sites. The stable, high-quality observing conditions.

A. Harper described plans for operation of a 24-inch telescope for nighttime infrared work. He offered the possibility of collaboration to use this facility in the daytime for solar work as well as additional nighttime stellar research. J. Lynch described his initiative for international Antarctic stations, one to be located near S82.5°, E66° at an elevation of 4.2 km. This should offer outstanding infrared performance and possibly good seeing at the expense of a diurnal cycle. The latter is such that even at the beginning of austral winter there would be no nights without some astronomical twilight. For comparison, the South Pole is without twilight for about 2 1/2 months each year. This disadvantage might be offset by a reduction in auroral interference at the proposed site.

A number of possible helioseismology experiments were described by A. Cacciani, D. Deming, J. Fontenla and E. Fossat. D. Deming also emphasized the diagnostic advantages of the 12µm solar emission lines (first discovered in South Pole observations), for magnetic field measurements using the Zeeman effect. An infrared telescope located at the South

Pole and equipped with a suitable instrument would allow studies of the evolution of magnetic fields which cannot now be done at single observatories.

M. Mumma pointed out the value of the South Pole for research concerning planets, comets and the search for extra-solar planetary systems. For planets, the South Pole offers the opportunity to do synoptic studies of time varying phenomena such as clouds on Jupiter and volcanic activity of Io and variations of its torus. New wavelength regimes may be available on the ground only at the South Pole. Seismology of Jupiter has been studied theoretically and a couple of observational attempts have been made. It is clear that the advantages of the South Pole for helioseismology would apply as well to giant planet seismology. Ideally, comet activity should be studied with continuous observations from a single site. This is particularly difficult at low-latitude observatories because active comets are usually close to the sun and are thus visible only for a short period of time each night. At the South Pole, the proximity problem would still occur with the result that the comet would, if visible at all, be low in the sky but would be visible for a long period of time. The possibility that the night sky at 2 to 2.5 µm might be 200 times darker at the South Pole than at mid-latitude observatories offers some very interesting opportunities for the study of protostellar and protoplanetary systems.

SCIENCE OPPORTUNITIES AND NEEDS

In this section some possible science projects involving the sun, stars and planets are sketched along with implied instrumental needs. These are presented in the form of three tables organized by the specific advantages offered by an Antarctic site such as the South Pole. The contents of the tables developed from a strawman list followed by discussion within the working group.

Table I. Wavelength coverage is the significant advantage.

SCIENCE	NEED
sun	
solar disk spectrum >20µm	high resolution spectrometer
sunspot spectrum >4µm	high resolution spectrometer
coronal spectrum >1µm	+ coronagraph
stars and planets	
detection of extra-solar planetary systems and brown dwarfs	>1 m (?) telescope with mid-IR imaging at high angular resolution

Table II. High-quality, stable observing conditions are the advantages.

SCIENCE	NEED
sun	
structure and dynamics of photosphere and chromosphere	50 cm VIS/near IR telescope + narrowband filters & polarimetry
proper motions ('heliometry')	10 cm VIS telescope + filters
stars/planets	
star formation planetary system formation brown dwarfs	telescope with good capability in dark window at 2-3 µm and 0.25 arc sec resolution

Table III. Time scale is the significant advantage.

SCIENCE	NEED
Seismology	
sun	modest sized telescope +
low-degree measurements	new detector package
high-degree measurements	under construction
new excitation theory	observations at 2 heights
higher duty cycle	2 instruments at 2 Antarctic stations
study thermodynamics, energetics	observations at 10-30 μm
stars	1 meter class telescope +
	fiber-fed echelle spectrograph
planets	precise spectrophotometric imaging
Activity	
sun	IR: ~2 m class telescope +
	VIS: 10-50 cm class telescope +
evolution of active regions	narrow band imaging + polarimetry
evolution of convective patterns	narrow band imaging
stars	
active binaries	40 cm spectrophotometric telescope
short period variables	40 cm spectrophotometric telescope
evolution/migration of starspots	1 m telescope +
	fiber-fed echelle spectrograph
planets and comets	VIS & IR broad and narrowband imaging

Reference is made to the original contributions made at the working group meeting and at other working group meetings for more details than can be presented here. Many of these reports are included in these proceedings. In addition to the necessarily brief and incomplete lists in this summary, there are no doubt many overlooked opportunities for solar, stellar and planetary research in Antarctica. Two possibilities are optical interferometry and radio observations. I consulted with one expert in each of these fields and received only a cool reaction to the possibilities of Antarctic observations in these fields.

CONCLUSIONS AND RECOMMENDATIONS

From the discussions within the working group it is possible to draw four broad conclusions in no order of significance.

1. Telescopes in the ~0.5 m (solar) and ~ 1.0 m (stellar) size class offer many new opportunities to do important stellar, planetary and solar research.
2. Day and night sharing of such telescopes seems feasible and desirable.
3. There is a potentially large payoff in solar and stellar research from a 2-m class IR telescope equipped with high resolution spectrometers, narrow and broadband imaging and polarimetric capabilities.
4. The South Pole has a continuing role to play in helioseismology with modest second and third generation experiments and the site could play an important role in the development of

stellar and planetary seismology.

Two recommendations pertaining to the South Pole emerged from the discussions.
1. Obtaining good 'site survey' data is a critical need. This can be obtained by new observations and, in part, from an astronomically-oriented reduction of existing meteorological and auroral data. Required information includes cloudiness, seeing, frequency of ice crystal storms, extinction, precipitable water vapor, sky brightness, frequency and magnitude of auroral interference, etc.
2. The support infrastructure needs power and communication improvements to help enhance the quality of the research environment.

It is interesting that the first of these recommendations is quite similar to the recommendations made in 1970.[3]

REFERENCES

1. T. J. Janssens, *Solar Phys.* **11**, 222 (1970).
2. E. H. Rogers, *Solar Phys.* **13**, 57 (1970).
3. A. Wyller, *Polar Research - A Survey* (National Academy of Sciences, Washington, D.C.), Ch. 8, pp. 170-177 (1970).
4. J. Harvey, M. Pomerantz and T. Duvall, Jr., *Sky and Telescope* **64**, 520 (1982).
5. D. H. Smith, *Sky and Telescope* **77**, 598 (1989).
6. F. J. Murcray, A. Goldman, F. H. Murcray, C. M. Bradford, D. G. Murcray, M. T. Coffey and W. G. Mankin, *Astrophys. J.* **247**, L97 (1981).
7. M. Taylor, *Sky and Telescope* **76**, 351 (1988).
8. *U. S. Research in Antarctica in 2000 A. D. and Beyond - A Preliminary Assessment* (National Academy Press, Washington D. C.), (1986).

Antarctic Environment and Experience

A Proposal for an International Station
in Antarctica

John T. Lynch
Program Manager
Polar Aeronomy and Astrophysics
Division of Polar Programs
National Science Foundation

Introduction

Many people have drawn an analogy between the manned exploration of the solar system and scientific stations in Antarctica. Even Wernher von Braun discussed the similarities between Antarctic exploration and the Apollo program. NSF and NASA have jointly sponsored a workshop on the subject and are also jointly funding a committee of the National Academy of Science's Polar Research Board (PRB) to study the parallels. Similarly the European Space Agency (ESA) has issued a report entitled "Study of Analogies Between Living Conditions at an Antarctic Scientific Base and on a Space Station".

The present would, for several reasons, be a portentous time to begin planning a truly international station in Antarctica. First of all, it would be a demonstration to the world that international scientific cooperation is alive, and has been growing and prospering after nearly thirty years under the Antarctic Treaty. Furthermore, 1992 has been declared International Space Year, it is the 500th anniversary of Columbus' first voyage to the New World and is scheduled to be the year of the economic unification of Europe. It would be most appropriate for the nations of the world to band together to make use of the last great un-nationalized continent to prepare themselves for the joint, peaceful, exploration of space. Clearly a joint venture of this nature could be important in promoting continued world peace.

In the summer of 1988 President Gorbachev suggested to President Reagan that the US and the USSR should begin to lay plans for a joint manned mission to Mars. That proposal was rejected, mainly due to the extreme cost of such an ambitious project. However, we could begin to prepare for eventual international space missions at a tiny fraction of the cost of such a mission by building an analogue station in Antarctica. In many ways the problems to be solved in space are very similar to those encountered in Antarctica.

Some of the Space/Antarctic parallels are quite obvious, such as the necessity to select small groups of highly trained individuals who can work together in isolation for extended periods, or in the case of the Moon/Antarctic comparison, the long day/night cycle. However, the parallel can be carried considerably further to include the types of science to be done, and, in some cases, there is even a strong similarity in environmental conditions.

Politically Antarctica is the perfect place to build an internationally owned and operated laboratory. The Antarctic Treaty, to which most of the major nations subscribe, strongly encourages cooperative projects for peaceful purposes, and cooperation has been a hallmark of relations between nations in Antarctica up to the present. For example, for a number of years there was always an American wintering over at one of the Soviet stations and a Soviet citizen spending a year at a US station. Also, the US and New Zealand programs are closely entwined in many important areas of logistics. An international project in Antarctica is greatly simplified by the facts that no nation can enforce sovereignty nor may the citizens of one nation be subjected to the laws of another nation. Another strong parallel is that the Outer Space Treaty of 1967 and the unratified Moon Treaty of 1979 both share much in common with the Antarctic Treaty, and to great extent are modelled on it.

Specifically, I recommend that the United States attempt to interest other Antarctic treaty nations in building at least one new Antarctic station that would be completely international and managed in such a way that no single nation would be dominant. It might be desireable to use the planned International Antarctic Center in Christchurch, New Zealand as the headquarters for the project.

Two Specific Proposed Stations

I would like to suggest two very specific projects for consideration. The first would be a wave injection facility (WIF) that would do science similar to that done at the recently closed Siple Station. The second would be a high altitude, high (geomagnetic) latitude station in East Antarctica. In this case the science would be similar to that done at Amundsen-Scott South Pole Station or Vostok Station, but the site would be chosen to optimize the station for science.

Wave Injection Facility

This station would have to be located near the base of the Antarctic Peninsula, within a few hundred kilometers of Siple Station. The invariant geomagnetic latitude would be about 60° an the geographical coordinates would be about 75°S, 90°W. This site would offer unique science opportunities for magnetospheric wave particle interaction studies. The thick ice (2000 m) allows the construction of reasonably efficient long wire very low frequency transmitting antennas to launch waves that can travel out into the magneosphere along the plasmapause. It is important that the magnetically conjugate point is accessible in Canada.

The general environmental characteristics of the area are well known. The altitude would be about 1000 meters, the snow accumulation is high at about 2 meters per year and if it is much like Siple there would be episodic fog. It is possible that the site could be logistically supported using blue ice runways in the Patriot Hills and overland traverse from there, or by using the planned hard runway at Rothera Station.

High Altitude, High Latitude Station

This station would be located on the East Antarctic ice sheet at an altitude above 4000 meters. The ice sheet thickness in this area is about 2,500 meters. The invariant geomagnetic latitude would be about 78°S and the geographical coordinates would be about 82°S, 45°E. The wintertime pressure altitude would be over 5,000 meters (16,000 to 17,500 ft), so high that it would be desirable, and possibly even necessary, to pressurize the station. This would certainly be the coldest and lowest water vapor observatory on Earth, which would make it highly desireable for infrared, submillimeter and millimeter astronomical measurements. The region is magnetically connected to a portion of the magnetosphere that will be a focus of attention of the world's solar-terrestrial physics community for the next decade or two. The site would be nearly magnetically conjugate to Sondre Stromfjord, Greenland.

The environmental conditions can be reliably extrapolated from other sites in East Antarctica such as South Pole, Vostok and Dome C. The precipitation would be very low (1-5 cm/yr?) with generally clear skies and consequently an abundance of solar energy in the sunlit period. The wind speed would be very low, averaging about

3 meters per second (7 mph) for the windiest month. The coldest temperatures would be near -90°C (-130°F). This site would be nearly equidistant (about 800 km) from South Pole and Vostok and could be logistically supported by ski equipped aircraft or overland traverse from McMurdo, South Pole or Vostok.

Clearly this station could closely emulate a lunar base. It should be pressurized, make use of solar power and the science would focus on astrophysics and space physics, which would certainly be major activities on the moon. The low precipitation and low average wind speed should minimize problems with accumulation and drifting, especially if the highest point is chosen for the station.

Conclusion

The Antarctic Treaty nations should consider the joint construction of one or more stations in Antarctica. The shared logistical effort could greatly enhance the scientific yield of Antarctica for all of the participating countries and make better use of the unique physical assets to be found there. I have tried to present a case for two such stations which could offer science opportunities that can be found nowhere else on Earth, but which are quite similar to the opportunities to be found in space.

In addition, Antarctica can serve as a logical stepping stone to the manned exploration of the solar system. The physical, physiological and psychological conditions that exist in Antarctica can be made good use of for this purpose. In particular, a high altitude, high latitude station on the polar plateau can be built at low cost to very closely emulate a lunar base. Of course, it would be possible even there to take a low technology approach, using diesel generators rather than solar power and sewer outfalls rather than recycled waste water, but the possibility of building a nearly isolated, self contained, habitat remains a viable and desirable option.

THE ROLE OF THE WINTER-OVER SCIENTIST IN ANTARCTIC ASTROPHYSICS

W. B. Gail
The Aerospace Corporation, Los Angeles, Ca. 90009

ABSTRACT

Winter-over scientists have traditionally played a central role in the success of astrophysics research in Antarctica. The unique access of the winter-over scientist to the data as they are collected provides an insight into the phenomenology which is often difficult for principal investigators to recreate subsequently. The current trend toward automatic data collection and the improved ability to communicate data to the home laboratory during the Antarctic winter have redefined the role of the winter-over scientist. Nevertheless, future astrophysics work can benefit greatly from the design of experiments which incorporate the potential contribution of the winter-over scientist.

INTRODUCTION

From the earliest days of Antarctic exploration, winter-over scientists have played a fundamental role in the progress of geophysics and astrophysics research in Antarctica. Using often inadequate or unreliable equipment in makeshift laboratories, the early scientists investigated with great success phenomena such as cosmic rays, geomagnetic variations, and atmospheric radio noise. These pioneers established the terminology and developed many of the experimental techniques that are still in use today. Living conditions have improved considerably since the early days of exploration and experimental equipment is now far more reliable. Communications advances have even made it possible to conduct experiments interactively from home laboratories. Nevertheless, the winter-over scientist continues to play a central role in the success of Antarctic research.

THE CURRENT ROLE OF THE WINTER-OVER SCIENTIST

The importance of the winter-over scientist results directly from the isolation of the Antarctic winter. The winter-over is in constant physical proximity to the experiments and has no other professional and personal distractions. These characteristics combine to focus the winter-over's activities almost exclusively on the scientific duties. The resulting intimacy with the experiments provides an irreplaceable working knowledge of the equipment and operations. This working knowledge is essential in the Antarctic because the effect of the harsh operating conditions is difficult to predict when designing experiments. Continous monitoring over a full year allows for identification of any conditions (e.g., high winds, low temperatures, ice fog, station power surges, static discharges) which cause problems for a particular experiment. This information can be used to fix the equipment, identify periods of unreliable data, or provide guidelines for future experimental modifications. In addition, the experience of the winter-over can be important for identifying otherwise inexplicable data problems (for example, interference with optical experiments caused by lights from the Caterpillar used to fill the snow melter).

Perhaps even more important, the constant exposure to the data produces an insight into the phenomology which is often difficult to recreate subsequently by principal investigators. The winter-over observes the phenomena in their natural temporal progression with instant access to related phenomena measured by other instruments. This long-term multi-instrument familiarity results in a unique knowledge of what constitutes normal phemenology, what the long-term trends are, and which periods or events are unusual or interesting. Much of the initial knowledge of geomagnetism, VLF emissions, and auroral effects was a direct result of personal observations by winter-over scientists. Recent examples of winter-over contributions include the first identification of correlations between particle precipitation, optical emissions, and VLF wave bursts, made by J. H. Doolittle and W. C. Armstrong while wintering at Siple Station[1], and the discovery of "Trimpi events", perturbations in sub-ionospherically propagating VLF transmitter signals caused by whistler-induced electron precipitation, by M. L. Trimpi during a winter at Eights Station[2]. In both of these cases, the phenomena observed were isolated events which could have been easily overlooked during subsequent analysis if they had not been identified by the winter-overs.

Despite the potential contributions of the winter-over scientist, the lack of continuity resulting from the yearly turnover can be a disincentive to reliance upon winter-overs. The need for annual training of new personnel may seem a burden when equipment can be designed to run autonomously. Moreover, an experiment which is dependent upon a winter-over scientist for its success is, in effect, hostage to the winter-over for the full year; winter-overs who do not respond well to the psychological and sociological stresses introduced by the isolation can degrade or even ruin an entire year's data. To the credit of previous winter-overs and the people who selected and trained them, such instances have been rare. Even with these limitations, however, the yearly turnover has many positive attributes. Newcomers often see novel relationships and ask questions that may have been missed by more experienced people. Winter-overs also have the time and interest to pursue occasionally fruitful avenues of research that may have been rejected initially by principal investigators as unpromising. In addition, winter-over programs, such as those that have been run for many years by Bartol Research Institute and Stanford University, have proven to be excellent training grounds for students entering geophysics and astrophysics research.

When designing and operating an experiment, proper consideration of the role of the winter-over can have considerable impact on the success of the experiment. Winter-overs tend to be independent and self-motivated, but the level of interest shown by a winter-over in a particular experiment is strongly influenced by the interest level displayed by the principal investigator. Experiments which are carelessly designed or tested with the expectation that the winter-over will fix them suggest a lack of concern for data quality that is passed along to the winter-over. A similar lack of interest is implied by principal investigators of ongoing experiments who assume interaction is not neccessary. Given competing responsibilities, the strongest support will in general be provided for those experiments in which principal investigators have demonstrated active interest and concern for quality. Similarly, the level of responsibility expected of the winter-over is in general more than returned by the amount of responsibility shown. To best take advantage of the potential contributions of the winter-over, he/she should be closely involved with the science. It may be tempting to design automated experiments, but removing the winter-over from the data analysis process reduces the opportunity for unexpected discoveries, eliminates the capability to modify or enhance the experiment except on an annual basis, and decreases the overall understanding of the phenomena being studied.

THE FUTURE OF THE WINTER-OVER SCIENTIST

The role of the future winter-over scientist will be strongly affected by technology and policy changes that are taking place now. Analog electronics and data recording are being replaced to a large extent by digital systems. The increasing reliablity of the equipment should result in a diminishing maintenance role, although this could be compensated for by an increase in the number of experiments. The reduction in size and cost of computational equipment provides excellent opportunities to increase the extent of on-site data analysis done by winter-overs. On-site analysis would be particularly beneficial to experiments requiring collection of large amounts of data which cannot be shipped to home laboratories during the winter. Such preliminary analysis, along with ongoing improvements in communications, provides opportunities for more productive interactions with principal investigators.

Given these trends, the role of the winter-over scientist may be expected to evolve in one of two directions. One possiblity is that the winter-over could become a caretaker, responsible for operating and maintaining experiments but not contributing directly to the science. Such a role would minimize the variability in data quality and reduce the level of interaction required of principal investigators. The other possibility is that the winter-over could take an increasingly important role in preliminary data analysis and in interactive experiments requiring collaboration with home facilities. For all of the above-mentioned reasons, the second approach provides the best utilization of the unique attributes of the winter-over.

The direction toward which the winter-over position evolves depends on the extent to which experienced principal investigators promote the need for involved winter-overs and the degree to which new principal investigators recognize this need. With a relatively large number of principal investigators associated with one or several winter-overs, the responsiblity for making such decisions is not clearly defined. The establishment of a Science and Technology Center would provide an excellent solution to this problem. Winter-overs could be hired through the Center and trained both at the Center and at each of the principal investigator's home institutions. The Center would establish a focus for continuity in the winter-over program and provide a much closer association with the science than is possible when winter-over scientists are hired by a commercial contractor.

REFERENCES

1. R. A. Helliwell, S. B. Mende, J. H. Doolittle, W. C. Armstrong, D. L. Carpenter, J. Geophys. Res. 85, 3376 (1980).
2. R. A. Helliwell, J. P. Katsufrakis, M. L. Trimpi, J. Geophys. Res. 78, 4679 (1973).

FIELD OPERATIONS IN ANTARCTICA FROM THE WINTER OVER SCIENTIST'S PERSPECTIVE

David P. Clements
Bartol Research Institute
University of Delaware
Newark, DE 19716

ABSTRACT

Having wintered at Amundsen-Scott South Pole Station in 1983-84 and at McMurdo Station in 1985-86 I will present my observations of the differences between wintering at the two stations, and the differences between the summer season and the winter isolation period. I was an employee of the Bartol Research Institute in charge of their cosmic ray stations and other upper and lower atmosphere experiments while in Antarctica, but I had no student status during my time "on the ice." However, I will also try to describe the conditions there from the point of view of doing graduate studies while wintering over.

INTRODUCTION

After a crowded and hectic summer season at the Pole, I and many of my fellow winter overs considered the winter to be the reward for surviving the summer. There was of course some hesitation about entering the eight and a half month isolation period there (mid-February to early November) but we were all anxious to see the end of water rationing and the general overcrowding that characterize all summer operations in Antarctica. At McMurdo, the isolation period is only about seven months long (late February to early October) and the transition between summer and winter is more gradual, though there is the same relief at seeing the summer crowds depart. At both stations, my job changed from summer to winter as dramatically as the change between the six month day and six month night. Summertime meant deploying new experiments or refurbishing existing ones with a host of Principle Investigators who appeared throughout the season for varying lengths of time. Wintertime saw this part of the job taper off, but there was a corresponding increase in the load as many of the aurora experiments came on line for the dark period.

SOUTH POLE VERSUS McMURDO

Working and living conditions at the two stations differ dramatically. The winter crew at South Pole is made up entirely of civilians, and totals about 20 people. The year that I was there I was one of seven science personnel out of a total of 19, the other twelve being support contractor employees. These numbers vary slightly from year to year but give a good idea of the relative importance of science at the Pole. In contrast, at McMurdo I was one of five science personnel out of a total of 136 winterers. The rest of the crew was made up of about 90 U.S. Navy personnel and 40 civilian contractor employees. The number of people and the more extensive facilities at McMurdo made it much easier to winter there from a psychological point of view. There was plenty to do (both work and recreation) at South Pole also, but the pace of life was slower and more effort had to be made to keep yourself occupied there. Recreation especially was much more diverse and organized at McMurdo, and the scenery and wildlife made the whole year quite enjoyable.

The working conditions were much worse at McMurdo however. While the absolute temperature is much lower at South Pole (-90°F is not uncommon in the winter at Pole, while -50°F is about as cold as McMurdo gets) the rest of the weather characteristics fall in South Pole's favor. McMurdo is much stormier throughout the winter, with sustained winds of over 70 mph not unusual away from the sheltered bowl of the main station complex. South Pole, by comparison, is several hundred miles away from the ocean and its storms and the highest gust ever recorded is only about 50 mph. The worse weather conditions combined with the fact that McMurdo is more spread out and requires a lot more outdoor travel makes it much more difficult to work at McMurdo. At the South Pole, everything is within walking distance, usually in sheltered archways. At McMurdo, the Cosmic Ray laboratory is about a mile outside of town, and the new Arrival Heights laboratory is two miles out in a different direction. A two mile commute to work may not seem formidable to many people in a temperate climate, but it is no picnic in perpetual darkness and blowing snow on a road that consists of the path of least resistance across volcanic terrain. The trip is not as bad in the summertime, assuming you are able to procure a running vehicle from the shortage plagued motor pool, which runs on a first come, first served basis.

THE EXPERIMENTAL ENVIRONMENT

The conditions mentioned above are not only hard on vehicles and personnel, but also on any equipment deployed. Dealing with the extremely low temperatures usually means having as few moving parts as possible. For experiments that need to be tended outside during the winter, one should design things with the thought in mind that access needs to be gained by someone with three or more layers of gloves on. The environment is not only very cold, but also very dry. Static electricity is a big problem, due to the dryness and the difficulty in getting a good earth ground. Station power blackouts, brownouts, and glitches caused by generator changes are not uncommon, so adequate battery backup of critical components is essential. Blowing snow is a big problem at South Pole, accumulating on any exposed surface, particularly on the optical experiments. The snow is joined by blowing volcanic grit at McMurdo, which finds it way into almost any container, no matter how well sealed. The entry vestibule and one of the windows at the Arrival Heights lab are located on the windward side of the building. The vestibule had to be shoveled out regularly during the winter just to allow access to the inner door, and no matter how tightly I shut and taped the window a pile of grit always accumulated on the inner sill. Any remote experiments are exposed to similar conditions.

POSSIBILITIES FOR GRADUATE STUDIES

As for graduate studies in Antarctica, as a current third year graduate student at Bartol my reaction is that conditions for leaving a student with an experiment for the winter are highly unfavorable at present. Much would depend on the specific situation though, and it would be far too general to give a blanket yes or no. Communication with the outside world during the austral winter is slow as best, and dependent on the whims of both Mother Nature and the communications personnel on station. My own experience at South Pole was that by the middle of the winter the communications coordinator was unwilling to use up any significant portion of our propagation window in the transmission of long data messages. Talking with prior and later winter overs has confirmed that this was not a problem isolated to my year. The worst problems of nonexistent mid-winter propagation have been alleviated by the installation of satellite links at both South Pole and McMurdo (the year after I left in each case) but the communications situation is far from ideal and could seriously hamper a graduate student in need of advice or information exchange during the winter.

There is also the question of the availability of computing and study facilities. The "Science Study Area" at South Pole has turned into a de facto station lounge and occasional overflow berthing area, and all pretense of use as a quiet study area has long since gone. The station library is well stocked with fiction and VCR tapes but other than a few old textbooks on station there are almost no usable references and journals. The library is unavailable as a legitimate study area as it is usually in use as a movie theatre or pool hall. McMurdo is a bit better off in this regard, as the current biology laboratory has a small study area with a small selection of mostly biology and geology references. Communications at McMurdo are somewhat better both in terms of number of operating personnel (thus avoiding single point failure when personality clashes occur) and available propagation time. At both stations, for viable graduate studies one would need either a large number of journals on station, which would be expensive and a logistical nightmare to keep updated, or very good electronic access to stateside computers and databases, which would put undue strain on the existing communication links and computing facilities.

SUMMARY

Anyone contemplating the deployment of an experiment in the Antarctic should send down well tested equipment that can withstand extreme temperatures and that is immune to large amounts of blowing and drifting snow. Adequate numbers of spare parts and detailed documentation must also accompany the equipment, as "highly automated" experiments rarely are that for the first few years, no matter how much testing occurs in a stateside environmental chamber. One also should make no assumptions about being able to get any parts or replacements into the stations during the winter. Ease of repair and maintenance will gain you great standing with the winter overs in charge of your equipment. You will find a very high correlation between how much you seem to care about your experiment (measured by quickness and helpfulness of response to winter over questions) and how much time the winter overs care to put in on your equipment. Regarding graduate studies, the way things currently stand spending a year in Antarctica, even running an experiment directly related to a thesis, cannot be equated to spending a year in graduate school in terms of progress one would be able to make towards a degree. While I think that there are plenty of areas that could use improvement, this is not meant to discourage anyone from going to the Antarctic, but quite the opposite, as I thoroughly enjoyed both winters and found the whole thing a tremendous experience.

ASTROPHYSICAL EXPERIMENTATION IN ANTARCTICA: A WINTER-OVER SCIENTIST'S VIEW

Laura E. Kay
Lick Observatory
University of California, Santa Cruz, CA 95064

ABSTRACT

Designing experiments to work under the extreme environmental conditions that exist at the Amundsen Scott South Pole base in Antarctica requires considerable advance planning. During the year I spent at the station there were many serious equipment malfunctions due to station power brownouts and blackouts, static, ice crystal precipitation, blowing snow, and poor thermal control. Future researchers are urged to consult with previous South Pole Principle Investigators to get more information about the common recurring problems.

INTRODUCTION

During the 1984-1985 season at the South Pole, my partner Cyril Lance and myself were responsible for the operation and maintenance of thirteen experiments in cosmic ray physics, upper atmosphere physics, helioseismology, auroral physics, and astronomy for scientists from ten universities and research labs. In the summer many of the Principal Investigators visited the station to check on their established experiments or to deploy new ones for the winter. In the winter we attempted to keep these experiments functioning despite the difficulties mentioned below. The following notes on some of the technical realities of working at the South Pole, especially during the winter season, come from our year-end reports to NSF Polar Programs and to the Principal Investigators at the conclusion of our winter-over in November, 1985.[1]

EXPERIMENTAL DIFFICULTIES

Power brownouts or short blackouts occurred frequently during our Antarctic season, for example in the winter during the generator rotations every ten days we would often get at least a brownout. Some of the experimental equipment, as well as the computer, was on an Uninterruptible Power Supply backup system and thus had no difficulties during brownouts or blackouts. For other experiments we had to reset clocks and timers, replace tapes, reset electronics, or make other minor adjustments. For more complex experiments however a loss in power was more serious, eg. a cryostat lost temperature and vacuum, necessitating long pumpdowns outside in very cold weather. Most blackouts did not last for long, therefore a UPS system attached to all complicated experiments would solve the problem. (New generators planned for the South Pole station may improve the situation). A related point is that the South Pole site is one of the driest on Earth; this combined with the lack of a real 'ground' meant that static often affected electronics, resulting in system crashes, altered times on clocks, etc.

Ice crystal precipitation on optical surfaces was another major headache. The domes on the roof of Skylab for the auroral experiments (all-sky cameras and zenith photometers) had a history of fogging and snow buildup. During my winter season we wiped snow off of the outside of each dome after every windstorm, and used blowing heaters on the inside of the domes once they fogged up. Dry nitrogen gas was blown through two of the domes about once a week for 5-10 minutes, and the base of these domes was kept heated; this seemed to prevent fogging problems. Limiting access to the room with these experiments also helped, as exhalations from nearby people were sometimes enough to fog up the inside of the dome if the experiment had been removed for repairs. For an outdoor experiment utilizing a mirror placed on the roof of the Clean Air facility, thick layers of ice crystals would build up in a few hours during windstorms. Even during light winds the mirror would become optically opaque from a fine layer of ice crystals after about 30 hours. This problem was temporarily solved by blasting the mirror with cooled compressed air several times a day.

Blowing snow created additional difficulties for the outdoor 'automated' experiments. Unmarked buried cables were in danger of being run over by people or tractors. Blowing snow entered cracks in boxes or buildings, as well as in holes in tubes or pipes, often creating problems in air intake and exhaust systems, internal thermal control of the box or building, etc. Equipment was then damaged from the resultant excessively high or low temperatures. The realities of fine blowing snow must be considered when designing plumbing systems for new experiments. The cold temperatures at the South Pole also affected the operation of some of the electronics, motors, power supplies, etc. Of special importance to astronomers perhaps is the observation that during my winter season the longer periods of clear skies correlated with the coldest weather, thus an experiment would have to work its best at these -85° to -105°F temperatures.

Even though summer conditions are less extreme than in the winter, during our year four new experiments failed before or a few days after the departure of the summer field team for that project. We thus recommended that the field teams go down to Antarctica relatively early in the summer season so that there is more time to do shakedown testing of all apparatus once it has been deployed, and that at least one field team member remain on site for a few weeks to monitor their instrument. It is crucially important for experimentors to talk with other researchers who have deployed outdoor 'automated' wintertime experiments at the South Pole in previous seasons in order to get more information about logistics and the feasiblity of specific systems. We spent a lot of aggravating time in our winter season trying to salvage 'automated' experiments that were supposed to run on their own. (The same was true for our predecessors). Thus we also recommended that a person who is well acquainted with the apparatus should winter-over for the first season of any new outdoor complex experiment.

In any case systems should be designed with backups as much as possible and configured so that separate components can be brought inside to be worked on in a warm environment. It it very hard to debug equipment outside; use of even simple tools is difficult during the cold and dark winter months. Obviously complete system documentation and a generous supply of spare parts (including a complete inventory) should be left at the station.

SITE TESTING

Astronomers may also want to conduct some further site tests before planning large observatories. A recent paper [2] suggests that nighttime average total sky coverage at the South Pole has been underestimated over the last decade by a factor of 1.5. Specifically the authors note that thin and scattered sky coverage is often difficult to detect during the two week period of moonless skies, and thus the times the skies were actually clear was overestimated when the moon was below the horizon. Ice crystals in the air may limit the seeing fairly often, and the light from the aurora may interfere with some projects (during a good auroral storm, 60% of the sky can become covered in five minutes). There may also be a recent increase in cloudiness, perhaps due to the number of polar stratospheric clouds in the sky in the late winter. Wide angle all-sky camera and zenith photometer data indicating cloud coverage, auroral coverage, and auroral brightness is available for many recent years from the Upper Atmosphere scientists; this data should be analyzed with the interests of astronomers in mind.

STATION FACILITIES

Beginning with the 1984-85 Antarctic season, a communications link was established with the continental United States using various satellites. It thus became possible to forward larger amounts of data back to the principal investigators. However there were still problems in determining priorities for data transmission. Thus experimentors requiring large amounts of data transference via satellite should discuss in advance their requirements with NSF and with the communications personnel in order to get an idea as to how much is feasible.

For the South Pole to become a major observatory some improvements to the library, computer, and communications facilities are needed. At least as of a few years ago there was no astronomy library available on site; useful books and some current journals should be sent there. The present station is also short on quiet office space and study areas. Adequate computer facilities to accommodate an increase in scientific projects, as well as easy and automated access to electronic mail should be provided.

Other papers have indicated why the South Pole may prove to be an excellent site for certain types of astronomical research. It is important however to keep in mind the specifics of the Polar environment when designing these experiments. In some ways preparing for this environment is more difficult than planning for space, since at least then there are established guidelines for designing and specific tests for debugging experiments prior to launch (and in space there is no blowing snow!). My year at the South Pole was a great experience, but I will say that a major frustration there was dealing with equipment that was poorly designed, inadequately tested, and insufficiently documented and supplied. The winter-over person in charge of new projects (and probably the NSF) will be much happier if new experimentors find out all they can about possible problems from previous researchers and plan accordingly.

REFERENCES

1. L. Kay and C. Lance, Year End Reports for NSF Polar and Upper Atmospheric Physics Program Investigators, (1985)
2. G. Schneider, P. Paluzzi, and J. P. Oliver, Journal of Climate, **2**, 295 (1989)

A GRADUATE STUDENT'S APPRECIATION OF THE 'WINTER-OVER' EXPERIENCE.

N.J.T.Smith.
University of Leeds, Leeds, LS2 9JT, England.

ABSTRACT

A discussion is given on the problems experienced by the author during the austral winter of 1988 whilst running the SPASE experiment at the USAP Amundsen-Scott South Pole research station as a winter-over scientist. This paper is based on a talk given to the workshop on ground based gamma-ray astronomy at this conference, although it is hoped that the information given is applicable to all fields of astronomy which utilise experiments run throughout the winter period.

INTRODUCTION

The SPASE experiment is a ground based ultra high energy gamma ray telescope consisting of sixteen scintillation detectors constructed at the USAP Amundsen-Scott South Pole research station during the austral summer of 1987 / 1988 [1]. The design of the detector and recording electronics is based on that used at a similar experiment, the GREX experiment [2], run by a group from the University of Leeds at Haverah Park in England under Professor Alan Watson, Antarctic deployment being made possible through a collaboration with the Bartol Research Institute, under Dr. Martin Pomerantz. The incentives for basing an array at the South Pole are outlined by Hillas [3] and the problems faced in the initial construction are discussed by Perrett [4]. The author was involved in the initial construction of the SPASE array during the 1987 austral summer, and subsequently ran the experiment during the 1988 austral winter as the sole winter-over operator. Comments received from Alistair Walker, the current winter-over scientist, support the views expessed here.

GRADUATE STUDENTS AS 'WINTER-OVERS'

Prior to his involvement with the Antarctic research program, the author had been working for two years as a graduate student on the collection, calibration and analysis of data from the GREX experiment. This experience was found to be invaluable in the subsequent involvement with the SPASE experiment. Once the summer field team departs the South Pole, communications to the home institutions can become slow and unreliable. As an example the average message turnaround time to the group in Leeds was about two days, longer if specific questions required investigation at the home institute. The communications link is also susceptible to noise from wind borne snow causing static on the antennae, solar interference at sunset and sunrise, networking computer outages and such unforeseen circumstances as the destruction of the satellite tracking station in Florida by brush fire (Feb 1989). These breaks in communication mean that the winter-over operator must have a full understanding of the experimental systems and the ramifications of any changes he / she implements, not just for the experimental hardware, but also for subsequent data analysis. Such knowledge should be possessed by a graduate student whose doctoral thesis is to based on that experiment.

A graduate student also has a vested interest in the data collected from the experiment and is already considered part of the investigative team before Antarctic deployment. This latter point is important from the psychological aspect of winter-over life as caring about the data and feeling part of a group helps the winter-over through the inevitable 'down' periods during the year.

An added bonus of using a graduate student as a winter-over, who then returns to the home institution to complete his doctoral work, is the continuity this allows between the operators. Recent policy on many experiments has been to hire the winter-over operator on a one year contract as a technician and have his experiences passed on through a 'turnover report'. It is inevitable that this document will not be exhaustive, and the general 'feel' that is gained by working with an experiment for a year at the South Pole is lost. Use of a graduate student will preserve this experience within the group, ensuring knowledge remains about running the experiment under the winter conditions.

During the year at the South Pole however a graduate student will lose some contact with his field. A more extensive publications library is needed, and some technique such as FAX for transmitting recent publications, before the year can profitably be used in writing a thesis. The intellectual stimuli existing within a university group working on a project is missing once the winter-over becomes isolated. This isolation also causes a general lack of enthusiasm and initiative that must be allowed for, especially in projects that commence well into the winter period, such as optical telescopes. Some facility for analysing and checking the integrity of data collected must exist to allow the winter-over graduate to retain a sense of motivation. This facility must also exist due to the current problems in transmitting large quantities of data back to home institutions for analysis, especially as the quantity of data collected from experiments is increasing.

ENVIRONMENTAL PROBLEMS

The most obvious problems associated with any experiment run at the South Pole are due to environmental constraints. A high altitude (10 000 feet pressure equivalent), low temperatures (reaching -80°C), and low humidity make the South Pole a harsh environment for an experiment. A continual 5 - 10 knot wind during the winter also compounds the problems. Drifting snow builds up around any ground level obstruction, the bow wave around the SPASE control centre reached five feet over the winter period. Snow also accumulated onto the vertical surfaces of the unheated, plywood, detector modules, several inches on the lee side, less than an inch onto the windward and top surfaces.

The wind borne snow particles can cause static spikes on any raised power cables, and any earth planes that use exposed metal. Earthing is a major problem at the South Pole where bedrock is under 9 000 feet of ice. Experimental power should be fed through an Uninterruptable Power Supply which totally isolates the load from the input power, as the station power was found to fluctuate in frequency and voltage and was prone to occasional 'brown-outs' and voltage spikes during maintenance on the generators.

Electronic components are also susceptible to static damage. Before adequate anti-static protection was installed in the control centre operator-induced static discharges could cause a disruption of the control computer.

For optical type experiments the added problem of frosting of the exterior windows exists. This frosting was found to occur on several types of material currently in use at the South Pole: solutions to the frosting incorporating dry, heated air or nitrogen blown across the window surface. Another problem for optical experiments is light pollution from the station, the moon and aurora. A detailed site study is needed to address these problems.

For the winter-over scientist the environment is most noticeable when working on equipment outside as even a light wind can have a drastic effect on the length of time the operator can remain outside. All external equipment should be designed for use with two pairs of mittens and a face mask, with all external tasks being as short as possible in duration.

Beyond the problems caused by the low temperatures affecting the outside apparatus, such as brittle cabling, the rarefied and dry atmosphere causes heat dissipation and overheating problems in electronics - power ratings are of the order of 50% greater than at sea level. Maintaining building temperatures can also cause problems as wide fluctuations arise due to solar heating and wind cooling of small huts.

EXPERIMENTAL DESIGN

Beyond the environmental considerations discussed above it should also be noted that once the station is isolated external contact is minimal (mid-winter airdrops have occurred in previous seasons). All experiments should thus be well equipped in spares and test equipment, with adequate documentation and reference manuals for the winter-over operator to overcome unforeseen problems. Redundancy and flexibility in the experimental systems allows a problem to be addressed in a number of ways.

Prior testing of all components is important, especially components exposed to the outside temperatures, and testing of the complete system before deployment is also a necessity. To facilitate this testing, and also the training of the winter-over personnel, duplication of the system at the home institute is desirable as far as is possible. This duplication also aids fault diagnosis through the year, as has been experienced with the SPASE and GREX experiments. Computer compatibility with the home institute not only allows for the development and revision of software for experiment control and data analysis at the home institute, but also allows programs developed by winter-over operators to be transported back, an important consideration for graduate students developing analysis techniques.

CONCLUSIONS

From the point of view and experiences of the author as the winter-over operator of the SPASE experiment it was found that prior system design and testing was of paramount importance. To this end the experience and suggestions of the Bartol staff was found to be invaluable. The training and knowledge gained as a graduate student on a similar experiment was found to be of great assistance in diagnosing and rectifying faults as they occurred, with a knowledge of how changes would affect the calibre and style of data collected. The frequent and, as far as possible, prompt communications from the home institutes not only aided rapid solution of problems but also helped in general morale throughout the year. All these factors allowed an on-time of 90% to be achieved throughout the winter period, a value comparable to similar, complex experiments in more temperate and accessible regions.

ACKNOWLEDGEMENTS

I would like to take this opportunity to thank all those involved in the SPASE project who supported me throughout my winter-over period, especially the initial setup team of Alan Watson, Jay Perrett, Paul Ogden and Martin Pomerantz. I would also like to thank all the other South Pole winter-over crew of 1988 for making that period so memorable and entertaining and for making a sole Briton so welcome.

REFERENCES

1. N.J.T.Smith et al. Nucl. Instr. Meth. in Phys. Res. A276: 622-7 (1989)
2. G.Brooke et al. Proc. 19th Int. Cosmic Ray Conf. La Jolla. Vol 3: 426 (1985)
3. A.M.Hillas. This conference.
4. J.C.Perrett. This conference.

AIP Conference Proceedings

		L.C. Number	ISBN
No. 151	Neural Networks for Computing (Snowbird, UT, 1986)	86-72481	0-88318-351-X
No. 152	Heavy Ion Inertial Fusion (Washington, DC, 1986)	86-73185	0-88318-352-8
No. 153	Physics of Particle Accelerators (SLAC Summer School, 1985) (Fermilab Summer School, 1984)	87-70103	0-88318-353-6
No. 154	Physics and Chemistry of Porous Media—II (Ridge Field, CT, 1986)	83-73640	0-88318-354-4
No. 155	The Galactic Center: Proceedings of the Symposium Honoring C. H. Townes (Berkeley, CA, 1986)	86-73186	0-88318-355-2
No. 156	Advanced Accelerator Concepts (Madison, WI, 1986)	87-70635	0-88318-358-0
No. 157	Stability of Amorphous Silicon Alloy Materials and Devices (Palo Alto, CA, 1987)	87-70990	0-88318-359-9
No. 158	Production and Neutralization of Negative Ions and Beams (Brookhaven, NY, 1986)	87-71695	0-88318-358-7
No. 159	Applications of Radio-Frequency Power to Plasma: Seventh Topical Conference (Kissimmee, FL, 1987)	87-71812	0-88318-359-5
No. 160	Advances in Laser Science–II (Seattle, WA, 1986)	87-71962	0-88318-360-9
No. 161	Electron Scattering in Nuclear and Particle Science: In Commemoration of the 35th Anniversary of the Lyman-Hanson-Scott Experiment (Urbana, IL, 1986)	87-72403	0-88318-361-7
No. 162	Few-Body Systems and Multiparticle Dynamics (Crystal City, VA, 1987)	87-72594	0-88318-362-5
No. 163	Pion–Nucleus Physics: Future Directions and New Facilities at LAMPF (Los Alamos, NM, 1987)	87-72961	0-88318-363-3
No. 164	Nuclei Far from Stability: Fifth International Conference (Rosseau Lake, ON, 1987)	87-73214	0-88318-364-1
No. 165	Thin Film Processing and Characterization of High-Temperature Superconductors	87-73420	0-88318-365-X
No. 166	Photovoltaic Safety (Denver, CO, 1988)	88-42854	0-88318-366-8

No. 167	Deposition and Growth: Limits for Microelectronics (Anaheim, CA, 1987)	88-71432	0-88318-367-6
No. 168	Atomic Processes in Plasmas (Santa Fe, NM, 1987)	88-71273	0-88318-368-4
No. 169	Modern Physics in America: A Michelson-Morley Centennial Symposium (Cleveland, OH, 1987)	88-71348	0-88318-369-2
No. 170	Nuclear Spectroscopy of Astrophysical Sources (Washington, D.C., 1987)	88-71625	0-88318-370-6
No. 171	Vacuum Design of Advanced and Compact Synchrotron Light Sources (Upton, NY, 1988)	88-71824	0-88318-371-4
No. 172	Advances in Laser Science–III: Proceedings of the International Laser Science Conference (Atlantic City, NJ, 1987)	88-71879	0-88318-372-2
No. 173	Cooperative Networks in Physics Education (Oaxtepec, Mexico 1987)	88-72091	0-88318-373-0
No. 174	Radio Wave Scattering in the Interstellar Medium (San Diego, CA 1988)	88-72092	0-88318-374-9
No. 175	Non-neutral Plasma Physics (Washington, DC 1988)	88-72275	0-88318-375-7
No. 176	Intersections Between Particle Land Nuclear Physics (Third International Conference) (Rockport, ME 1988)	88-62535	0-88318-376-5
No. 177	Linear Accelerator and Beam Optics Codes (La Jolla, CA 1988)	88-46074	0-88318-377-3
No. 178	Nuclear Arms Technologies in the 1990s (Washington, DC 1988)	88-83262	0-88318-378-1
No. 179	The Michelson Era in American Science: 1870–1930 (Cleveland, OH 1987)	88-83369	0-88318-379-X
No. 180	Frontiers in Science: International Symposium (Urbana, IL, 1987)	88-83526	0-88318-380-3
No. 181	Muon-Catalyzed Fusion (Sanibel Island, FL, 1988)	88-83636	0-88318-381-1
No. 182	High T_c Superconducting Thin Films, Devices, and Application (Atlanta, GA 1988)	88-03947	0-88318-382-X
No. 183	Cosmic Abundances of Matter (Minneapolis, MN 1988)	89-80147	0-88318-383-8

No. 184	Physics of Particle Accelerators (Ithaca, NY 1988)	89-83575	0-88318-384-6
No. 185	Glueballs, Hybrids, and Exotic Hadrons (Upton, NY 1988)	89-83513	0-88318-385-4
No. 186	High-Energy Radiation Background in Space (Sanibel Island, FL 1987)	89-83833	0-88318-386-2
No. 187	High-Energy Spin Physics (Minneapolis, MN 1988)	89-83948	0-88318-387-0
No. 188	International Symposium on Electron Beam Ion Sources and their Applications (Upton, NY 1988)	89-84343	0-88318-388-9
No. 189	Relativistic, Quantum Electrodynamic, and Weak Interaction Effects in Atoms (Santa Barbara, CA 1988)	89-84431	0-88318-389-7
No. 190	Radio-frequency Power in Plasmas (Irvine, CA 1989)	89-45805	0-88318-397-8
No. 191	Advances in Laser Science–IV (Atlanta, GA 1988)	89-85595	0-88318-391-9
No. 192	Vacuum Mechatronics (First International Workshop) (Santa Barbara, CA 1989)	89-45905	0-88318-394-3
No. 193	Advanced Accelerator Concepts (Lake Arrowhead, CA 1989)	89-45914	0-88318-393-5
No. 194	Quantum Fluids and Solids—1989 (Gainesville, FL, 1989)	89-81079	0-88318-395-1
No. 195	Dense Z-Pinches (Laguna Beach, CA, 1989)	89-46212	0-88318-396-X
No. 196	Heavy Quark Physics (Ithaca, NY, 1989)	89-81583	0-88318-644-6
No. 197	Drops and Bubbles (Monterey, CA, 1988)	89-46360	0-88318-392-7